T5-CQZ-156

Advances in

ECOLOGICAL RESEARCH

VOLUME 37

Advances in Ecological Research

Series Editor: HAL CASWELL

Biology Department
Woods Hole Oceanographic Institution
Woods Hole, Massachusetts

Advances in ECOLOGICAL RESEARCH

VOLUME 37

Population Dynamics and Laboratory Ecology

Edited by
ROBERT A. DESHARNAIS

Department of Biological Sciences
California State University
Los Angeles, California

2005

ELSEVIER
ACADEMIC
PRESS

AMSTERDAM • BOSTON • HEIDELBERG • LONDON
NEW YORK • OXFORD • PARIS • SAN DIEGO
SAN FRANCISCO • SINGAPORE • SYDNEY • TOKYO

Elsevier Academic Press
525 B Street, Suite 1900, San Diego, California 92101-4495, USA
84 Theobald's Road, London WC1X 8RR, UK

This book is printed on acid-free paper. 

For all information on all Elsevier Academic Press publications visit our Web site at www.books.elsevier.com

ISBN-13: 978-0-12-013937-8
ISBN-10: 0-12-013937-5

PRINTED IN THE UNITED STATES OF AMERICA
05 06 07 08 09 9 8 7 6 5 4 3 2 1

Contributors to Volume 37

ANDREW P. BECKERMAN, *Department of Animal and Plant Sciences, University of Sheffield, Western Bank, Sheffield S10 2TN, United Kingdom.*

TIM G. BENTON, *Earth and Biosphere Institute, School of Biology, University of Leeds, Leeds LS2 9JT, United Kingdom.*

BRENDAN J.M. BOHANNAN, *Department of Biological Sciences, Stanford University, Stanford, California 94305.*

MICHAEL B. BONSALL, *Department of Biological Sciences, Imperial College London, Ascot, Berkshire SL5 7PY, United Kingdom.*

MARC W. CADOTTE, *Department of Ecology and Evolutionary Biology, University of Tennessee, Knoxville, Tennessee 37996.*

ROBERT F. COSTANTINO, *Department of Ecology and Evolutionary Biology, University of Arizona, Tucson, Arizona 85721.*

JIM M. CUSHING, *Department of Mathematics, University of Arizona, Tucson, Arizona 85721-0089.*

BRIAN DENNIS, *Department of Fish and Wildlife Resources, University of Idaho, Moscow, Idaho 83844.*

ROBERT A. DESHARNAIS, *Department of Biological Sciences, California State University, Los Angeles, Los Angeles, California 90032.*

ODO DIEKMANN, *Department of Mathematics, Utrecht University, 3508 TA Utrecht, The Netherlands.*

JAMES A. DRAKE, *Department of Ecology and Evolutionary Biology, University of Tennessee, Knoxville, Tennessee 37996.*

STEPHEN P. ELLNER, *Department of Ecology and Evolutionary Biology, Cornell University, Ithaca, New York 14853.*

SAMANTHA E. FORDE, *Department of Biological Sciences, Stanford University, Stanford, California 94305-5020.*

TADASHI FUKAMI, *Landcare Research, Lincoln, New Zealand.*

GREGOR F. FUSSMAN, *Biology Department, McGill University, Montreal, Quebec H3A 1B1, Canada.*

NELSON G. HAIRSTON, JR., *Department of Ecology and Evolutionary Biology, Cornell University, Ithaca, New York 14853.*

MICHAEL P. HASSELL, *Department of Biological Sciences, Imperial College London, Ascot, Berkshire SL5 7PY, United Kingdom.*

SHANDELLE M. HENSON, *Department of Mathematics, Andrews University, Berrien Springs, Michigan 49104.*

MARCEL HOLYOAK, *Department of Environmental Science and Policy, University of California, Davis, California 95615.*

ARNE JANSSEN, *Institute for Biodiversity and Ecosystem Dynamics, University of Amsterdam, 1098 SM Amsterdam, The Netherlands.*

VINCENT A.A. JANSEN, *School of Biological Sciences, Royal Holloway, University of London, Egham, Surrey TW20 0EX, United Kingdom.*

CHRISTINE M. JESSUP, *Department of Biological Sciences, Stanford University, Stanford, California 94305.*

LAURA E. JONES, *Department of Ecology and Evolutionary Biology, Cornell University, Ithaca, New York 14853.*

AARON A. KING, *Ecology and Evolutionary Biology, University of Tennessee, Knoxville, Tennessee 37996.*

SHARON P. LAWLER, *Department of Entomology, University of California, Davis, California 95616.*

LAWRENCE D. MUELLER, *Department of Ecology and Evolutionary Biology, University of California, Irvine, California 92697.*

CASANDRA L. RAUSER, *Department of Ecology and Evolutionary Biology, University of California, Irvine, California 92697.*

MICHAEL R. ROSE, *Department of Ecology and Evolutionary Biology, University of California, Irvine, California 92697.*

MAURICE W. SABELIS, *Institute for Biodiversity and Ecosystem Dynamics, Universiteit van Amsterdam, 1090 GB Amsterdam, The Netherlands.*

KYLE W. SHERTZER, *Center for Coastal Fisheries and Habitat Research, National Oceanic and Atmospheric Administration, Beaufort, North Carolina 28516.*

MASAKAZU SHIMADA, *Department of Systems Sciences (Biology), University of Tokyo, Meguro, Tokyo 153-8902, Japan.*

MIDORI TUDA, *Institute of Biological Control, Faculty of Agriculture, Kyushu University, Fukuoka 812-8581, Japan.*

PAUL TURNER, *Department of Ecology and Evolutionary Biology, Yale University, New Haven, Connecticut 06520.*

MINUS VAN BAALEN, *Laboratoire de'Ecologie - UMR 7625, Ecole Normale Supérieure, 75230 Paris Cedex 05, France.*

ERIK VAN GOOL, *Institute for Biodiversity and Ecosystem Dynamics, University of Amsterdam, 1098 SM Amsterdam, The Netherlands.*

TAKEHITO YOSHIDA, *Department of Ecology and Evolutionary Biology, Cornell University, Ithaca, New York 14853.*

Preface

The role of laboratory studies in population ecology has been both formative and controversial. For example, the early experiments of Gause (1934), Nicholson and Bailey (1935), and Park (1939, 1948) played a central role in ecological discussions of density-dependent population growth, competitive exclusion, predatory-prey coexistence, and niche theory. On the other hand, regarding laboratory studies, ecologists are often warned that "any extrapolation to nature should be done with the utmost caution" (Andrewartha & Birch 1954, p. 433). MacArthur (1972) suggested that Gause's early laboratory studies on competition "misled ecologists for forty years." Although he was critical of these early "bottle experiments," MacArthur did call for further laboratory studies with more heterogeneous microcosms along the lines of the work by Crombie (1946) and Huffaker (1958). Studies, referring to them pejoratively as "bottle experiments." Kareiva (1989), lamenting the influence of MacArthur's devaluation of laboratory studies and its effect on theoretical ecology, advised that, "Ecologists gave up on bottle experiments too soon." In 1996, the controversy over laboratory studies motivated a special feature in the journal *Ecology* on "The Role of Microcosms in Ecological Research" (Daehler and Strong, 1996). There one can read a range of opinions from "the utility and power of microcosm analyses to provide insight into ecological systems is limited only by imagination and creativity" (Drake *et al.*, 1996) to "there is a cognitive danger that the microcosm (rather than the ecological system) will become the object of study, leading to needless confusion as results are overinterpreted and overextended" (Carpenter, 1996). The tension between laboratory and field ecology remains.

Laboratory studies have led to numerous advances in ecology within last few years. For example, Costantino *et al.* (1997) provided the first experimental demonstration of chaos in population ecology. McCauley *et al.* (1999) provided convincing evidence for the coexistence of multiple stable cyclic attractors in experimental populations of *Daphnia* and algae. Turner and Chao (1999) showed experimentally how game theory can be used to understand the dynamics of viruses infecting their hosts. Petchey *et al.* (1999) showed how environmental warming affects food web structure in experimental aquatic microcosms. Fussman *et al.* (2000) were able to induce a "bifurcation" from stable equilibria to predator-prey cycles in experimental populations of rotifers and algae. Using bacterial microcosms, Kassen *et al.* (2000) showed that selection for specialization in heterogeneous environments produces an unimodal relationship between diversity and productivity. Ellner *et al.* (2001) identified the mechanisms behind spatially-mediated coexistence in an acarine predator-prey system. Bjornstad *et al.* (2001) demonstrated experimentally that strong coupling between a predator and prey can lead to an increase in the dimensionality of the prey's population dynamics. Henson *et al.* (2001) documented the importance of "lattice effects" in population dynamics which are a consequence of the discreteness of animals numbers. Kerr *et al.* (2002) used bacterial populations to show how spatially localized species interactions can lead to increased biodiversity. Yoshida *et al.* (2003) demonstrated how prey evolution can alter predator-prey dynamics in and algal-rotifer system. Fukami and Morin (2003) used aquatic microbial communities to demonstrate that the history of community assembly can affect productivity-diversity patterns. All of these examples appeared in the journals *Science* or *Nature*. Many other recent examples have also appeared in the pages of prominent ecological journals. Despite their controversial history, one could argue that laboratory studies are still demonstrating their value to the discipline of ecology as a whole.

In the Winter of 2000 my "Beetle Team" colleagues and I were approached by Professor Hal Caswell regarding our interest in guest editing a volume of *Advances in Ecological Research* which would be focused on the role of laboratory studies in ecology. We excitedly agreed, but as often happens with busy people, the project languished for nearly three years as we attended to other priorities. After several gentle prods from Hal, my colleagues and I decided that the only way we would make progress on this project was if one of us would to take the lead. I agreed and started the process of contacting potential authors regarding their interest in contributing a chapter. Invitations went out in the Spring of 2003 and the response was enthusiastic.

The title of this special volume might be somewhat misleading. The words "population dynamics" have been interpreted broadly to include evolutionary dynamics, metapopulation studies, multi-species interactions, and community ecology. The broad range of topics addressed in this special volume is a reflection of the diverse ways in which experimental microcosms are being used to study ecological problems.

The main purpose of this book is to highlight some of the contributions laboratory studies are making to our understanding of the dynamics of ecological systems. The table of contents is not a comprehensive list of the various systems of laboratory microcosms being employed in ecological studies, but it is a good representative sample. More importantly, these chapters provide excellent case studies for the rationale of using laboratory studies to investigate ecological principles.

The book begins with a chapter by Mike Bonsall and Mike Hassell on the role of laboratory systems for understanding ecological concepts. Chapters 2–10 review progress made using a variety of laboratory systems including bruchid beetles and parasitoids, *Drosophila*, flour beetles, soil mites, acarine predators and prey, algal-rotifer chemostats, protozoans, bacteria, and viruses. The final chapter by Marc Cadotte, James Drake and Tadashi Fukami returns to the theme of laboratory systems as ecological tools.

This book addresses fundamental questions that are of interest to ecologists and evolutionary biologists whether they work in the laboratory or field or whether they are primarily empiricists or theorists. While theoretical models appear in several of the chapters, the mathematics and statistics are at a level where the book should be accessible to biologists with broad training. Given the range of opinions among ecologists surrounding the utility of laboratory microcosms, this book might also be a good focus for a graduate seminar in ecology.

This special volume has benefited from the contributions of a large number of individuals. I would like to give a special thanks to my "Beetle Team" colleagues, Bob Costantino, Jim Cushing, Brian Dennis, Shandelle Henson, and Aaron King. They were involved in the early planning stages of this book and have been very supportive of this project. I would also like to give a special thanks to Hal Caswell who extended the initial invitation to the Beetle Team and refused to let the project fade to oblivion. The authors of the contributed chapters and I owe our gratitude to numerous colleagues who agreed to review the original manuscripts and provide suggestions for improvement: Tim Benton, Ottar Bjornstadt, Cherie Briggs, Adam Chippindale, Bob Costantino, Santiago Elena, Jeremy Fox, James Fry, Kevin Higgins, Marcel Holyoak, Jennifer Hughes, Aaron King, Sharon Lawler, Peter Morin, Owen Petchey, Susi Remold, Pejman Rohani, Steve Sait, Karen-Beth Scholthof, Philip Warren and John Wertz. Finally, I would like to acknowledge the U.S. National Science Foundation for their grant support.

REFERENCES

Andrewartha, H.G. and Birch, L.C. (1954) *The Distribution and Abundance of Animals*. The University of Chicago Press, Chicago, Illinois.

Bjornstad, O.N., Sait, S.M., Stenseth, N.C., Thompson, D.J. and Begon, M. (2001) Coupling and the impact of specialised enemies on the dimensionality of prey dynamics. *Nature* **401**, 1001–1006.

Carpenter, S.R. (1996) Microcosm experiments have limited relevance for community and ecosystem ecology. *Ecology* **77**, 667–680.

Costantino, R.F., Desharnais, R.A., Cushing, J.M. and Dennis, B. (1997) Chaotic dynamics in an insect population. *Science* **275**, 389–391.

Crombie, A.C. (1946) Further experiments on insect competition. *Proc. Royal Society (London) B* **133**, 76–109.
Daehler, C.C. and Strong, D.R. (1996) Can you bottle nature? The roles of microcosms in ecological research. *Ecology* **77**, 663–664.
Drake, J.A., Huxel, G.R. and Hewitt, C.L. (1996) Microcosms as models for testing and generating community theory. *Ecology* **77**, 670–677.
Ellner, S.P., McCauley, E., Kendall, B.E., Briggs, C.J., Hosseini, P.R., Wood, S.N., Janssen, A. and Sabelis, M.W. (2001) Habitat structure and population persistence in an experimental community. *Nature* **412**, 538–542.
Fukami, T. and Morin, P.J. (2003) Productivity-biodiversity relationships depend on the history of community assembly. *Nature* **424**, 423–426.
Fussmann, G.F., Ellner, S.P., Shertzer, K.W. and Hairston, N.G., Jr. (2000) Crossing the Hopf bifurcation in a live predator-prey system. *Science* **290**, 1358–1360.
Gause, G.F. (1934) *The Struggle for Existence*. Williams and Wilkins, Baltimore. Reprint 1971, Dover, New York.
Henson, S.M., Costantino, R.F., Cushing, J.M., Desharnais, R.A. and Dennis, B. (2001) Lattice effects observed in chaotic dynamics of experimental populations. *Science* **294**, 602–605.
Huffaker, C.B. (1958) Experimenal studies on predation. *Hilgardia* **27**, 343–383.
Karevia, P. (1989) Renewing the dialogue between theory and experiments in population ecology. In: *Perspectives in Ecological Theory* (Ed. by J. Roughgarden, R.M. May and S.A. Levin), pp. 68–88. Princeton University Press, Princeton, New Jersey, USA.
Kassen, R., Buckling, A., Bell, G. and Rainey, P.B. (2000) Diversity peaks at intermediate productivity in a laboratory microcosm. *Nature* **406**, 508–512.
Kerr, B., Riley, M., Feldman, M. and Bohannan, B.J.M. (2002) Local dispersal promotes biodiversity in a real life game of rock-paper-scissors. *Nature* **418**, 171–174.
MacArthur, R. (1972) *Geographical Ecology*. Harper and Row, New York, USA.
McCauley, E., Nisbet, R.M., Murdoch, W.W., DeRoos, A.M. and Gurney, W.S.C. (1999) Large-amplitude cycles of *Daphnia* and its algal prey in enriched environments. *Nature* **402**, 653–656.
Nicholson, A.J. and Bailey, V.A. (1935) The balance of animal populations, Part I. *Proceed. Zool. Soc. London* **3**, 551–598.
Park, T. (1939) Analytical population studies in relation to general ecology. *American Midland Naturalist* **21**, 235–255.
Park, T. (1948) Experimental studies of interspecies competition. I. Competition between populations of the flour beetles, *Tribolium confusum* Duvall and *Tribolium castaneum* Herbst. *Ecological Monographs* **18**, 267–307.
Petchey, O.L., McPhearson, P.T., Casey, T.M. and Morin, P.J. (1999) Environmental warming alters food-web structure and ecosystem function. *Nature* **402**, 69–72.
Turner, P.E. and Chao, L. (1999) Prisoner's dilemma in an RNA virus. *Nature* **398**, 441–443.
Yoshida, T., Jones, L.E., Ellner, S.P., Fussmann, G.F. and Hairston, N.G., Jr. (2003) Rapid evolution drives ecological dynamics in a predator–prey system. *Nature* **424**, 303–306.

Robert A. Desharnais
California State University, Los Angeles

Foreword: Message in a Bottle

I work at an institution devoted to the study of the oceans. One edge of one of those oceans is across the street from my office. From there, it goes on to cover three quarters of the planet. Big as it is, there are people here who study ocean circulation in laboratory experiments smaller than a bathtub. They get away with this because the laws governing the behavior of a drop of water are the same no matter where that drop happens to be, and because there exist well-defined scaling relationships that relate the results of those laws in a bathtub with their results in an ocean.

If you can study the ocean in a bathtub, what about studying populations in bottles (or cages, aquaria, greenhouses, chemostats, climatrons, ecotrons, etc.)? Although the scaling problem is more challenging, it is not impossible. The dynamics of such a bottle population are determined by births, deaths, immigration, and emigration, just as for any population. Inside the lab or out of it, the birth, death, immigration, and emigration of individuals depend on age or developmental stage, and thus reflect the life cycle of the species. Inside the lab or out of it, these processes are set by the interaction of the life cycle with the biotic environment (resources, competitors, predators, parasites) and the physical environment (including both physical conditions like temperature and aspects of the physical layout, such as size, shape, and whether containers are open or closed).

What makes an experiment possible is the ability, in the lab, to control aspects of the biotic and physical environment. What makes an experiment *interesting* is the investigator's cleverness in setting up the environment to address interesting ecological questions. Robert MacArthur (1972), in an oft-quoted but usually misinterpreted passage, criticized the tradition of bottle experiments on competition not because they were bottle experiments, but because he felt that they had missed the interesting questions. Instead of more documented instances of competitive exclusion, he wanted experimentalists to study coexistence. Instead of asking whether the competitive exclusion principle held in yet another homogeneous environment, he called for studies of how species could coexist in heterogeneous environments. Studies that addressed this (e.g., Crombie, 1946; Huffaker, 1958; Pimentel *et al.*, 1963) have rightly become classics.

Ecologists seem beset with epistemological insecurities that, all too often lead to attacks on one or another approach to science, followed by defenses and counter-attacks. I was preparing to follow this tradition and write a defense of laboratory population experiments. But that seems a waste of time. It is *obvious* that field observations, comparative observations (so-called "natural experiments"), field experiments, lab experiments, and mathematical models (and others I haven't thought of right now) are all capable, in any given circumstance, of providing insight and understanding. Some will, in any given circumstance, be more powerful than others.

Theoretical models have important parallels with experiments (in the lab or the field; Caswell, 1988). Both approaches rely on artificiality, on exclusion of some factors to focus on others, and a purposeful simplification of reality. An experiment that tries to manipulate every factor, or a model that tries to include every variable, are equally worthless.

The ability to control the environment of laboratory populations gives them an important ability to discern relationships that are obscured in observational studies. The demographer Nathan Keyfitz once wrote a chapter on "how we know the facts of demography" (it can be found in Keyfitz and Caswell, 2005). He considered as an example the question of how the rate of increase affects the proportion of a human population over 65 years of age. (Everyone knows, as he points out, that the correlation is negative.) An observational approach, plotting proportion over 65 against the rate of increase, r, for 18 Latin American countries gave a negative correlation, but one that leaves much variance unexplained. On the other hand, an analysis of the renewal equation for the stable age distribution shows clearly that the proportion over 65 decreases exponentially with increases in r, at a rate that depends on the mean and variance of the ages of individuals over and under 65. Why the difference between the empirical correlation and the theoretical result? Because the observations do not show the relation between r and the proportion over 65, even though that is what is plotted in the figure. Instead, they show the projection, onto the r – proportion plane, of the response of the proportion over 65 to changes in r and in a host of other unmeasured factors, varying in whatever way they happened to vary among the 18 countries examined. The calculation, in contrast, shows the effect of r on the proportion, holding all else constant.[1]

Experiments can accomplish the same thing; by holding other factors constant (or as constant as the ingenuity of the experimenter will allow)

---

[1]This is not to say that the observational pattern may not be interesting in its own right. Such a correlation can answer the question "out of the variability in proportion over 65 among these 18 Latin American populations, how much is due to variation in r?" This is a different question from "all else being equal, how will the proportion over 65 respond to a change in r?" The first is a retrospective question, the second a prospective question (sensu Caswell, 2000). Both are important in the proper context.

they show the results of the manipulated factors alone. This process of discovering relationships is one of the great benefits of lab experiments, one that is quite distinct from hypothesis testing. The most dramatically satisfying experimental programs feature a close integration of mathematical models, experimental design and manipulation, and statistical analysis. The work on *Tribolium* population dynamics by the editor of this volume and his colleagues (Cushing *et al.*, 2003) sets the standard for that kind of integration. The chapters in this volume provide even more examples. This book is important because it collects accounts of some of the best laboratory experiments in population dynamics, on a range of taxa, using a variety of experimental approaches. The messages in these bottles show that laboratory population experiments can ask and answer interesting questions about population dynamics. Laboratory population ecology has come a long way from the early days of Pearl, Park, Gause, and Utida. They would be excited by this book.

REFERENCES

Caswell, H. (1988) Theory and models in ecology: A different perspective. *Ecological Modelling* **43**, 33–44.

Caswell, H. (2000) Prospective and retrospective perturbation analyses and their use in conservation biology. *Ecology* **81**, 619–627.

Crombie, A.C. (1946) Further experiments on insect competition. *Proceedings of the Royal Society London* **B133**, 76–109.

Cushing, J.M., Costantino, R.F., Dennis, B., Desharnais, R.A. and Henson, S.M. (2003) *Chaos in ecology*. Academic Press, New York, New York, USA.

Huffaker, C.B. (1958) Experimental studies on predation: Dispersion factors and predator-prey oscillations. *Hilgardia* **27**, 343–383.

Keyfitz, N. and Caswell, H. (2005) *Applied Mathematical Demography* (3rd edition). Springer-Verlag, New York, NY.

MacArthur, R.H. (1972) *Geographical Ecology*. Harper and Row, New York, New York, USA.

Pimentel, D., Nagel, W.P. and Madden, J.L. (1963) Space-time structure of the environment and the survival of parasite-host systems. *American Naturalist* **97**, 141–167.

Hal Caswell
Biology Department MS-34
Woods Hole Oceanographic Institution
Woods Hole, MA 02543

Contents

Complexity, Evolution, and Persistence in Host-Parasitoid Experimental Systems With *Callosobruchus* Beetles as the Host

MIDORI TUDA AND MASAKAZU SHIMADA

Population Dynamics, Life History, and Demography: Lessons From *Drosophila*

LAURENCE D. MUELLER, CASANDRA L. RAUSER AND MICHAEL R. ROSE

Nonlinear Stochastic Population Dynamics: The Flour Beetle *Tribolium* as an Effective Tool of Discovery

ROBERT F. COSTANTINO, ROBERT A. DESHARNAIS, JIM M. CUSHING, BRIAN DENNIS, SHANDELLE M. HENSON AND AARON A. KING

Population Dynamics in a Noisy World: Lessons From a Mite Experimental System

TIM G. BENTON AND ANDREW P. BECKERMAN

Global Persistence Despite Local Extinction in Acarine Predator-Prey Systems: Lessons From Experimental and Mathematical Exercises

MAURICE W. SABELIS, ARNE JANSSEN, ODO DIEKMANN, VINCENT A.A. JANSEN, ERIK VAN GOOL AND MINUS VAN BAALEN

Ecological and Evolutionary Dynamics of Experimental Plankton Communities

GREGOR F. FUSSMANN, STEPHEN P. ELLNER, NELSON G. HAIRSTON, JR., LAURA E. JONES, KYLE W. SHERTZER AND TAKEHITO YOSHIDA

The Contribution of Laboratory Experiments on Protists to Understanding Population and Metapopulation Dynamics

MARCEL HOLYOAK AND SHARON P. LAWLER

Microbial Experimental Systems in Ecology

CHRISTINE M. JESSUP, SAMANTHA E. FORDE AND BRENDAN J.M. BOHANNAN

Parasitism Between Co-Infecting Bacteriophages

PAUL E. TURNER

Constructing Nature: Laboratory Models as Necessary Tools for Investigating Complex Ecological Communities

MARC W. CADOTTE, JAMES A. DRAKE AND TADASHI FUKAMI

Understanding Ecological Concepts: The Role of Laboratory Systems

MICHAEL B. BONSALL AND MICHAEL P. HASSELL

I. INTRODUCTION

Ecology is a disparate discipline with studies ranging from the highly abstract to the highly applied. Understanding ecological concepts often involves the use of theoretical models, laboratory studies, and/or field studies, all of which exist on a continuum, with laboratory experiments often considered to be somewhere between abstract theoretical models and the field studies of the natural world. Laboratory systems have many appealing features such as ease of culturing, manipulation, minimal sampling difficulties, repeatability, and quantifiable resource supply, but there are more fundamental reasons for using laboratory microcosms.

ADVANCES IN ECOLOGICAL RESEARCH VOL. 37 0065-2504/05 $35.00

DOI: 10.1016/S0065-2504(04)37001-7

Long-term population (time series) studies are the cornerstone for testing hypotheses about the processes and mechanisms of population regulation, limitation, and persistence. However, field studies of population ecology suffer from a double jeopardy. While ecological time series are at the heart of population ecological research, they are often of insufficient length and are unreplicated which makes statistical inferences often weak or even impossible. Laboratory experiments provide one solution to this jeopardy by allowing well-replicated, long-term experiments to be done under controlled environmental conditions. The use of laboratory or model systems in ecology has been widely advocated as a plausible and realistic approach to understanding the processes and mechanisms underpinning ecological systems (Kareiva, 1989; Lawton, 1995).

In this review, we discuss the use of laboratory microcosms in understanding ecological concepts focusing particularly on the ideas central to population ecology. However, our rationale and approach is equally applicable to evolutionary and genetic experiments, and can be more widely applied in the design and implementation of field-based ecological experiments. We begin by outlining the philosophy and rationale for laboratory microcosm experiments. We emphasize the role of appropriate model selection and choice as the underlying approach for integrating ecological theory and experiments. Following this, we review the historical contributions that laboratory experiments have made to our understanding of a number of core ecological concepts. In the final section, we conclude by examining the prospects for the use of laboratory systems in ecological research.

II. RATIONALE

There are some pertinent philosophical prerequisites placed on the rationale for the use of laboratory microcosm experiments. For inference and belief in an ecological phenomenon or particular result, it is essential from the outset that a distinction is made between proof and evidence. Proof shows that if x is so, then y is also so, while evidence shows that given x, y is probably so. The difference between proof and evidence rests in the accumulation of scientific beliefs which has important implications in ecological research. However, both proof and evidence require an x to start, and knowledge ultimately relies on some scientific truths that are known without proof or evidence.

One philosophical view is that we should derive ecological axioms from empiricism (the collation of ecological knowledge). Such empiricism is based on experience; by pooling evidence found through experience, patterns can be described and predicted. Under empiricism, scientific methods and rules are used to identify genuine from spurious patterns. Although the methods may be complex (e.g., analytical models, statistics), the rules are relatively straightforward. Hypotheses are formulated, data are collected and predictions are made. Acceptance is the verdict of further experiences that is eventually generalized to scientific theorems or the derivation of axioms. However, this approach may be too simplistic (even with the concepts of modern mathematics and statistics) as it is more geared towards predictive success than telling the genuine ecological patterns from spurious ones. A second view is that of rationalism, which involves explaining the ecological pattern and finding the mechanistic cause. Such an approach to ecological research often focuses on pragmatic reasons for asking a particular question.

However, it may be preferable to take a different approach with ecological studies and reject the assumptions of both rationalism and empiricism. Rather than passively recording ecological patterns, an active judgment is required to reject or amend a hypothesis. The ideal then becomes one of model evaluation using data and assessing how much a model is 'worth'. For example, how relevant is the model at explaining the ecological phenomenon under scrutiny? Is it simple enough? Is it realistic? By taking this approach of model selection and choice, any discoveries made are only those which reason (defined in terms of statistical power or inference) will finally certify. If laboratory experiments are to contribute to the falsification of hypotheses in strict Popperian fashion, then it is of paramount importance in ecological research that the 'worth of the model' is clearly known. While it is obvious that these philosophical approaches (rationalism, empiricism, Popperian) are not mutually exclusive, it is clear that the use of laboratory microcosms systems or organisms requires a sensible frame of reference against which to base a scientific rationale.

A. Model Choice

One frame of reference is through the use and selection of the most appropriate model for data. This is the model that best approximates the data and allows inference on the most likely ecological effects (model parameters) supported by the data. Models are an approximation, and understanding the methods of model choice and selection are of overriding importance. While advocated as a robust alternative to standard Popperian hypothesis testing (Burnham and Anderson, 2002), the application of model selection approaches (e.g., through the use of information-theoretic criteria) remains relatively under-utilized in ecology. In this chapter, we advocate fitting different explanatory models under the principles of parsimony (that is, the model should be as simple as possible) and place less emphasis on merely testing null hypotheses. Given the advancement in ecological theory (May, 1981; Gurney and Nisbet, 1998; McGlade, 1999; Murdoch *et al.*, 2003a), mathematical models abound in ecology and, in particular, in population

Table 1 Candidate models used to describe density-dependent processes in single-species populations (May and Oster, 1976, Bellows, 1981). Models are fitted using a Gaussian likelihood (equation 5) to time-series data. The most parsimonious description of the dynamics is based on the log-likelihood values by evaluating AIC (equation 1). Model comparison is made by computing the difference between the most parsimonious model (model 2) and the ith model

	Model	AIC	$\Delta_i\ (AIC_i - AIC_{min})$
1	$dN/dt = r \cdot N_t - b \cdot log(N_t) \cdot N_t$	2566.438	5.516
2	$dN/dt = r \cdot N_t - log(1 + a \cdot N_t) \cdot N_t$	2560.922	-
3	$dN/dt = r \cdot N_t - a \cdot N_t \cdot N_t$	2611.352	50.43
4	$dN/dt = r \cdot N_t - (a \cdot N_t^b) \cdot N_t$	2566.952	6.03
5	$dN/dt = r \cdot N_t - b \cdot log(1 + a \cdot N_t) \cdot N_t$	2582.214	21.292
6	$dN/dt = r \cdot N_t - log(1 + (a \cdot N_t)^b) \cdot N_t$	2570.676	9.754

ecology. As an illustration of this, a range of continuous-time models for density dependence are listed in Table 1. These represent a variety of ecological processes and mechanisms that may underpin the dynamics of a single-species population. However, the link between these candidate models and empirical data remains relatively weak. Although much is known about these models and laboratory experiments on density dependence (see section titled Density Dependence), there has been little development in assessing which model (density-dependent processs) may best describe an ecological system.

In attempting to find the model that best describes or approximates an ecological problem (data), then in some sense it is most appropriate to consider the model that minimizes information loss (Burnham and Anderson, 2002). This is most straightforwardly achieved through the use of information-theoretic criteria, such as the Akaike Information Criterion (AIC) (Akaike, 1973, 1974). This criterion is based on the likelihood of each model given the data ($L(model|data)$) and is a bias-corrected (in terms of the number of parameters estimated from the data) maximum log-likelihood value:

$$AIC = -2 \cdot log\Big(L(model|data)\Big) + 2 \cdot np, \tag{1}$$

where np is the number of parameters estimated in the model. Of critical importance in model choice is obtaining a small set of candidate models that capture the ecological problem at hand. Once an appropriate set of models have been chosen, competing models (hypotheses) can be ranked by determining relative AIC values (based on the difference to the minimum AIC value). These methods of model choice and selection using information-theoretic approaches are an extension of simple parameter estimation optimization problems (e.g., minimizing the log-likelihood) (Burnham and Anderson, 2002) and are amenable to development within the scope of ecological research problems. Within these schemes, methods now exist that allow selection, uncertainty and inference from multiple models to be considered. It is often the case that more that one theoretical ecological model adequately explains a problem. Within the information-theoretic paradigm, approaches exist for examining multiple models and making formal inference from more than one process. Determining the weighted evidence of one model over another (similar to likelihood ratio tests) has been widely advocated as a robust, simple, and useful approach for providing information on contrasting (ecological) effects (Burnham and Anderson, 2002).

To illustrate the rationale of model selection as a plausible approach for inferring ecological processes and mechanisms, we consider the density-dependent models in Table 1 as the candidate models for the population dynamics of *Drosophila ananassae* illustrated in Fig. 1A. To explain the approach, we take a straightforward description of a density-dependent interaction:

$$\frac{dN}{dt} = r \cdot N_t \cdot \left(1 - \frac{N_t}{K}\right) - d \cdot N_{t,} \tag{2}$$

where r is the population rate of increase, K is the population carrying capacity, and d is the natural mortality rate. To determine the expected number of individuals at the next census point ($E_{n[i]}$) in the time series, this equation is solved over a fixed time period ($t \rightarrow \mathrm{t} + \tau$) corresponding to the census interval (τ):

$$E_{n[i]} = \int_t^{t+\tau} \left[r \cdot N_t \cdot \left(1 - \frac{N_t}{K}\right) - d \cdot N_t \right] \cdot dx. \tag{3}$$

In order to contrast different density-dependent mechanisms, we use a maximum likelihood-based approach. In order to implement this, we need to give explicit consideration to the role of stochasticity on the dynamics. Given that the abundances of *D. ananassae* are estimates of adult flies alive at each census point, we assume that the variability in population size is due to a general environmental component of noise (rather than any particular demographic process). This is a straightforward way to capture a variety of noise processes and the different forms of stochasticity and their statistical implementation are outlined more fully elsewhere (Bonsall and Hastings, 2004). Under environmental noise, changes in population size between successive time intervals occur due to the deterministic processes of births and deaths, and the effects of random noise (acting additively on a log-scale) (Dennis *et al.*, 1995). The stochastic version of the logistic density-dependent model (equation 3) under environmental noise is:

$$E_{n[i]} = exp(v) \cdot \int_t^{t+\tau} \left[r \cdot N_t \cdot \left(1 - \frac{N_t}{K}\right) - d \cdot N_t \right] \cdot dx, \tag{4}$$

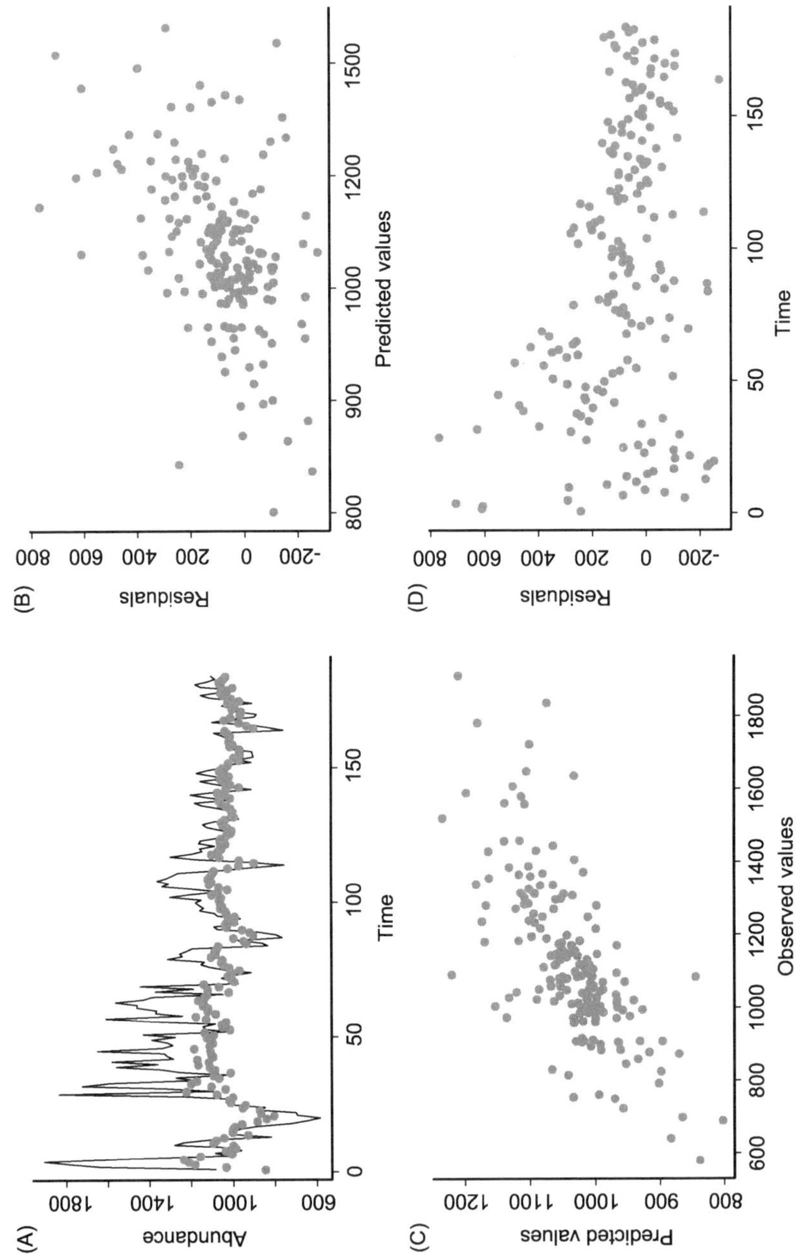
(A)
Abundance
1800
1400
1000
600
0
50
100
150
Time
(B)
Residuals
800
600
400
200
0
-200
800
900
1000
1200
1500
Predicted values
(C)
Predicted values
1200
1100
1000
900
800
600
800
1000
1200
1400
1600
1800
Observed values
(D)
Residuals
800
600
400
200
0
-200
0
50
100
150
Time

where v is an independent, identically distributed random variable with mean 0 and variance σ^2. It is assumed that the autocovariances between successive time points are small (autocorrelations are relatively weak) and as such are assumed to be uncorrelated. The appropriate likelihood is based on a univariate Gaussian probability distribution (Johnson *et al.*, 1994) and is of the form:

$$L(F) = \prod_{i=1}^{z} \frac{1}{\sqrt{2\pi}} \cdot exp\left[-\frac{1}{2}\left(\frac{N_{[i]} - E_{n[i]}}{\sigma}\right)^2\right] \tag{5}$$

where $N_{[i]}$ is the observed numbers of individuals at each census point i and $E_{n[i]}$ is the expected number of individuals (determined from the integrated logistic model, equation 3) at each census point i. Given that the autocovariances are small, the overall likelihood is simply the product over all census points (z). Maximum likelihood estimates of the parameter set (F) are those that minimize the negative log-likelihood of equation 5. To compute this likelihood, an optimization method incorporating a numerical integration routine is implemented (Bonsall and Hastings, 2004).

As outlined in the previous text, models for the population dynamics can be contrasted and ranked by rescaling AIC values by simple differences (Burnham and Anderson, 2002). For this time series (Fig. 1A), Skellam's model (Skellam, 1951) has the most support for explaining the dynamics. Two models (Varley and Gradwell, 1960; Bellows, 1981) with intermediate AIC values ($\Delta_i = 4 - 9$) have some support while three models (Ricker, 1954; Maynard-Smith and Slatkin, 1973; Hassell, 1975) have no support for explaining the population dynamics (with $\Delta_i > 10$). Skellam's model is of the form:

$$\frac{dN}{dt} = r \cdot N_t - log(1 + a \cdot N_t) \cdot N_t \tag{6}$$

where r is the population growth rate and $1/a$ is a measure of the threshold above which competition for resources occurs. This form of intraspecific

Figure 1 Population dynamics of *Drosophila ananassae*. *D. ananassae* has a life-cycle of about 8 days, populations estimates were sampled every 3.5 days and, in chronological order, 4 resource patches (apple-based media) out of 20 were renewed every week. Model fitting and selection criteria reveal that the dynamics of *D. ananassae* are influenced by a density-dependent process and information-theoretic model selection criteria reveal that the functional form for the density dependence is $\mu(N) = log(1 + 0.001 \cdot N_t)$. (A) Population dynamics (solid line) and one step-ahead predictions (solid circles) from the best-fitting model. (B to D) Post-model selection diagnostics show the (B) predicted values versus residuals, (C) correlation between observed and predicted abundances ($\rho = 0.719$), and (D) residuals versus time. These diagnostics confirm that the assumptions about the error structures are appropriate.

competition suggests that the ecological mechanism underpinning the dynamics of *D. ananassae* is determined by a density-dependent process operating above a threshold density of about 1,000 flies. To ensure that the assumptions about the errors are appropriate, goodness of fit criteria (standardized residual plots, model one-step ahead prediction) are estimated for the likelihood function (equation 5). For the most parsimonious model of *D. ananassae* dynamics (equation 6), model selection diagnostics are illustrated in Fig. 1B to D.

Model choice (based on a maximum-likelihood approach, information-theoretic criteria, and post-selection diagnostics) motivates a departure from simple hypotheses testing for contrasting ecological effects. This holds the potential for exciting developments in population ecology through the combined, iterative approach of theoretical model predictions, manipulative and observational experiments, and robust statistical assessment and inference.

B. Simplicity

Laboratory experiments and microcosms are often cited as simple analogs of nature where the ability to control environmental variables allows key theories to be tested. However, simplicity is too often equated with lack of relevance. Ecological laboratory systems have clear relevance and are not simply caricatures of nature. One way in which complexity can be incorporated into laboratory systems is by developing microcosms in which modules (small number of species interactions) can be assembled into larger groups of interacting species (Neill, 1975; Naeem *et al.*, 1994; Weatherby *et al.*, 1998). Such systems are not simple. Complicated environments, the availability of resource patches, and spatial heterogeneity all introduce levels of complexity into laboratory experiments (e.g., Huffaker *et al.*, 1963) that allow the ecological detail of nature to be approximated. Even one of the central practicalities, the control over environmental variables, can be manipulated in laboratory experiments to test the complex role of stochasticity and abiotic flucations on species interactions.

Laboratory experiments are not simple analogs of nature nor are they simple extensions of theoretical models. They provide a robust and rational approach for ecological research, provided it is accepted that they cannot be conducted on inappropriate scales of space or time, that one does infer from them experimental response of manipulating entire ecosystems (Carpenter *et al.*, 1995) and that they are not simply extrapolated to the field.

C. Risk and Reward

Advancement in scientific understanding involves combining risk and reward. Risk can be defined as the likelihood that the objectives or aims of a study will fall short of expectations. Such failure might arise because of the inappropriateness of the methods: poor selection of model, organisms, or question. Risk may also be expressed in terms of the (ecological) costs of the experiment. Such costs might arise through the manipulation of rare or endangered habitats. While the risks associated with ecological experiments are real, laboratory experiments allow risks to be minimized and reward maximized through the ability to manipulate, replicate, and infer on particular research questions.

For example, whole ecosystem experiments are often undertaken to assess a response rather than to falsify hypotheses. Such perturbations could be considered to have a high element of risk. The lack of replication limits the power that such experiments have in the comparison of models or competing hypotheses. Laboratory systems might circumvent some of these problems by allowing appropriate models and hypotheses to be tested and/or compared. However, even laboratory microcosms are not without risk. For example, laboratory-based systems are often used to explore the long-term dynamics of populations (e.g., Nicholson, 1954; Costantino *et al.*, 1995, 1997) and while the level of replication and manipulation extend the statistical power of such experiments, problems of methodology and inference still remain. For instance, statistical inference from population time series is often based on the measures of covariance such as autocorrelation and partial autocorrelation (e.g., Turchin, 1990). It is argued that patterns in these correlations are indicative of exogenous or endogenous dynamics and provide ideas about the period of oscillations (Nisbet and Gurney, 1982). However, poor experimental design (e.g., resource renewal regimes on a period equivalent to the generation time of the organisms) or weak statistical inferences can bias the interpretation of such data and introduces elements of risk into laboratory systems. Risk and failure through the poor selection of models, organisms, or questions have clear implications for how we approach ecological research questions with laboratory microcosms. In the next section, we review how a number of ecological concepts have been advanced by judicious choice and implementation of particular laboratory systems.

III. CONCEPTS IN POPULATION ECOLOGY

Several concepts in population ecology have been shaped by the interplay between theoretical model development and experiments. In this section, we give an eclectic history of four broad ecological concepts (density dependence, competition, predation, and stability-complexity). For each concept, we introduce the core ecological theory and highlight how laboratory experiments and microcosms have been used to advance each area. In general, we illustrate how laboratory microcosms have been used in testing hypotheses rather than comparing contrasting models. However, we use examples from a recent series of experiments to illustrate how these two approaches can be combined to provide a more rigorous understanding of a particular ecological problem and, more generally, demonstrate the rationale for laboratory experiments in ecological research.

A. Density Dependence

Five broad concepts can be considered to be at the center of population ecology. These are: 1) population growth, 2) population equilibrium, 3) limitation, 4) regulation, and 5) persistence. Population growth is defined as the change in population size from one generation to the next. A population is at equilibrium if it does not change in size over time. Limitation is the process that sets the equilibrium, and regulation is the process by which a population returns to its equilibrium. Regulatory processes act in a density-dependent manner, and long-term population persistence requires that density-dependent processes operate. It is clearly evident that density-dependent processes can affect each of these concepts. Quite simply, density dependence is an increase in mortality or reduction in natality as population density increases.

B. Nicholson's Concepts

Using studies from laboratory populations on *Drosophila* (Pearl, 1928) and *Tribolium* (Chapman, 1928; Holdaway, 1932), Nicholson (1933) conceptualized the idea of regulation through density-dependent processes. Particularly influential in Nicholson's ideas of regulation was the work on *Tribolium* by Holdaway (1932). Holdaway illustrated that *Tribolium* populations reached an equilibrium and persisted about this equilibrium. He argued that it was the interaction between the insects themselves that acted to maintain this persisting population, while the combination of physical environmental factors limited the population. Developing this theme, Nicholson lucidly argued that density dependence must be manifest through the effects of intraspecific competition since 'if the severity of its [competition] action against an average individual increases as the density of animals increases, the decreased chance of survival, or of producing offspring, is clearly brought about by the presence of more individuals of the same species in the vicinity' (Nicholson, 1933). This theme has been the focus of a wide range of laboratory population-dynamic experiments to explore the ideas of density dependence and population regulation.

To corroborate the conceptual ideas of density dependence, Nicholson conducted a series of laboratory experiments using *Lucilia cuprina*

(Nicholson, 1954, 1957). By defining concepts such as 'scramble' and 'contest' as various forms of intraspecific competition, Nicholson highlighted that these two mechanisms might have different consequences for the dynamics of a population. *Lucilia cuprina* was shown in a separate experiment to experience scramble competition such that some, and at times all, of the resources were insufficient to support the individuals in the population. Nicholson's population dynamic experiments (under different feeding regimes) showed that large amplitude cycles occurred rather than regulation to an equilibrium through density-dependent processes. In contrast to a recent criticism of this interpretation (Mueller and Joshi, 2000), these findings are quite succinctly defined in terms of Nicholson's mechanisms of intraspecific competition. Observations on the relatively high fecundity of *L. cuprina*, the time-delay in the effects of adult recruitment, and the fact that adult recruitment was highly non-linear, may actually be more precursory to developments in non-linear population dynamics rather than clarifying the theory associated with regulation and density dependence. Advances in non-linear population dynamics have been motivated by detailed analysis of laboratory microcosm systems. For example, in an investigation of non-linear population dynamics, Hassell *et al.* (1976a) examined the role of density dependence in the regulation of 28 insect populations. Of these, several were derived from laboratory studies. Using non-linear least-squares methods to estimate parameters from a modified Ricker model for population growth (Hassell, 1975), it was demonstrated that most insect populations were predicted to show stable dynamics. As a test of this model, Hassell *et al.* (1976a) used data from laboratory populations of *Callosobruchus chinensis* showing monotonic, damped oscillations, or two-point limit cycles to validate their predictions. Both the monotonic and damped populations appeared in the appropriate regions of a stability plane while the cycling population was on the margins of the appropriate stability zone. More interestingly, by using Nicholson's blowfly data, Hassell *et al.* (1976a) predicted that the model parameters for this population should show chaotic (seemingly random) population fluctuations.

C. Chaos

Chaos is a phenomenon that has attracted a wealth of interest in population ecology (May, 1974; Schaffer, 1984; Hastings *et al.*, 1993). Biologists have shown that the strong non-linear negative feedback (due to density-dependent processes) in ecological populations have the potential to generate chaotic (apparently random) dynamics. For example, in a number of theoretical models it has been demonstrated that single populations show seemingly chaotic dynamics (May, 1974, 1976; May and Oster, 1976).

Laboratory experiments have been truly instrumental in shaping our understanding of these non-linear processes. Analysis of Nicholson's blowfly data has been worked and re-worked as a test of chaotic or limit-cycle dynamics (May, 1973; Gurney *et al.*, 1980; Ellner and Turchin, 1995). Currently, with the availability of a range of non-linear statistical techniques, most laboratory populations have been shown not to have chaotic dynamics (Ellner and Turchin, 1995). However, the dynamics of Nicholson's blowflies remain a moot point: a realistic population model of these dynamics has a positive Lyapunov exponent (a statistic that asks whether in the long run two initially nearby points converge or diverge). This is indicative of chaotic dynamics (Ellner and Turchin, 1995). These dynamics of *L. cuprina* first studied by Nicholson to investigate the nature of equilibrium dynamics and regulation remain a focus of contemporary research in non-linear population dynamics.

More extensive laboratory experiments have been undertaken to explore the nature of non-linear dynamics and chaos using *Tribolium castaneum*. Desharnais, Costantino, Dennis, and colleagues have explored the non-linear dynamics of these *Tribolium* populations (Desharnais and Liu, 1987; Costantino *et al.*, 1995, 1997; Dennis *et al.*, 1995, 2001) by developing a single theoretical model of the age-structured beetle populations, using robust statistical analogs of this model and long-term laboratory experiments. By manipulating key demographic processes (e.g., adult survival) of the beetles (predicted to change the population dynamics from analysis of the age-structured model), the dynamics of *Tribolium* were monitored. In an initial series of replicated experiments, manipulation of adult death rate changed the population dynamics from stable dynamics to periodic cycles (Costantino *et al.*, 1995). In a further set of replicated experiments, increments in adult recruitment were shown to shift the dynamics of the populations from equilibrium to cycles to chaos (Costantino *et al.*, 1997). Based on the initial approach, Hassell *et al.* (1976a) and more recent developments of fitting single models to ecological data (Turchin and Taylor, 1992; Dennis and Taper, 1994; Dennis *et al.*, 1995, 2001), there remains the potential for extending the rationale of model choice, fitting, and selection to understanding the dynamics of density-dependent feedback for a range of species under different ecological scenarios.

D. Competition

1. Interspecific Competition

Originally formulated as a theoretical problem (Lokta, !925; Volterra, 1926), the central ideas of interspecific competition have principally been shaped by laboratory experiments. The effects of interspecific competition were first illustrated by Gause (1933) who used yeast cells. It is clear that Gause chose the confines of the laboratory to focus on the ecological nature of the interaction between competitors in a controlled environment. In a series of controlled experiments using yeast cell populations (*Saccharomyces cerevisiae* and *Schizosaccharomyces kephir*), Gause investigated the mechanisms of both intra- and interspecific competition by combining a series of theoretical ideas (logistic population growth, Lokta-Volterra competition) with a set of carefully designed experiments. By estimating the coefficients of interspecific competition between *S. cerevisiae* and *S. kephir*, Gause demonstrated that the effects of interspecific competition can, as theory predicted, inhibit the population growth of the two species. Further, from these experiments, it was shown that the effects of interspecific competition is remarkably asymmetric: *S. kephir* has much more of a dramatic effect on the growth of *S. cerevisiae* than the reciprocal effect. In fact, *S. kephir* decreases the opportunity for growth of *S. cerevisiae* by a factor of three. This phenomenon of asymmetries in competitive interactions has wide appeal and has been observed in a number of additional laboratory and field experiments (Lawton and Hassell, 1981) and, more recently, has been extended to more complex species interactions (Chaneton and Bonsall, 2000).

In a separate series of experiments, Gause extended these ideas of interspecific competition using Protozoa. In particular, experiments were designed to mimic the renewal of energy in a competitive system. Although experiments on Protozoa had preceded Gause's work (e.g., Woodruff, 1914; Eddy, 1928), by designing carefully controlled experiments, Gause was to bring rigor to experimental population ecology by reducing a system to a number of understandable components. First, Gause corroborated that the logistic growth patterns also underpinned the population growth of the Protozoa populations. Second, the effects of interspecific competition were clearly demonstrated in a system where the energy inputs were carefully controlled. Finally, the amount of energy obtained by each species was evaluated and used to predict the outcome of the competitive interaction. From these experiments, theories have been developed and extended into the quantitative rules and predictions on the outcome of competition, such as the competitive exclusion hypothesis (Hardin, 1960) and energy utilization (R*) rules (Tilman, 1982, 1988).

Using a similar series of laboratory experiments, Park (1948) demonstrated that the effects of interspecific competition can be manifest in flour beetle (*Tribolium*) cultures. Using cultures of *Tribolium castaneum* and *Tribolim confusum* the population dynamics of single and interspecific competitive interactions were studied over a four-year period. Single-species populations established equilibrium densities for the duration of the study. In competition, one of the species of *Tribolium* was driven to extinction. Typically, *T. confusum* was lost from the interaction. Park demonstrated that the outcome of competition can also be mediated by the effects of a non-discriminant parasite. In the presence of the microsporidia, *Adelina tribolii*, the outcome of the competitive interaction was switched and typically *T. castaneum* was driven to extinction. This effect of a third species on two potential competitors is a class of indirect interactions that involves the population dynamics of multiple species.

2. Apparent Competition

More specifically, the class of indirect interactions in which two species that do not compete for resources share a common natural enemy is known as apparent competition. Theoretically, it is known that this indirect interaction can lead to the loss of one species due to the numerical and functional response of the natural enemy. For example, consider the consequences of the invasion of a second prey species into an established predator-prey interaction. The availability of alternative prey in the diet of a predator leads to an increase in the size of the predator population. With a larger predator population there is potential for more attacks on both prey species and if sustained over time, the species that suffers the higher number of attacks or has the lower growth rate is eliminated. The natural enemy is then a dynamic monophage (Holt and Lawton, 1993): through the dynamical interaction with its prey, the predator persists with a single prey due to the consequences of apparent competition. Although the theoretical consequences of apparent competition are relatively well-established (Holt, 1977, 1984; Holt and Lawton, 1993, 1994; Bonsall, 2003; Bonsall and Holt, 2003), recent laboratory experiments have demonstrated that apparent competition has the potential to shape the structure of multispecies predator-prey interactions.

Although the single-generation cohort effects of apparent competition have been largely understood in predator-prey systems through laboratory and field observations (Holt and Lawton, 1994), only relatively recently have laboratory experiments taken a more central role and been used to explore the long-term transgenerational effects of apparent competition. Using a laboratory system of a simple insect host-parasitoid assemblage (*Plodia interpunctella, Ephestia kuehniella, Venturia canescens*), Bonsall and Hassell (1997, 1998) demonstrated that the parasitoid, *V. canescens* can induce apparent competition between two of its hosts (*Plodia and Ephestia*). Replicated time series were established in which the separate pairwise predator-prey interactions (*P. interpunctella-V. canescens* and *E. kuehniella-V. canescens*) were shown to always persist. Time series analysis revealed that the dynamics of these pairwise interactions were described by simple linear statistical models (Bonsall and Hassell, 1998). The full three-species interaction, in which the two hosts were not allowed to compete for resources and parasitoid foraging on both species was not restricted, did not persist. *E. kuehniella* was always lost from the interaction and *V. canescens* became a dynamic monophage through its population dynamic actions on the availability of multiple prey. Although the empirical tests of competition and apparent competition have been motivated by theory, testing contrasting concepts for these patterns remains a challenge for this area of ecology.

E. Predation

Initially our understanding of the components of predation were shaped by the theoretical models of Lokta (1925), Volterra (1926), Nicholson and Bailey (1935), and Thompson (1924). In testing the ideas that predation can lead to oscillatory or diverging dynamics, the use and application of controlled environment experiments or observations have been exceptionally influential. Nicholson and Bailey (1935) illustrated theoretically that the interaction between an exponentially growing host and a randomly searching parasitoid leads to non-persistent, diverging dynamics due to overexploitation of the host by the natural enemy. Several empirical studies have confirmed that this combination of factors can lead to extinction of predators and/or prey (Burnett, 1958; Huffaker, 1958; May *et al.*, 1974; Bonsall *et al.*, 2002). Understanding the ecological factors that allow predator-prey interactions to persist has been a dominant area of ecological research utilizing a wide range of models and model systems (e.g., Hassell, 1978, 2002; Murdoch *et al.*, 2003a).

1. Components of Predation

One of the fundamental components of predation is the functional response. This is the relationship between the number of prey attacked and prey density (Solomon, 1949). For the three types of functional response (linear, asymptotic, sigmoid) laboratory experiments have had an influential role in determining how these responses affect the population dynamics of predator-prey interactions. In particular, in a series of experiments and model analysis, Lawton and colleagues explored how the components of arthropod predation are shaped by the prey death rate, the effects of prey density on predator development and survival, and the effects of prey density on predator fecundity (Lawton *et al.*, 1975; Beddington *et al.*, 1976; Hassell *et al.*, 1976b). In a similar vein, laboratory experiments have been used to explore the detailed foraging activities of insect predators. In particular,

Cook and Hubbard (1977) examined how insect parasitoids might be expected to forage optimally for hosts. They predicted that the expected rate of prey an insect predator will encounter in a patchy environment should be reduced to the same level in all patches. By developing an optimal foraging model they showed that the search strategy for *Nemeritis (Venturia) canescens* should alter as the density of the parasitic wasp is changed. In an earlier set of experiments, Hassell and Varley (1969) had shown that the behavior of insect predators and in particular parasitic wasps can be clearly influenced by the presence of conspecifics. As predators aggregate around patches of high prey density it is increasingly likely that they will interact while searching for prey. Such an effect has been termed 'mutual interference' and has been observed in the behavior of a number of predators such as *Diaeretiella rapae* (Hassell, 1978), *Trybliographa rapae* (Visser *et al.*, 1999), *Nemeritis canescens* (Hassell, 1971), and *Typhlodromus longipilus* (Kuchlein, 1966). The generic effect of mutual interference is to reduce the available search time in direct proportion to the frequency of encounters. However, not all search behaviors by predators in the presence of patchily distributed prey can be ascribed to mutual interference. Free *et al.* (1977) argued that aggregation of predators to patchily distributed prey can lead to a phenomenon known as 'pseudointerference'. Pseudointerference arises due to the differential exploitation of host patches. Initial aggregation to high-density patches is profitable as a higher proportion of hosts are attacked than would be expected under a random search strategy. However, such behavioral tendencies introduce a density-dependent reduction in searching efficiency that causes an apparent interference relationship. This is now known as pseudointerference. Although a number of empirical studies have highlighted the different role of behavioral interference (Visser *et al.*, 1999), interference is essentially the phenomenon of optimal foraging (Charnov, 1976) extended to predator-prey interactions (Cook and Hubbard, 1977) but with a more direct connection to the role of spatial heterogeneity in predator-prey interactions.

2. *Spatial Heterogeneity*

The effect of heterogeneity where some individuals are more at risk of attack than others is predominant in predator-prey interactions. Heterogeneity can be manifest at a number of different levels such as at the individual level through physiological or genetic differences, at the population level through differences in the local spatial distribution of prey or at the community level by the availability of different prey or habitats. Nevertheless, it is the role of spatial heterogeneity (patchily distributed prey within a habitat) that has attracted considerable attention, and laboratory experiments have been instrumental in shaping our understanding of the role of this ecological process on the persistence of predator-prey interactions.

Most influential in the study of spatial heterogeneity has been the work on host-parasitoid interactions (Bailey *et al.*, 1962; Hassell and May, 1973, 1974; May, 1978). These early theoretical models demonstrated that aggregation by parasitoids to patches of high-host density was potentially a major contributing factor to the temporal stability of the interaction. More recently this has been re-examined in terms of heterogeneity in the distribution of parasitism which, if sufficient, can be strongly stabilizing (Chesson and Murdoch, 1986; Pacala *et al.*, 1990; Hassell *et al.*, 1991a). Such variability in parasitism can arise through behavioral responses by the parasitoid to host density (positively or negatively dependent on host density) or responses unrelated to host density (host density-independent). This variability gives rise to (positive, negative, or independent) spatial patterns of parasitism. It has been argued that these spatial patterns of parasitism are indicative of the temporal stability of the predator-prey interaction (Hassell and May, 1973, 1974); however, rather than the precise pattern, it is the distribution and risk of parasitism between host patches that is more influential in determining the stability and persistence of the predator-prey interaction (Pacala and Hassell, 1991; Hassell *et al.*, 1991a).

Many empirical studies have sought to identify the role of these spatial patterns of parasitism. Of the studies listed in the reviews by Lessells (1985), Stiling (1987), Walde and Murdoch (1988), and Hassell and Pacala (1990), 29% show positive (direct) density-dependent patterns of parasitism, 26% show inverse patterns, and 45% show patterns of parasitism uncorrelated with host density. Discussions on how these patterns might arise has focused on the role of the functional response (Hassell, 1982), parasitoid biology (Lessells, 1985), and the spatial scale of the interaction (Heads and Lawton, 1983). Originally highlighted by Reeve *et al.* (1989) and more recently explored by Gross and Ives (1999), inferring temporal stability from spatial patterns of parasitism is complicated by a number of difficulties, including the way parasitism is distributed within and between patches (Gross and Ives, 1999). A more robust way to test the ideas of spatial heterogeneity on the persistence of host-parasitoid interactions is to examine the population-dynamic consequences of heterogeneity from a series of well-replicated laboratory experiments. Here, we present evidence for the role of spatial heterogeneity on the population-dynamic interaction between *Drosophila ananassae* and its parasitic wasp *Leptopilina victorae*.

In particular, we contrast two different types of patchy environments: resource patches that are equally available for fly and wasp oviposition and resource patches that provide a refuge for the fly from wasp oviposition. Replicated population dynamics of these two different host-parasitoid interactions are illustrated in Figs. 2 and 3. For these population-level data, ecological models describing the predator-prey interaction between *D. ananassae* and *L. victorae* were contrasted (Table 2). In these patchy resource systems, the most appropriate description for the dynamics of the host-parasitoid interaction in both environments is a model incorporating density dependence acting on the parasitoid:

$$\frac{dH}{dt} = r \cdot H_t - k \cdot log\left[1 + \frac{\alpha \cdot P_t}{k}\right] \cdot H_t \tag{7}$$

$$\frac{dP}{dt} = k \cdot log\left[1 + \frac{\alpha \cdot P_t}{k}\right] \cdot H_t - d \cdot P_t \tag{8}$$

where r is the intrinsic rate of increase of the host population (H), k is a measure of the dependence of parasitoid efficiency on parasitoid density such that as k gets smaller, the magnitude of the density-dependent effect increases, α is the parasitoid attack rate, and d is the parasitoid (P) death rate. This form of parasitism introduces density dependence to the population interaction through the processes of pseudointerference and allows persistence of the host-parasitoid interaction. In the no-refuge treatment, parameter values for the dynamics (Fig. 2), and in particular, the strength of parasitoid density dependence predicts that the dynamics of the interaction will be unstable ($k = 4.31$, 95% $CI = 0.095$). In contrast, in the refuge treatment, the dynamics are predicted to be stable since $k < 1$ (May, 1978; Hassell, 2000) ($k = 0.622$, 95% $CI = 0.0004$) (Fig. 3).

3. Predator-Prey Metapopulations

The role of space and spatial dynamics are central to the patterns of the distribution and abundance of species. Although the explicit inclusion of space into population ecological theory is a relatively recent development (Gilpin and Hanski, 1991; Hanski and Gilpin, 1997), a number of early studies postulated that space might affect the persistence of ecological systems (Nicholson and Bailey, 1935; Wright, 1940; Andrewartha and Birch, 1954). That extinction-prone species interactions can persist as a result of metapopulation processes has been frequently demonstrated in models (Allen, 1975; Hassell *et al.*, 1991b, 1994; Comins and Hassell, 1996; Bonsall and Hassell, 2000). However, only rarely has this effect of increased persistence due to spatial processes been observed in experimental microcosms (Huffaker, 1958; Holyoak and Lawler, 1996; Ellner *et al.*, 2001; Bonsall *et al.*, 2002) or in the field (Hanski, 1999).

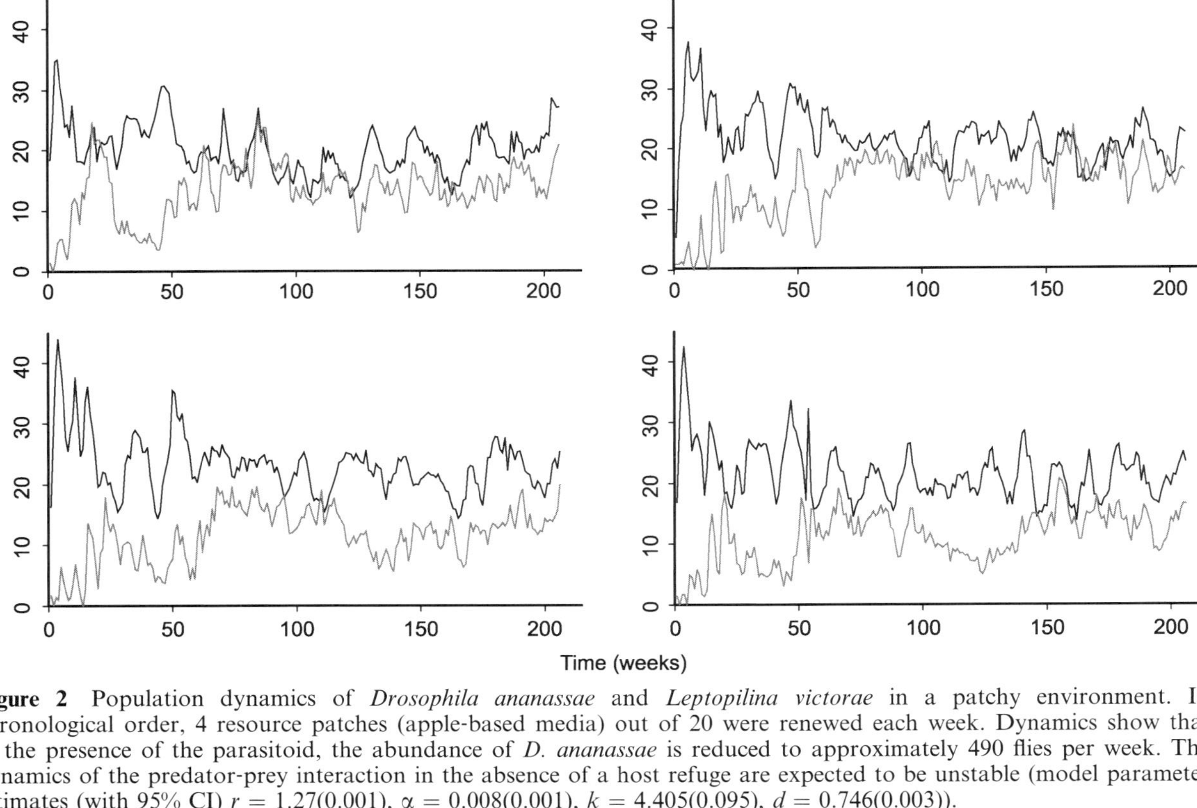

Figure 2 Population dynamics of *Drosophila ananassae* and *Leptopilina victorae* in a patchy environment. In chronological order, 4 resource patches (apple-based media) out of 20 were renewed each week. Dynamics show that in the presence of the parasitoid, the abundance of *D. ananassae* is reduced to approximately 490 flies per week. The dynamics of the predator-prey interaction in the absence of a host refuge are expected to be unstable (model parameter estimates (with 95% CI) $r = 1.27(0.001)$, $\alpha = 0.008(0.001)$, $k = 4.405(0.095)$, $d = 0.746(0.003)$).

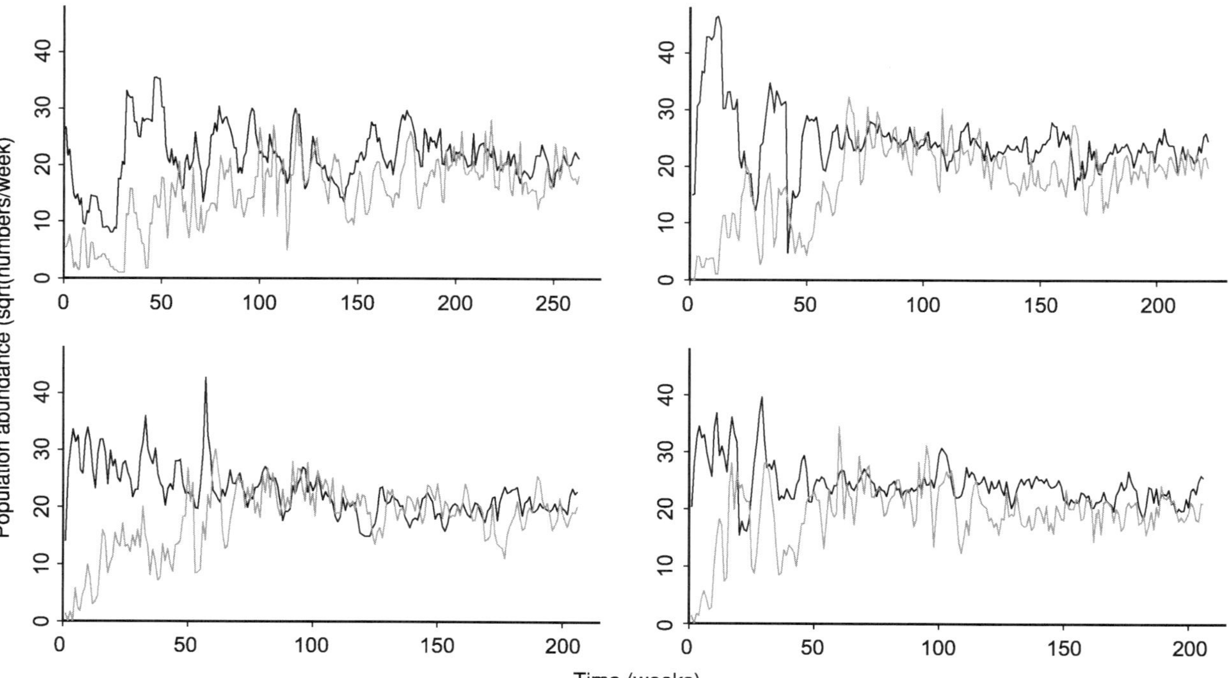

Figure 3 Population dynamics of *Drosophila ananassae* and *Leptopilina victorae* in a patchy environment with a (70%) refuge for the host on all patches. In chronological order, 4 resource patches (apple-based media) out of 20 were renewed each week. Dynamics show that in the presence of the parasitoid, the abundance of *D. ananassae* is reduced to approximately 560 flies per week. The dynamics of this predator-prey interaction in the presence of a host refuge are predicted to show stable dynamics (model parameter estimates (with 95% CI) $r = 0.371(0.0004)$, $\alpha = 0.001(0.001)$, $k = 0.622(0.0004)$, $d = 0.864(0.001)$).

Table 2 Information theoretic (AIC) criteria for three candidate models for describing the overall predator-prey dynamics between *Drosophila ananassae* and *Leptopilina victorae* in two different patchy environments (no refuges, available refuges) (all replicates pooled). Model 1 is the standard Lokta-Volterra model of the form $\frac{dx}{dt} = rx - \alpha xy, \frac{dy}{dx} = c\alpha xy - dy$. Model 2 is a Lokta-Volterra model incorporating a type II functional response $(\frac{dx}{dt} = rx - \frac{\alpha xy}{1+\beta x}, \frac{dy}{dt} = \frac{\alpha xy}{1+\beta x} - dy)$. Model 3 is a Lotka-Volterra model replacing the type II functional response by an expression for density dependence acting on the parasitoid $(k \cdot log[1 + \frac{\alpha y}{k}])$ (May, 1978). Models were fitted using a bivariate Gaussian likelihood to time series data. Comparison of the models for the dynamics of *D. ananassae* and *L. victorae* in two different environments (refuges, no refuges) reveals that the most parsimonious description is the model incorporating parasitoid density dependence (Model 3)

Host-parasitoid Interaction		
	No refuges	Refuges available
Model 1	23,302.442	28,406.886
Model 2	23,342,220	26,430.148
Model 3	21,316.096	24,437.878

In a classic series of experiments, Huffaker (1958; Huffaker and Kennet, 1956; Huffaker *et al.*, 1963) investigated the effects of spatial structure on a laboratory mite predator-prey interaction. Huffaker designed a suite of experiments using oranges as patches in various arrangements (fully exposed oranges, partially exposed oranges) to test ideas of the effects of spatial structure on the persistence of predator-prey interactions. First, Huffaker (1958) showed that the predator-prey interaction could not survive in a homogenous environment without dispersal. Second, to test the role of migration, Huffaker elegantly manipulated the spatial environment (wooden posts to aid dispersal, petroleum jelly to restrict dispersal) and demonstrated an increased persistence time for the system. Finally, in an extension to this work, the experimental system was expanded to include more patches and more environmental complexity (Huffaker *et al.*, 1963). From this series of experiments Huffaker and colleagues showed that increased persistence of the predator-prey interaction was a consequence of the spatial structure and complexity of the system. In a similar series of experiments, Pimental *et al.*, (1963) showed that the persistence of the interaction between *Musca domestica* and *Nasonia vitripennis* and *Lucilia sericata* and *Nasonia vitripennis* could be enhanced by linking patches by dispersal (via tubes). In a series of population-level experiments, it was observed that when host densities were low, wasp dispersal increased. However, when parasitoid dispersal was restricted (via baffles covering the dispersal tubes), persistence time was increased. In contrast to the conceptual predictions of both Andrewartha and Birch (1954) and Nicholson and Bailey (1935), as dispersal of the natural enemy was reduced, the densities of both the wasp and its host increased and the amplitude of the predator-prey fluctuations declined. Pimental *et al.* (1963) argued that the original Nicholson-Bailey model was predicted to captured predator-prey dynamics in simple systems and that their designed laboratory systems may not be simple due to the inclusion of dispersal and baffles.

Recently, the role of metapopulation structures on the persistence of predator-prey interactions have been more extensively tested using laboratory systems (Holyoak and Lawler, 1996; Ellner *et al.*, 2001; Bonsall *et al.*, 2002). Theoretical models predict that non-persistent predator-prey interactions may persist due to the effects of limited dispersal (Hassell *et al.*, 1991b; Comins *et al.*, 1992). The emerging consensus from a series of laboratory experiments on a wide range of organisms is that limited dispersal clearly has a major influence on the persistence of populations. For example, recent work on the interaction between the bruchid beetle *Callosobruchus chinensis* and its parasitic wasp *Anisopteromalus calandrae* has shown that persistence of this extinction-prone host-parasitoid interaction is enhanced by a metapopulation structure (Bonsall *et al.*, 2002). Through a series of replicated experiments the persistence of host-parasitoid interactions in metapopulations with limited and unlimited dispersal was explored. By controlling for the effects of any increased persistence due to the effects of the availability of more resource in the system, persistence of this host-parasitoid interaction was shown to be principally driven by coupling habitats with limited dispersal (Fig. 4). In comparison to other studies on predator-prey metapopulation dynamics, Bonsall *et al.* (2002) demonstrated that spatial structure has a primary influence on the persistence of trophic interactions. However, in a series of controlled experiments using a mite system (*Tetranychus urticae-Phytoseiulus persimilis*), Ellner *et al.* (2001) have shown that habitat structure per se has a limited role in the persistence of a predator-prey metapopulation and it is the reduced probability of attack by the predator that allows the system to persist. These two examples show that by using replicated, designed experiments, laboratory systems permit a number of models on spatial dynamics and persistence of predator-prey metapopulations to be evaluated. For instance, it is now feasible, using the *C. chinensis-A. calandrae* system to explore how the effects of local patch processes (demographic stochasticity) affect the regional dynamics of metapopulations (Bonsall and Hastings, 2004).

F. Stability and Complexity in Ecosystems

The stability-complexity debate was initially sparked by the question whether assemblages with more species are more or less persistent than species-poor assemblages (Elton, 1958; May 1971, 1973). The debate was initially

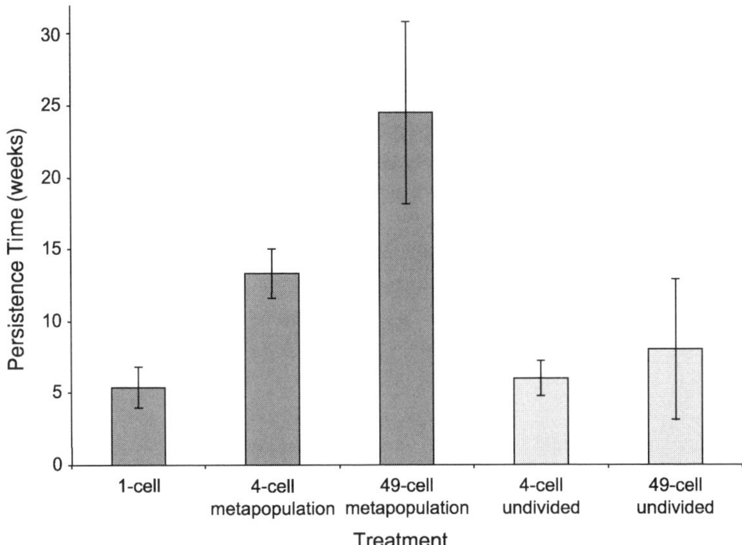

Figure 4 Predator-prey persistence times in two different metapopulation environments (4-cell, 49-cell systems). Persistence time for the interaction between *Callosobruchus chinensis-Anisopteromalus calandre* increases as the metapopulation size increases. This is due to the effects of limited dispersal (barriers to dispersal open for 2 hours each week) rather than the increased availability of resources in the metapopulations (Bonsall *et al.*, 2002).

focused on the population dynamics of assemblages, and theoretical models predicted that assemblages with more species may not be more stable (in the strict population dynamic sense) than assemblages with fewer species. In particular, May (1973) showed, theoretically, that multispecies communities in which the individual populations of each species are stable could be unstable if the food-web is highly connected or the average interaction strength is too large. That is, from an assessment of the local stability of a community of n species, the system is stable if:

$$s(nC)^{1/2} < 1, \tag{9}$$

where s is the average interaction strength and C is the connectance of the food-web. More recently, McCann *et al.* (1998) have corroborated May's (1971, 1973) findings by showing that weak links allow the persistence of simple three and four species assemblages. This general result that complexity does not necessarily beget stability has been tempered by discussion on the definitions of 'stable' and 'persistent' (Pimm, 1984), the role of energy flow through ecosystems (Pimm and Lawton, 1977), and more recently on the role of ecosystem function and services (e.g., decomposition, facilitation) provided by species in communities (Kinzig *et al.*, 2002).

While the theory surrounding this concept is well developed, laboratory experiments on the stability-complexity debate have only recently been undertaken. For instance, Fox and McGrady-Steed (2002) tested the prediction that species connected randomly by strong interactions are unlikely to exhibit stable, feasible equilibria using experimental communities of bacteria, protists, and small metazoans. Analysis of long-term data from experimental communities that varied initially in species richness (n), connectance (C), and composition showed that the probability of stability declined with complexity. Moreover, Fox and McGrady-Steed (2002) demonstrated that communities converged in connectance over time so that the fraction of interspecific interactions was independent of species richness. In a similar series of replicated microcosm experiments, Weatherby *et al.* (1998) established 63 different microcosms from a pool of six protist species (three bacterivores, two predators and one omnivore). Forty-seven of these showed a repeatable collapse to one of eight communities, and persistent communities could be reached from several different starting communities. In a more detailed study of the effects of species richness, Lawler (1993) tested the hypothesis that the degree of species richness affected the persistence, abundance and temporal variability of protist communities. In general, species food-web composition affects the persistence and mean abundance of several species: species abundance declines in species-rich food-webs and temporal variability is increased as the number of predator-prey pairs increases.

Laboratory microcosms, and in particular the use of aquatic communities have been exceedingly beneficial to our understanding of and in prediction of the effects of food-web architecture on population dynamics (Lawler and Morin, 1993; Morin and Lawler, 1995), the effects of multispecies competition on community composition and resource utilization (Neill, 1975), the effects of diversity on ecosystem function (McGrady-Steed *et al.*, 1997), the effects of productivity on food-chain dynamics (Kaunzinger and Morin, 1998, Fukami and Morin, 2003), and the effects of environmental warming on food-web structure and ecosystem function (Petchey *et al.*, 1999). However, it is the effects of species diversity on ecosystem function that has attracted the most recent interest. Naeem *et al.* (1994) clearly demonstrated using replicated terrestrial microcosms that the loss of biodiversity can alter the performance and function of ecosystems. By manipulating levels of biodiversity, ecosystem performance and community responses, such as respiration, decomposition, nutrient retention, plant productivity, and water retention, are affected. More recently, based on the findings from these controlled microcosm experiments the effects of species diversity on ecosystem performance has been extended to field studies (Hector *et al.*, 1999). However, there remains little rigorous comparison of models of community dynamics with experimental data. With the availability of a diverse wealth of community biology theories (e.g., Morin, 1999; Tokeshi, 1999), it is appropriate to integrate more fully ecological experiments on community and ecosystem dynamics with the theoretical developments in this area of ecology.

IV. FUTURE PERSPECTIVES

Our principal argument for the rationale for the use of laboratory microcosms is to move away from simple significance tests towards a more pluralistic approach to understanding ecological processes and mechanisms. Selecting an a priori set of candidate models, computing a criterion based on information loss for each of these possible models, and accepting the one that minimized the information loss for the ecological problem remains a simple and compelling concept. Such an approach has formal, theoretical roots in information theory and provides a powerful alternative to the conventional approach to ecological research. With the availability of theoretical ecological models, this rationale coupled with laboratory experiments is appealing and holds the hope of some exciting developments in ecology.

A number of broad ecological problems still necessitate the design and implementation of laboratory microcosms. While by no means complete, this list includes the use of laboratory experiments to understand space, noise, transient dynamics, and the rules of species assembly. In this final section, we highlight five areas where laboratory experimental microcosms are likely to have an important impact on our understanding of the underlying ecological principles.

A. Population Dynamics

One of the obvious ways that laboratory experiments can be integrated with the rationale of model choice is through an analysis of the time series of the population dynamics of species interactions. While time series are observational and often inference of mechanism from such data is difficult (Ellner *et al.*, 1997), identifying the ecological mechanism from ecological population data is a direct application of the rationale of model selection and choice. While previous studies have highlighted the integration of mechanistic models and time series analysis to interpreting population-level phenomena such as cyclic dynamics (e.g., Kendall *et al.*, 1999) or the dynamics of specific systems (Dennis *et al.*, 1995, 2001), a fuller approach to testing candidate models and assessing contrasting ecological effects remains a straightforward but novel approach to problems on the dynamics of populations. For example, recent findings have highlighted that the population dynamics of generalist consumers can be described by simple single-species models (Murdoch *et al.*, 2002). This result was obtained from 112 time series from a variety of sources. That multispecies interactions can be collapsed to single-species interactions is appealing but not without controversy (Rohani *et al.*, 2003; Murdoch *et al.*, 2003b). However, a set of manipulative experiments could be designed to test this idea. Generalist natural enemies such as some parasitic wasps are readily cultured and multispecies laboratory systems are clearly amenable to resolving this problem through the testing and comparison of competing single and multispecies models. Moreover, the maxim that mechanism cannot be inferred from population time series is now in doubt (Table 1). The combination of laboratory experiments and appropriate model choice holds the possibility that mechanistic understanding can be gained from ecological time series.

B. Role of Productivity in Ecosystems

While laboratory experiments may not be appropriate for testing ecosystem responses to perturbations, such microcosm experiments can be used successfully in testing specific ecosystem processes. One such process that warrants further study is the role of productivity. Productivity is the amount of locally available energy to a particular ecosystem and is known, theoretically, to influence species diversity (Rosenzweig and MacArthur, 1963), predator-prey interactions (Rosenzweig, 1971), and more complex multispecies assemblages (Diehl and Feissel, 2000; Bonsall and Holt, 2003). While the theory surrounding the role of productivity on species interactions is well-established and the broad pattern of the relationship between species diversity and productivity is widely documented (e.g., Tilman and Pacala, 1993), empirical support for the ecological effects (model parameters) remains relatively scarce. Models of productivity are known to take a variety of contrasting forms (Waide *et al.*, 1999) and comparing models that describe montonic relationships ($diversity = m_1 \cdot productivity + c$), humped relationships ($diversity = m_1 \cdot productivity + m_2 \cdot productivity^2 + c$), or more general (non-parametric) relationships ($diversity = f(productivity) + c$) between productivity and diversity would allow temporally consistent patterns between different species interactions to be explored. There is clearly a scope for developing robust, well-replicated laboratory microcosm experiments to contrast a range of concepts (models) on the role of ecosystem productivity on species diversity and interactions.

C. Role of Weak Links in Ecosystems

Theoretical models have indicated that multispecies interactions with strong interaction strengths are not compatible with the persistence of speciose and diverse food-webs (May, 1971; McCann *et al.*, 1998). While this has been the focus of considerable attention (Pimm, 1982; King and Pimm, 1983; Pimm, 1984), several testable predictions remain unanswered. These include: Do weak interactions predominant in persisting multispecies assemblages? Are species-poor food-webs more likely to show oscillatory dynamics compared to more species-diverse food-webs? What is the relative effect of more strong interactions over weak interactions on multispecies dynamics? Understanding how these different rules of assembly affect the dynamics and persistence of multispecies interactions rests on an appropriate measure of species interaction strengths (the magnitude of one species on another) (Laska and Wootton, 1998). While several measures of interaction exist such as estimates based on the community matrix (Levins, 1968), the Jacobian matrix (May, 1973), the inverted Jacobian matrix (Bender *et al.*, 1984), per capita effects (Paine, 1992), or population-dynamic effects (Laska and Wootton, 1998), ecological interaction strengths are traditionally thought to be indeterminate in their response to changes in biotic or abiotic perturbations (Bender *et al.*, 1984). Recent work has highlighted that the species interactions are contingent, stochastic, and non-linear (Bonsall *et al.*, 2003). Species interactions may be misinterpreted for reasons associated with experimental design (Bender *et al.*, 1984) or non-linear effects (Bonsall *et al.*, 2003). However, it remains an attainable goal that a progressive understanding of species interaction strengths and their impact on multispecies assemblages can be achieved. Given that species interactions can be manipulated in laboratory experiments (through press or perturbation responses) and that complex food-webs can be constructed from community modules (Holt, 1997), the use of laboratory microcosms in understanding interaction strength effects holds the hope of exciting advancements in community ecology.

D. Role of Noise in Ecological Interactions

Although the effects of environmental noise and climatic variability (e.g., NAO, ENSO) on the dynamics of populations have been explored (Grenfell *et al.*, 1998; Pascual *et al.*, 2000), the effects of stochasticity on single and multispecies population dynamics still remain difficult to evaluate. For instance, it is not obvious whether stochasticity acts in an additive or multiplicative fashion to influence the dynamics of species interactions. That is, does noise just simply have a scaling effect on dynamics or is it that noise interacts with endogenous processes to induce qualitatively different dynamics? Theoretical models predict that noise can influence the dynamics in a more multiplicative way (Petchey *et al.*, 1997; Greenman and Benton, 2003). However, experimental tests of the effects of environmental and/or demographic stochasticity remain relatively rare (however, see

Petchey *et al.*, 1999, and Benton *et al.*, 2002). Parameter optimization and model selection requires a decision about the underlying noise processes (e. g., environmental or demographic), and laboratory experiments are well suited to testing the effects of noise on the dynamics (e.g., Dennis *et al.*, 2001). Manipulating environmental conditions have been shown to affect population dynamics (Beckerman *et al.*, 2002). However, contrasting different noise terms and differentiating between the additive and multiplicative effects of noise remains an area for future development. Extending our understanding of the effects of noise on multispecies interaction also remains unclear and understanding the population dynamic consequences through the use of laboratory systems and model choice remains a challenge for population ecological research.

E. Role of Space in Ecological Interactions

One of the most frequently raised questions in ecology is how do space and spatial processes affect species interactions. While the principle of space has been widely acknowledged (Nicholson and Bailey, 1935; Wright, 1940; Skellam, 1951; Andrewartha and Birch, 1954) and is fundamental to the understanding of the distribution and the abundance of species, experimental tests of spatial processes remain scarce. In particular, the concept of the metapopulation has become the central paradigm in spatial ecology (Hanski, 1999). However, designing well-replicated, spatially-explicit and spatially-relevant experiments in the field remains difficult if not impossible. The use of laboratory studies in understanding spatial processes is clearly of benefit. The original experiments by Huffaker (Huffaker *et al.*, 1963) and Pimental (Pimental *et al.*, 1963) highlighted how laboratory systems can be developed to tackle research problems on spatial ecology. More recently, laboratory microcosms examining metapopulation processes have been developed (Holyoak and Lawler, 1996; Bonsall *et al.*, 2002). However, several research problems that remain unanswered such as the effects of more complex habitat arrangements, the role of source-sink dynamics, and the existence of between-patch heterogeneity are all potential issues that impinge on our understanding of spatial dynamics. These can be realistically examined with designed laboratory experiments. Moreover, process- and individual-based models on metapopulation dynamics and interactions abound in the literature (Allen, 1975; Hassell *et al.*, 1991b; DeAngelis and Gross, 1992; Durrett and Levin, 1994), and within our framework of model choice and selection, the comparison of different spatial concepts on the dynamics and persistence of ecological systems remains an exciting area for development.

V. CONCLUSIONS

We have argued that laboratory microcosm experiments have had a predominant role in conceptualizing ideas in ecology. If the aim in ecology is to find laws and axioms, then laboratory experiments will remain a valuable approach for collating evidence on ecological patterns and processes underpinning the distribution and abundance of organisms. In short, the detailed history of natural communities is seldom known and, therefore, it is often difficult to deduce the ecological patterns or mechanisms. To this end, laboratory experiments will remain essential for exploring the structure and function of ecological assemblages.

ACKNOWLEDGMENTS

Dr. Bonsall is a Royal Society University Research Fellow.

REFERENCES

Akaike, H. (1973) Information theory and an extension of the maximum likelihood principle. *2nd Int. Symp. Inf. Theor.*, 267–310.

Akaike, H. (1974) A new look at the statistical model identification. *IEEE Trans. Autom. Control.* **AC-19**, 716–723.

Allen, J.C. (1975) Mathematical models of species interactions in time and space. *Am. Nat.* **109**, 319–342.

Andrewartha, H.G. and Birch, L.C. (1954) *The Distribution and Abundance of Animals*. University of Chicago Press, Chicago.

Bailey, V.A., Nicholson, A.J. and Williams, E.J. (1962) Interactions between hosts and parasites where some host individuals are more difficult to find than others. *J. Theor. Biol.* **3**, 1–18.

Beckerman, A., Benton, T.G., Ranta, E., Kaitala, V. and Lundberg, P. (2002) Population dynamic consequences of delayed life-history effects. *Trends Ecol. Evol.* **17**, 263–269.

Beddington, J.R., Hassell, M.P. and Lawton, J.H. (1976) The components of arthropod predation. II. The predator rate of increase. *Journ. Anim. Ecol.* **45**, 165–185.

Bellows, T.S. (1981) The descriptive properties of some models for density dependence. *Journ. Anim. Ecol.* **50**, 139–156.

Bender, E.A., Case, T.J. and Gilpin, M.E. (1984) Perturbation experiments in community ecology: Theory and practice. *Ecology* **65**, 1–13.

Benton, T.G., Lapsley, C.T. and Beckerman, A.P. (2002) The population response to environmental noise: Population size, variance and correlation in an experimental system. *Journ. Anim. Ecol.* **71**, 320–332.

Bonsall, M.B. (2003) The role of variability and risk on the persistence of shared-enemy, predator-prey assemblages. *J. Theor. Biol.* **221**, 193–204.

Bonsall, M.B., French, D.A. and Hassell, M.P. (2002) Metapopulation structures affect persistence of predator-prey interactions. *Journ. Anim. Ecol.* **71**, 1075–1084.

Bonsall, M.B. and Hassell, M.P. (1997) Apparent competition structures ecological assemblages. *Nature* **388**, 371–373.

Bonsall, M.B. and Hassell, M.P. (1998) Population dynamics of apparent competition in a host-parasitoid assemblage. *Journ. Anim. Ecol.* **67**, 918–929.

Bonsall, M.B. and Hassell, M.P. (2000) The effects of metapopulation structures on indirect effects in host-parasitoid assemblages. *Proc. Roy. Soc. Lond.* **267**, 2207–2212.

Bonsall, M.B. and Hastings, A. (2004) Demographic and environmental stochasticity in predator-prey metapopulations. *Journ. Anim. Ecol.* **73**, 1043–1055.

Bonsall, M.B. and Holt, R.D. (2003) The effects of enrichment on the dynamics of apparent competitive interactions in stage-structured systems. *Am. Nat.* **162**, 780–795.

Bonsall, M.B., van der Meijden, E. and Crawley, M.J. (2003) Contrasting dynamics in the same plant-herbivore system. *Proc. Natl. Acad. Sci.* **100**, 14932–14936.

Burnham, K.P. and Anderson, D.R. (2002) *Model Selection and Multi-Model Inference: A Practical Information-Theoretic Approach*. Springer-Verlag, New York.

Burnett, T. (1958) A model of host-parasite interaction. *Proc. 10th Int. Congr. Ent.* **2**, 679–686.

Carpenter, S.R., Chisholm, S.W., Krebs, C.J., Schindler, D.W. and Wright, R.F. (1995) Ecosystem experiments. *Science* **269**, 324–327.

Chaneton, E.J. and Bonsall, M.B. (2000) Enemy-mediated apparent competition: Empirical patterns and the evidence. *Oikos* **88**, 380–394.

Chapman, R.N. (1928) The quantitative analysis of environmental factors. *Ecology* **9**, 111–122.

Charnov, E.L. (1976) Optimal foraging: The marginal value theorem. *Theor. Popul. Biol.* **9**, 129–136.

Chesson, P.L. and Murdoch, W.W. (1986) Aggregation of risk: Relationships among host-parasitoid models. *Am. Nat.* **127**, 696–715.

Comins, H.N. and Hassell, M.P. (1996) Persistence of multispecies host-parasitoid interactions in spatially distributed models with local dispersal. *J. Theor. Biol.* **183**, 19–28.

Comins, H.N., Hassell, M.P. and May, R.M. (1992) The spatial dynamics of host-parasitoid systems. *Journ. Anim. Ecol.* **61**, 735–748.

Cook, R.M. and Hubbard, S.F. (1977) Adaptive searching strategies in insect parasites. *Journ. Anim. Ecol.* **46**, 115–125.

Costantino, R.F., Cushing, J.M., Dennis, B. and Desharnais, R.A. (1995) Experimentally-induced transitions in the dynamic behaviour of insect populations. *Nature* **375**, 227–230.

Costantino, R.F., Desharnais, R.A., Cushing, J.M. and Dennis, B. (1997) Chaotic dynamics in an insect population. *Science* **275**, 389–391.

DeAngelis, D.L. and Gross, L.J. (1992) *Individual-Based Models and Approaches in Ecology*. Chapman and Hall, London.

Dennis, B., Desharnais, R.A., Cushing, J.M. and Costantino, R.F. (1995) Non-linear demographic dynamics—mathematical models, statistical methods, and biological experiments. *Ecol. Monog.* **65**, 261–281.

Dennis, B., Desharnais, R.A., Cushing, J.M., Henson, S.M. and Costantino, R.F. (2001) Estimating chaos and complex dynamics in an insect population. *Ecol. Monog.* **71**, 277–303.

Dennis, B. and Taper, M. (1994) Density-dependence in time series observations of natural population—estimation and testing. *Ecol. Monog.* **64**, 205–244.

Desharnais, R.A. and Liu, L. (1987) Stable demographic limit cycles in laboratory populations of *Tribolium castaneum*. *Journ. Anim. Ecol.* **56**, 885–906.
Diehl, S. and Feissel, M. (2000) Effects of enrichment on three-level food chains with omnivory. *Am. Nat.* **155**, 200–218.
Durrett, R. and Levin, S.A. (1994) Stochastic spatial models—a user's guide to ecological applications. *Phil. Trans. Roy. Soc. Lond. B* **343**, 329–350.
Eddy, S. (1928) Succession of Protozoa in cultures under controlled conditions. *Trans. Am. Micr. Soc.* **47**, 283.
Ellner, S.P., Kendall, B.E., Wood, S.N., McCauley, E. and Briggs, C.J. (1997) Inferring mechanism from time-series data: Delay-differential equations. *Physica D* **110**, 182–194.
Ellner, S.P., McCauley, E., Kendall, B.E., Briggs, C.J., Hosseini, P.R., Wood, S.N., Janssen, A. and Sabelis, M.W. (2001) Habitat structure and population persistence in an experimental community. *Nature* **412**, 538–542.
Ellner, S.P. and Turchin, P. (1995) Chaos in a noisy world: New methods and evidence from time-series analysis. *Am. Nat.* **145**, 343–375.
Elton, C.S. (1958) *The Ecology of Invasions by Animals and Plants*. Methuen and Co, London.
Fox, J.W. and McGrady-Steed, J. (2002) Stability and complexity in microcosm communities. *Journ. Anim. Ecol.* **71**, 749–756.
Free, C.A., Beddington, J.R. and Lawton, J.H. (1977) On the inadequacy of simple models of mutual interference for parasitism and predation. *Journ. Anim. Ecol.* **46**, 543–554.
Fukami, T. and Morin, P.J. (2003) Productivity-biodiversity relationships depend on the history of community assembly. *Nature* **424**, 423–426.
Gause, G.F. (1933) *The Struggle for Existence*. Dover Publications, New York.
Greenman, J.V. and Benton, T.G. (2003) The amplification of environmental noise in population models: Causes and consequences. *Am. Nat.* **161**, 225–239.
Grenfell, B.T., Wilson, K., Finkenstadt, B.F., Coulson, T.N., Murray, S., Albon, S. D., Pemberton, J.M., Clutton-Brock, T.H. and Crawley, M.J. (1998) Noise and determinism in synchronized sheep dynamics. *Nature* **394**, 674–677.
Gilpin, M. and Hanski, I. (1991) *Metapopulation Dynamics: Empirical and Theoretical Investigations*. Academic Press, London.
Gross, K. and Ives, A.R. (1999) Inferring host-parasitoid stability from patterns of parasitism among patches. *Am. Nat.* **154**, 489–496.
Gurney, W.S.C., Blythe, S.P. and Nisbet, R.M. (1980) Nicholson's blowflies revisited. *Nature* **287**, 17–21.
Gurney, W.S.C. and Nisbet, R.M. (1998) *Ecological Dynamics*. Oxford University Press, Oxford.
Hanski, I. (1999) *Metapopulation Ecology*. Oxford University Press, Oxford.
Hanski, I.A. and Gilpin, M.E. (1997) *Metapopulation Biology. Ecology, Evolution and Genetics*. Academic Press, London.
Hassell, M.P. (1971) Mutual interference between searching insect parasites. *Journ. Anim. Ecol.* **40**, 473–486.
Hassell, M.P. (1975) Density dependence in single species populations. *Journ. Anim. Ecol.* **44**, 283–295.
Hassell, M.P. (1978) *The Dynamics of Arthropod Predator-prey Systems*. Princeton University Press, Princeton.
Hassell, M.P. (1982) Patterns of parasitism by insect parasitoids in patchy environments. *Ecol. Ent.* **7**, 365–377.

Hassell, M.P. (2000) *Spatial and Temporal Dynamics of Host-Parasitoid Interactions*. Oxford University Press, Oxford.

Hassell, M.P., Comins, H.N. and May, R.M. (1991b) Spatial structure and chaos in insect population dynamics. *Nature* **353**, 255–258.

Hassell, M.P., Comins, H.N. and May, R.M. (1994) Species coexistence and self-organizing spatial dynamics. *Nature* **370**, 290–292.

Hassell, M.P., Lawton, J.H. and May, R.M. (1976a) Patterns of dynamical behaviour in single species populations. *Journ. Anim. Ecol.* **45**, 471–486.

Hassell, M.P., Lawton, J.H. and Beddington, J.R. (1976b) The components of arthropod predation. I. The prey death rate. *Journ. Anim. Ecol.* **45**, 135–164.

Hassell, M.P. and May, R.M. (1973) Stability in insect host-parasite models. *Journ. Anim. Ecol.* **42**, 693–726.

Hassell, M.P. and May, R.M. (1974) Aggregation of predators and insect parasites and its effect on stability. *Journ. Anim. Ecol.* **43**, 567–594.

Hassell, M.P. and Pacala, S. (1990) Heterogeneity and the dynamics of host-parasitoid interactions. *Phil. Trans. Roy. Soc. Lond. B* **330**, 203–220.

Hassell, M.P., Pacala, S.W., May, R.M. and Chesson, P.L. (1991a) The persistence of host-parasitoid associations in patchy environments. I. A general criterion. *Am. Nat.* **138**, 568–583.

Hassell, M.P. and Varley, G.C. (1969) New inductive population model for insect parasites and its bearing on biological control. *Nature* **223**, 1133–1137.

Hardin, G. (1960) The competitive exclusion principle. *Science* **131**, 1292–1297.

Hastings, A., Hom, C.L., Ellner, S., Turchin, P. and Godfray, H.C.J. (1993) Chaos in ecology: Is mother nature a strange attractor? *Ann. Rev. Ecol. Syst.* **24**, 1–33.

Heads, P.A. and Lawton, J.H. (1983) Studies on the natural enemy complex of the holly leaf-miner—the effects of scale on the detection of aggregative responses and the implications for biological-control. *Oikos* **40**, 267–276.

Hector, A., Schmid, B., Beierkuhnlein, C., Caldeira, M.C., Diemer, M., Dimitrakopoulos, P.G., Finn, J.A., Freitas, H., Giller, P.S., Good, J., Harris, R., Hogberg, P., Huss-Danell, K., Joshi, J., Jumpponen, A., Korner, C., Leadley, P.W., Loreau, M., Minns, A., Mulder, C.P.H., O'Donovan, G., Otway, S.J., Pereira, J.S., Prinz, A., Read, D.J., Scherer-Lorenzen, M., Schulze, E.D., Siamantziouras, A.S.D., Spehn, E.M., Terry, A.C., Troumbis, A.Y., Woodward, F.I., Yachi, S. and Lawton, J.H. (1999) Plant diversity and productivity experiments in European grasslands. *Science* **286**, 1123–1127.

Holdaway, F.G. (1932) An experimental study of the growth of population of the flour beetle *Tribolium confusum* Duval, as affected by atmospheric moisture. *Ecol. Monog.* **2**, 262–304.

Holt, R.D. (1977) Predation, apparent competition and the structure of prey communities. *Theor. Popul. Biol.* **12**, 197–229.

Holt, R.D. (1984) Spatial heterogeneity, indirect interactions, and the coexistence of species. *Am. Nat.* **124**, 377–406.

Holt, R.D. (1997) Community modules. In: *Multitrophic Interactions in Terrestrial Systems* (Ed. by A.C. Gange and V.K. Brown), pp. 333–350. Blackwell Science, Oxford.

Holt, R.D. and Lawton, J.H. (1993) Apparent competition and enemy-free space in insect host-parasitoid communities. *Am. Nat.* **142**, 623–645.

Holt, R.D. and Lawton, J.H. (1994) The ecological consequences of shared natural enemies. *Ann. Rev. Ecol. Syst.* **25**, 495–520.

Holyoak, M. and Lawler, S.P. (1996) Persistence of an extinction-prone predator-prey interaction through metapopulation dynamics. *Ecology* **77**, 1867–1879.

Huffaker, C.B. (1958) Experimental studies on predation: Dispersion factors and predator-prey oscillations. *Hilgardia* **27**, 343–383.

Huffaker, C.B. and Kennet, C.E. (1956) Experimental studies on predation: Predation and cyclamen mite populations on strawberries in California. *Hilgardia* **24**, 191–222.

Huffaker, C.B., Shea, K.P. and Herman, S.G. (1963) Experimental studies on predation: Complex dispersion and levels of food in an acarine predator-prey interaction. *Hilgardia* **34**, 305–329.

Johnson, N.L., Kotz, S. and Balakrishnan (1994) *Continuous Univariate Distributions*, Volume 1. 2nd edition. John Wiley and Sons, New York.

Kareiva, P. (1989) Renewing the dialogue between theory and experiments in population ecology. In: *Perspectives in Ecological Theory*, Princeton University Press, Princeton68–88.

Kaunzinger, C.M.K. and Morin, P.J. (1998) Productivity controls food-chain properties in microbial communities. *Nature* **395**, 495–497.

Kendall, B.E., Briggs, C.J., Murdoch, W.W., Turchin, P., Ellner, S.P., McCauley, E., Nisbet, R.M. and Wood, S.N. (1999) Why do populations cycle? A synthesis of statistical and mechanistic modeling approaches. *Ecology* **80**, 1789–1805.

King, A.W. and Pimm, S.L. (1983) Complexity, diversity and stability—a reconciliation of theoretical and empirical results. *Am. Nat.* **122**, 229–239.

Kinzig, A.P., Pacala, S.W. and Tilman, D. (2002) *The Functional Consequences of Biodiversity: Empirical Progress and Theorectical Extensions*. Princeton University Press, Princeton.

Kuchlein, J.H. (1966) Mutual interfence among the predacious mite *Typhlodromus longipilius* Nesbitt (Acari, Phytoseiidae). I. Effects of predator density on oviposition rate and migration tendency. *Meded. Rijksfac. Lanb Wet. Genet.* **31**, 740–746.

Laska, M.S. and Wootton, J.T. (1998) Theoretical concepts and empirical approaches to measuring interaction strength. *Ecology* **79**, 461–476.

Lawler, S.P. (1993) Species richness, species composition and population dynamics of protists in experimental microcosms. *Journ. Anim. Ecol.* **62**, 711–719.

Lawler, S.P. and Morin, P.J. (1993) Food-web architecture and population dynamics in laboratory microcosms of protists. *Am. Nat.* **141**, 675–686.

Lawton, J.H. (1995) Ecological experiments with model systems. *Science* **269**, 328–331.

Lawton, J.H. and Hassell, M.P. (1981) Asymmetrical competition in insects. *Nature* **289**, 793–795.

Lawton, J.H., Hassell, M.P. and Beddington, J.R. (1975) Prey death rates and rates of increase of arthropod predator populations. *Nature* **255**, 60–62.

Lessells, C.M. (1985) Parasitoid foraging: Should parasitism be density dependent? *Journ. Anim. Ecol.* **54**, 27–41.

Levins., R. (1968) *Evolution in Changing Environments*. Princeton University Press, Princeton.

Lokta, A. (1925) *Elements of Physical Biology*. Dover Publications, Baltimore.

May, R.M. (1971) Stability in multi-species community models. *Math. Biosci.* **12**, 59–79.

May, R.M. (1973) *Stability and Complexity in Model Ecosystems*. Princeton University Press, Princeton.

May, R.M. (1974) Biological populations with non-overlapping generations: Stable points, stable cycles and chaos. *Science* **186**, 645–647.

May, R.M. (1976) Simple mathematical models with very complicated dynamics. *Nature* **261**, 459–467.

May, R.M. (1978) Host-parasitoid systems in patchy environments: A phenomenological model. *Journ. Anim. Ecol.* **47**, 833–843.

May, R.M. (1981) *Theoretical Ecology*. Blackwell Scientific Publications, Oxford.

May, R.M., Conway, G.R., Hassell, M.P. and Southwood, T.R.E. (1974) Time delays, density-dependence and single species oscillations. *Journ. Anim. Ecol.* **43**, 747–770.

May, R.M. and Oster, G.F. (1976) Bifurcations and dynamic complexity in simple ecological models. *Am. Nat.* **110**, 573–600.

Maynard-Smith, J. and Slatkin, M. (1973) The stability of predator-prey systems. *Ecology* **54**, 384–391.

McCann, K., Hastings, A. and Huxel, G.R. (1998) Weak trophic interactions and the balance of nature. *Nature* **395**, 794–798.

McGlade, J. (1999) *Advanced Ecological Theory: Advances in Principles and Applications*. Blackwell Science, Oxford.

McGrady-Steed, J., Harris, P.M. and Morin, P.J. (1997) Biodiversity regulates ecosystem predictability. *Nature* **390**, 162–165.

Morin, P. (1999) *Community Ecology*. Oxford University Press, Oxford.

Morin, P.J. and Lawler, S.P. (1995) Food-web architecture and population dynamcis - theory and empirical evidence. *Ann. Rev. Ecol. Syst.* **26**, 505–529.

Murdoch, W.W., Briggs, C.J. and Nisbet, R.M. (2003a) *Consumer-Resource Dynamics*. Princeton University Press, Princeton.

Murdoch, W.W., Briggs, C.J., Nisbet, R.M., Kendall, B.E. and McCauley, E. (2003b) Natural enemy specialization and the period of population cycles—reply. *Ecol. Lett.* **6**, 384–387.

Murdoch, W.W., Kendall, B.E., Nisbet, R.M., Briggs, C.J., McCauley, E. and Bolser, R. (2002) Single-species models for many-species food webs. *Nature* **417**, 541–543.

Mueller, L.D. and Joshi, A. (2000) *Stability in Model Populations*. Princeton University Press, Princeton.

Naeem, S., Thompson, L.J., Lawler, S.P., Lawton, J.H. and Woodfin, R.M. (1994) Declining biodiversity can alter the performance of ecosystems. *Nature* **368**, 734–737.

Neill, W.E. (1975) Experimental studies of microcrustacean competition, community composition and efficiency of resource utilization. *Ecology* **56**, 809–826.

Nicholson, A.J. (1933) The balance of animal populations. *Journ. Anim. Ecol.* **2**, 131–178.

Nicholson, A.J. (1954) An outline of the dynamics of animal populations. *Austr. Journ. Zool.* **2**, 9–65.

Nicholson, A.J. (1957) The self-adjustment of populations to change. *Cold Spring Harbor Symp. Quant. Biol.* **22**, 153–173.

Nicholson, A.J. and Bailey, V.A. (1935) The balance of animal populations. Part I. *Proc. Zoo. Soc. Lond.* **3**, 551–598.

Nisbet, R.M. and Gurney, W.S.C. (1982) *Modelling Fluctuating Populations*. Wiley and Sons, Chichester.

Pacala, S.W. and Hassell, M.P. (1991) The persistence of host-parasitoid associations in patchy environments. II. Evaluation of field data. *Am. Nat.* **138**, 584–605.

Pacala, S.W., Hassell, M.P. and May, R.M. (1990) Host-parasitoid associations in patchy environments. *Nature* **344**, 150–153.

Paine, R.T. (1992) Food-web analysis through field measurement of per-capita interaction strength. *Nature* **355**, 73–75.

Park, T. (1948) Experimental studies of interspecific competition. I. Competition between populations of the flour beetles *Tribolium confusum* Duval and *Tribolium castaneum* Herbst. *Ecol. Monog.* **18**, 265–308.

Pascual, M., Rodo, X., Ellner, S.P., Colwell, R. and Bouma, M.J. (2000) Cholera dynamics and El Nino-southern oscillation. *Science* **289**, 1766–1769.

Pearl, R. (1928) *The Rate of Living*. University of London Press, London.

Petchey, O.L., Gonzalez, A. and Wilson, H.B. (1997) Effects on population persistence: the interaction between environmental noise colour, intraspecific competition and space. *Proc. Roy. Soc. Lond. B* **264**, 1841–1847.

Petchey, O.L., McPhearson, P.T., Casey, T.M. and Morin, P.J. (1999) Environmental warming alters food-web structure and ecosystem function. *Nature* **402**, 69–72.

Pimental, D., Nagel, W.P. and Madden, J.L. (1963) Space-time structure and the survival of parasite-host systems. *Am. Nat.* **97**, 141–167.

Pimm, S.L. (1982) *Food Webs*. Chapman and Hall, London.

Pimm, S.L. (1984) The complexity and stability of ecosystems. *Nature* **307**, 321–326.

Pimm, S.L. and Lawton, J.H. (1977) Number of trophic levels in ecological communities. *Nature* **268**, 329–331.

Reeve, J.D., Kerans, B.L. and Chesson, P.L. (1989) Combining different forms of parasitoid aggregation: Effects on stability and patterns of parasitism. *Oikos* **56**, 233–239.

Ricker, W.E. (1954) Stock and recruitment. *J. Fish. Res. Board Can.* **28**, 333–341.

Rohani, P., Wearing, H.J., Cameron, T. and Sait, S.M. (2003) Natural enemy specialization and the period of population cycles. *Ecol. Lett.* **6**, 381–384.

Rosenzweig, M.L. (1971) Paradox of enrichment: Destabilization of exploitative ecosystems in ecological time. *Science* **171**, 385–387.

Rosenzweig, M.L. and MacArthur, R.H. (1963) Graphical representation and stability conditions of predator-prey interactions. *Am. Nat.* **97**, 209–223.

Schaffer, W.M. (1984) Stretching and folding in lynx fur returns—evidence for a strange attractor in nature. *Am. Nat.* **124**, 798–820.

Skellam, J.G. (1951) Random dispersal in theoretical populations. *Biometrika* **38**, 196–218.

Solomon, M.E. (1949) The natural control of animal populations. *Journ. Anim. Ecol.* **18**, 1–35.

Stiling, P.D. (1987) The frequency of density dependence in insect host-parasitoid systems. *Ecology* **68**, 844–856.

Thompson, W.R. (1924) La théorie mathématique de l'action des parasites entomophages et le facteur du hasard. *Ann. Facul. Sci. Mars.* **2**, 69–89.

Tilman, D. (1982) *Resource Competition and Community Structure*. Princeton University Press, Princeton.

Tilman, D. (1988) *Plant Strategies and the Dynamics and Structure of Plant Communities*. Princeton University Press, Princeton.

Tilman, D. and Pacala, S. (1993) The maintenance of species richness in plant communities. In: *Species Diversity in Ecological Communities* (Ed. by R. Ricklefs and D. Schluter), pp. 13–25. The University of Chicago Press, Chicago.

Tokeshi, M. (1999) *Species Coexistence. Ecological and Evolutionary Perspectives*. Blackwell Science, Oxford.

Turchin, P. (1990) Rarity of density dependence or population regulation with lags. *Nature* **344**, 660–663.

Turchin, P. and Taylor, A.D. (1992) Complex dynamics in ecological time series. *Ecology* **73**, 289–305.

Varley, G.C. and Gradwell, G.R. (1960) Key factors in population studies. *Journ. Anim. Ecol.* **29**, 399–401.

Visser, M.E., Jones, T.H. and Driessen, G. (1999) Interference among insect parasitoids: A multi-patch experiment. *Journ. Anim. Ecol.* **68**, 108–120.

Volterra, V. (1926) Variazioni e fluttuazioni del numero d'individui in specie animali conviventi. *Mem. Acad. Lincei.* **2**, 31–113.

Waide, R.B., Willig, M.R., Steiner, C.F., Mittelbach, G., Gough, L., Dodson, S.I., Juday, G.P. and Parmenter, R. (1999) The relationship between productivity and species richness. *Ann. Rev. Ecol. Syst.* **30**, 257–300.

Walde, S.J. and Murdoch, W.W. (1988) Spatial density dependence in parasitoids. *Ann. Rev. Entomol.* **33**, 441–466.

Weatherby, A.J., Warren, P.H. and Law, R. (1998) Coexistence and collapse: An experimental investigation of the persistent communities of a protist species pool. *Journ. Anim. Ecol.* **67**, 554–566.

Woodruff, L.L. (1914) The effect of excretion products of Infusoria on the same and on different species, with special reference to the protozoan sequence in infusions. *Journ. Exp. Zool.* **14**, 575.

Wright, S. (1940) Breeding structures of populations in relation to speciation. *Am. Nat.* **74**, 232–248.

Complexity, Evolution, and Persistence in Host-Parasitoid Experimental Systems With *Callosobruchus* Beetles as the Host

MIDORI TUDA AND MASAKAZU SHIMADA

I. SUMMARY

Experimental laboratory systems of bruchid beetles, *Callosobruchus* in particular, and their parasitoids have been used as models to study population dynamics of single species and host-parasitoid interactions since the early 1940s. First, this paper reviews the recent advances in ecological studies on laboratory systems of bruchid hosts and their parasitoids as represented by bottom-up and top-down controls. Factors controlling the persistence of simple host-parasitoid systems that can be modified by an evolutionary change in a host beetle are demonstrated with reference to local carrying capacity, vulnerable time window of hosts to parasitism, and functional

ADVANCES IN ECOLOGICAL RESEARCH VOL. 37

0065-2504/05 $35.00
DOI: 10.1016/S0065-2504(04)37002-9

response of parasitoids. Second, we present experimental results on persistence of larger species assemblies analyzed in the light of the simple two-species host-parasitoid control factors. The most persistent association of species showed that both host and parasitoid control factors in the simple host-parasitoid system were consistently effective in the larger species complexes. There was also a general loss of persistence of host-parasitoid associations as species richness increased. Finally, at the interface between simple and complex assemblies, we asked how an addition of a third species to a simple host-parasitoid system affects resilience and duration of transients, with the *Callosobruchus* beetles as the host and two parasitoids (the pteromalid *Anisopteromalus calandrae* and braconid *Heterospilus prosopidis*). Semi-mechanistic models parameterized by fitting to the population data were constructed to help understand the driving forces that govern the behavior of interacting populations. The population dynamics of the three-species system was ascribed to cyclic/chaotic transient dynamics towards an attractor that has potential of not only a stable equilibrium but also a chaotic one. By comparing the three-species dynamics to the stable two-species (one host-one parasitoid) dynamics before *H. prosopidis* was introduced, the instability that leads to chaos was revealed to be induced by density-dependent host-feeding by *A. calandrae*. Although the destabilizing host-feeding was under the control of a stabilizing effect of mutual interference, the effect was estimated to be weakened by the introduction of the second parasitoid, *H. prosopidis*.

II. GENERAL INTRODUCTION

Following Pearl's (1927) seminal work, laboratory experiments under controlled environmental conditions have played a key role in testing ecological hypotheses. In the late 1930s, Syunro Utida was the first to initiate studies on laboratory experimental systems, using *Callosobruchus* beetles and their parasitoids; a series of publications followed in the 1940s (e.g., Utida 1943a,b, 1944a,b). Since then, the biology of intraspecific and interspecific density dependence and the consequences on dynamic behavior in the bruchid and parasitoid populations have provided a useful model for testing theories in ecology and for the development of new theoretical perspectives. Thus, these host-parasitoid model systems play a similar role to that of model organisms such as *Drosophila* (Pimentel and Stone, 1968) and *Plodia* Begon *et al.*, 1995) as the host.

Beetles of the genus *Callosobruchus* (Coleoptera: Chrysomeloidea: Bruchidae) and their parasitoids, *A. calandrae* (Hymenoptera: Pteromalidae) and *H. prosopidis* (Hymenoptera: Braconidae), are the model organisms we used in the series of experiments on population dynamics of host-parasitoid systems. Their basic ecology is well studied and described in Appendix 1. In this paper, we review recent advances in ecological studies of bruchid hosts and their parasitoids in laboratory systems. Bottom-up (competition over resource) and top-down (parasitism) controls on the persistence of simple host-parasitoid systems, which can be modified by the evolutionary change in resource competition in a host beetle, are reviewed. Particular reference is made to local carrying capacity, vulnerable time window of hosts and functional response of parasitoids as the control factors. We also present findings from experimental tests on persistence of larger species assemblies analyzed in the light of these control factors as well as number of species. Finally, at the interface between simple and complex assemblies, we asked how addition of a second parasitoid species to a simple host-parasitoid system affects stability. As a measure of stability, we used both persistence time and resilience.

III. PERSISTENCE OF A SIMPLE HOST-PARASITOID SYSTEM

To examine life history and behavioral characters that promote persistence of host-parasitoid associations, the simplest possible host-parasitoid systems were studied with different combinations of resource and host species, each assembled with the same species of parasitoid (Tuda, 1996b). The resources used were the mung bean, *Vigna radiata*, and azuki, *V. angularis*; the hosts were the cowpea beetle, *Callosobruchus maculatus* and *C. phaseoli*, and the parasitoid was *H. prosopidis. V. radiata* is more suitable for the development of the two bruchids, and the bruchids differ in the degree of scramble competition; *C. phaseoli* exhibits a higher degree of scramble competition (many individuals share a bean) than *C. maculatus*. It should be noted that *C. maculatus*, exhibits an extreme variation in larval competition between geographical populations; the population used here as well as the one in the following section was of the scramble type in that multiple individuals survive in a single bean but closer to the contest type (a single individual dominates a bean) in that survivors are close to one.

A. Bottom-Up Control Factors

1. Local Carrying Capacity or Larval Competition Type

Two life history characters have proven to be the causes of persistence of host-parasitoid systems: local/global carrying capacity and duration of vulnerability to parasitoid attacks (Tuda, 1996b). Local carrying capacity in bruchids on the bean scale is a direct result of density-dependent competition of larvae in a confined (bean) resource (Appendix 1); a population of contest type competitors, because of its higher resource requirement, has a lower local carrying capacity, and consequently a lower global carrying capacity at the population level.

Lower local/global carrying capacity of the host *C. maculatus* than that of *C. phaseoli* significantly promoted persistence of the host-parasitoid system (Fig. 1) (Tuda, 1996b). Simulation of a Nicholson-Bailey type model with age structure also supported the effect of local carrying capacity, with parameter values estimated independently of the long-term experiments. This result is consistent with the prediction of the paradox of enrichment, in which the increase in (global) carrying capacity reduces stability (Rosenzweig, 1971). On the contrary, when the global carrying capacity is increased by the addition of a greater resource for the host, longer coexistence was achieved (Tuda, 1999, unpublished data). Therefore, parallel increases in local and global carrying capacities or increases in global carrying capacity by changing the degree of density dependence (or

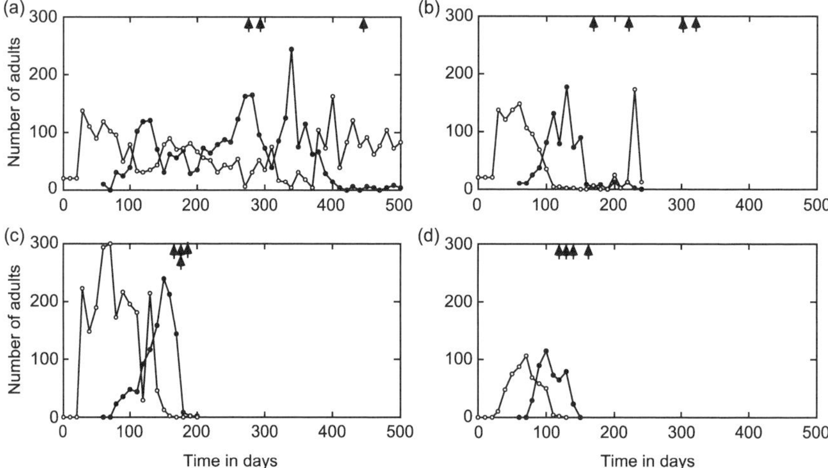

Figure 1 Population dynamics and persistence of host-parasitoid systems: (a) *V. radiata-C. maculatus-H. prosopidis*, (b) *V. angularis-C. maculatus-H. prosopidis*, (c) *V. radiata-C. phaseoli-H. prosopidis*, and (d) *V. angularis-C. phaseoli-H. prosopidis*. Both resource (*V. radiata* or *V. angularis*) and host (*C. maculatus* or *C. phaseoli*) effects are significant (Tuda, 1996b). The closed circles show the numbers of host adults emerged during the preceding 10 days, and the open circles show those of parasitoid adults emerged during the same periods as the hosts. The arrows indicate the times of extinction of either the host or the parasitoid in replicated systems. Modified from Tuda (1996b).

nonlinearity) rather than by increasing resource patches, is the key to the paradox of enrichment, which may often be implicitly violated in the field (Tuda, 1999, unpublished data).

In a competition system, the intraspecific contest-type of *C. maculatus* is also a superior interspecific competitor (Toquenaga and Fujii, 1991) that can competitively exclude other bruchid species in assemblies of bean resource and herbivores.

Bruchid beetles that utilize small beans of wild legumes are the contest type, that is, only a single individual that dominates (bean) resource can survive in each grain of beans, whereas pest species such as the azuki bean beetle, *Callosobruchus chinensis* and *Zabrotes subfasciatus*, are (or evolved to be) the scramble type, with multiple individuals sharing single grains of beans. This would predict that population dynamics of pest species, that are likely to have high carrying capacities because of resource sharing, can be destabilized by introduction of a parasitoid for control, which allows both outbreaks and extinction of the pests. Selection for either the scramble or contest competition types will be discussed further in the subsection on evolutionary change.

2. *Vulnerable Time Window*

It has been shown that the duration of the vulnerable developmental stages in the host to parasitism is one of the factors that have the greatest effect on the persistence of host-parasitoid dynamics (Tuda and Shimada, 1995; Tuda, 1996b). The longer the vulnerable period, the less persistent the host-parasitoid system, which was observed in host-parasitoid systems with azuki as the resource that elongates developmental time in both *C. maculatus* and *C. phaseoli* (Fig. 1). Feeding on poor-quality plants hinders development of herbivorous insects, increasing the time window of susceptibility to natural enemies and leads to higher mortality of the herbivores (Johnson and Gould, 1992; Benrey and Denno, 1997). However, this slow-growth-high-mortality relation is not always observed in nature because of confounding interactions (Clancy and Price, 1987; Benrey and Denno, 1997). Slow development can also be associated with low fecundity. Model simulations to test this possibility, however, showed that the effect of such low fecundity is minor compared to that of a vulnerable time window for a parasitoid with high searching efficiency as in *H. prosopidis* (Tuda, 1996b). Resource is not the only factor that affects the vulnerable time window of the host. Increase in temperature elongated the vulnerable time window of *C. chinensis* to *H. prosopidis*, which reduced persistence, or time to extinction, of the host-parasitoid system (Tuda and Shimada, 1995).

3. *Evolutionary Change*

We often assume characters that play an important role in determining persistence of host-parasitoid systems are constant over time. This assumption, however, can be violated as observed in the host *C. maculatus* during a long-term experiment of a host-parasitoid system. Larval competition of *C. maculatus* was initially the scramble type, and two to three individuals survived from a small *V. radiata* seed (Fig. 2a). When the experiment was terminated on day 800, only single adults emerged indicating the larval competition was of the contest type (Fig. 2a) (Tuda, 1998). This population of *C. maculatus* was brought into the laboratory relatively recently (two years prior to experimentation), unlike the other populations of bruchids and parasitoids in our laboratory. There was no difference in the attack rates between pre- and post-experiment of the parasitoid *H. prosopidis,* which has been maintained in the laboratory for about 20 years. The change in the population dynamics occurred approximately at 20 generations of the host (Fig. 2b), and the post-experimental host and parasitoid after 800 days of coexistence showed stable population dynamics unlike the initial oscillations when they were returned to initial densities (Fig. 2c). This confirms the change was not temporal.

With a parameterized game-theoretical model, Tuda and Iwasa (1998) showed that the evolutionary change in larval competition towards the contest type can induce a large shift in the population dynamics of host and parasitoid as observed in the experimental system. For such rapid evolutionary change to occur, 20 generations was shown to be sufficient by the model with the following assumptions: 1) a small fraction of initial population was a contest-type competitor, in which the phenotype is genetically determined; and 2) the contest competitor consumes a certain volume of bean and kills other individuals in that volume, enabling a scramble competitor to survive only when beans are large enough. It is worth noting that random parasitoid attacks did not alter the evolution towards contest type but can slow the evolution.

This explains the results of experiments on a single-species *C. maculatus* population by Toquenaga *et al.* (1994), in which competition type was estimated based on the resemblance of dynamical trajectories. Their study also indicated that the transition from scramble to contest type occurred on a time scale consistent with the prediction of the present model (i.e., 20 generations). The time scale required for such a change also corresponds to the one observed in Nicholson's laboratory blowfly population (Nicholson, 1957; Stokes *et al.*, 1988). This similarity may arise not only from the stable laboratory conditions but also from a common biological mechanism, e.g., density-dependent processes in the insects.

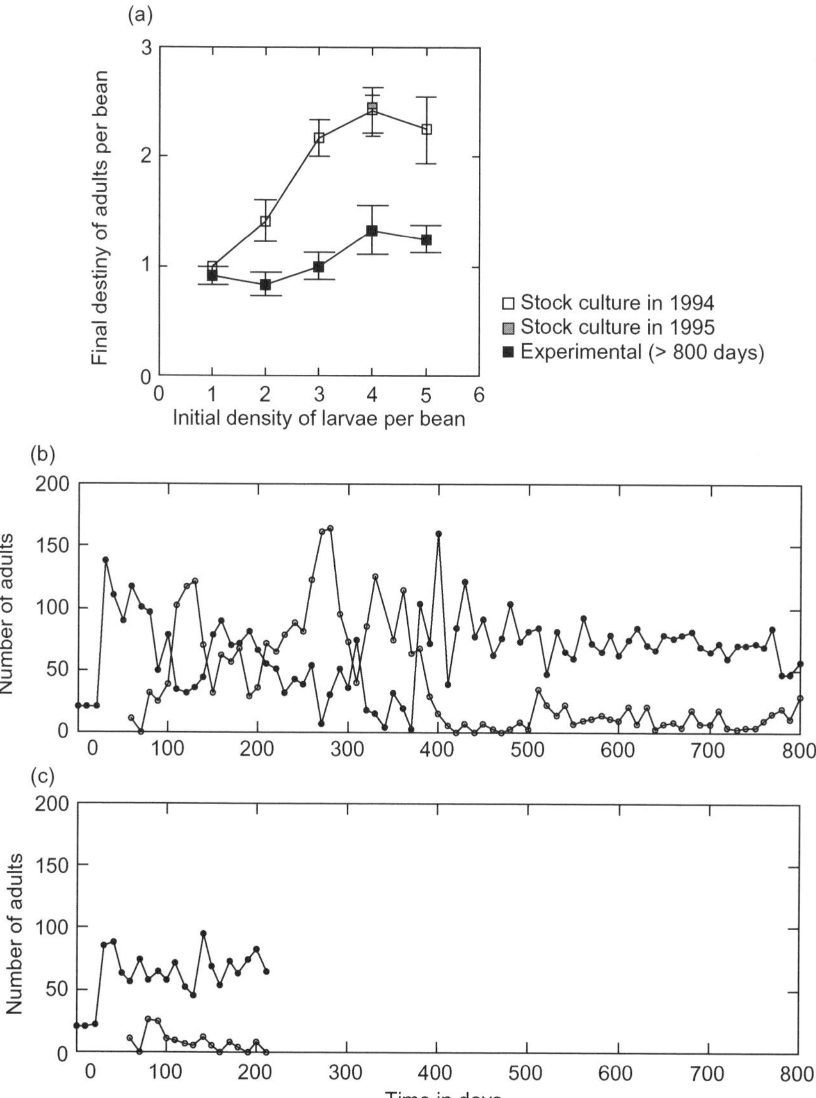

Figure 2 (a) Evolutionary change in the number of emerged adults that survived from larval competition of the host *C. maculatus* in the replicate shown in Figure 1a. Redrawn from Tuda (1998). (b) Temporal change observed in population dynamics of the *C. maculatus-H. prosopidis* host-parasitoid system (the replicate shown in Figure 1a). Redrawn from Tuda and Iwasa (1998). (c) A representative replicate of population dynamics of post-experimental host and parasitoid populations collected after 800 days from the replicate shown in Figure 2b. The open and closed circles as in Figure 1.

Because the contest type in the host is stabilizing in host-parasitoid dynamics, this bottom-up herbivore evolution enhanced the persistence of the host-parasitoid (*C. maculatus-H. prosopidis*) system. Genetic variability retained in the host probably facilitated the evolutionary change. Recently, the effect of genetic variability was directly tested on the persistence of the *C. chinensis-H. prosopidis* host-parasitoid system (Imura *et al.*, 2003). The system with an F_1 generation derived from a cross of two different strains of host as a founder persisted longer than those with single strains as founders. What is not known is how general this is and whether genetic variability in the parasitoid population has the same effect. The model prediction is that genetic variability that exhibits a trade-off in the parasitoid population does not have a critical effect, whereas that in the host population enables coexistence when the parasitoid attack rate is moderate which allows host-parasitoid coexistence in the early phase (Tuda and Bonsall, 1999).

Such evolutionary change in host-parasitoid dynamics may not be rare, especially when the insect populations are brought recently into the laboratory or a new environment. The population dynamics of *C. chinensis-H. prosopidis* system studied by Utida (1957a,b) appears to change over time (Utida, 1957a,b; Royama, 1992). Parameterization of a subset of the three-species model, by fitting to the time-series, indicated that there was indeed a decrease in the attack rate of the parasitoid as time windows for fitting are shifted gradually from the onset of the experiment to the end of the time series (Tuda, 2003, unpublished data). Evolutionary flexibility and phenotypic plasticity as a modification force on the food-web has been discussed theoretically and empirically (Pimentel, 1968; Pimentel and Stone, 1968; Thompson, 1998; Abrams, 2000; Agrawal, 2001; Kondoh, 2003). What we observed in the bruchid-parasitoid system illustrates how evolution can promote persistence of species coexistence that, in the first place, is attained ecologically. Although evolution can also reduce persistence of a host-parasitoid system, it is more difficult to capture once coexistence is terminated by extinction. For this reason, biological assemblies we see in nature may be biased examples of selection 'towards coexistence'. Experimental tests on the possibility of evolution 'against coexistence' are intriguing future challenges.

4. Spatial and Temporal Heterogeneity

Distribution of resources or plants for herbivores, which is often determined by environmental factors, is also crucial for host-parasitoid interactions to persist. Shimada (1999) showed that the degree of patchiness of resource for

C. chinensis affected variability of population dynamics of one of the two parasitoids, *A. calandrae* and *H. prosopidis*. The differential effect of patchiness between the two parasitoids is ascribed to their different searching behavior (Shimada, 1999). By comparing spatial to temporal heterogeneity of resources using two species of legumes, Mitsunaga and Fujii (1997) concluded that temporal heterogeneity or altering two resources in sequence contributed to the persistence of a two hosts-one parasitoid system. The conclusions of the two studies cannot be compared as they are because of the different structures of the assemblies. The host-parasitoid systems with different architectures have potential to be subjected to further testing for generality of the conclusions on the effect of heterogeneity, with a different combination of species.

B. Top-Down Control Factors

Functional response, or the number of attacked hosts as a function of host density, of parasitoid can be an important top-down control on host (Hassell, 1978). The attack rate of *A. calandrae* is characterized by the following: first, the number of attacked hosts increases gradually with increasing host density, specifically per-bean density, which itself is destabilizing (Utida, 1943b; Utida, 1957a; Kistler, 1985, unpublished data; Shimada and Fujii, 1985; Mitsunaga and Fujii, 1999); second, at a high density of *A. calandrae*, strong mutual interference stabilizes its population dynamics with the host. The *C. chinensis-A. calandrae* system with azuki beans (*Vigna angularis*), for example, exhibits mild oscillations with a small amplitude (Utida, 1948b, 1957a; Fujii, 1983). On the contrary, with blackeye bean, *V. unguiculata* as the resource, *A. calandrae* is able to attack more efficiently even when the density of host beetles is low, because the thin seed coat of the bean allows easier penetration of the ovipositor of the parasitoid. Eventually, host-parasitoid interaction ended with extinction probably because the destabilizing effect of host-feeding was greater than the stabilizing effect of mutual interference.

H. prosopidis has a high attack rate independent of host density but is limited by the number of eggs, which results in saturation in the number of attacked hosts as host density increases (Utida, 1957a). A *C. chinensis-H. prosopidis* system with azuki beans fluctuates with large amplitudes, sometimes showing population cycles of 100 days or longer (Utida, 1957a,b; Fujii, 1983; Tuda and Shimada, 1995). During the low-density phase, the system is likely to become extinct (Fujii, 1983; Tuda and Shimada, 1995), although it does persist for a fairly long time in some experimental replicates (Utida, 1957a,b).

IV. PERSISTENCE OF COMPLEX HOST-PARASITOID ASSEMBLIES

There is controversy over the relation between complexity and stability in a biological community. Random assembly models predict that complexity (in terms of species richness) reduces local stability and connectance (May, 1972), whereas Elton (1927) suggested that complexity enhances stability. In terms of global stability and permanence, which guarantee persistent coexistence of species, stability decreases with increasing complexity, which is consistent with May's prediction (Chen and Cohen, 2001b). Recent studies indicate that adaptive evolutionary changes in component species can either increase or decrease diversity (Abrams, 2000; Kondoh, 2003). Studies using micro-organisms supported May's prediction (Hairston *et al.*, 1968; Lawler, 1993; Lawler and Morin, 1993). Accumulation of sound empirical evidence is required for assemblies of higher organisms. Here, we review recent advances in empirical tests on the complexity-stability hypothesis in the experimental assembly of bruchid hosts and their parasitoids.

Tuda (1996a) and Tuda and Kondoh (2003, unpublished data) analyzed experimental results that tested the complexity-stability hypothesis by using bruchid beetles and their parasitoid (Ohdate, 1980; Fujii, 1981, 1994). The experimental design was as follows: each of the four compartments of a petri dish was filled with 5 g of azuki beans, then bruchids, followed by parasitoids. The maximum number of the initial component species was 5, i.e., three bruchid species (*C. maculatus*, *C. chinensis* and *Z. subfasciatus*) and two parasitoid species (*A. calandrae* and *H. prosopidis*), and the minimum was 2, i.e., two bruchid species (Fujii, 1994). Initial assemblies were replicated up to three times. As each species becomes extinct, different initial assemblies can end up with the same assembly, and replicated assemblies can lose different species. Each such transient assembly is treated as an additional replicate. Bean supply was continued until the assembly reached stable states, or stable species compositions. Persistence, or time to extinction of a species, of each assembly type was recorded. The original records were transformed to mean probability of extinction per generation of a component species (Tuda and Kondoh, 2003, unpublished data). The results indicate that extinction probability increased as the number of species increased (Tuda and Kondoh, 2003, unpublished data), and persistence decreased with the increasing number of species (Tuda, 1996a). While this supports May's complexity-instability relationship found from random combination of species, there was something beyond the prediction: a clear pattern in the composition of species in the final experimental assemblies (Ohdate, 1980; Fujii, 1983, 1994). The final assemblies were composed of either single bruchid, *C. maculatus*, or single host-single parasitoid, *C. maculatus-A. calandrae*. This shows that the bottom-up (contest competition or resultant low local carrying capacity of *C. maculatus*) and self-regulating top-down (mutual interference of *A. calandrae*) control factors proved critical in the previous section, stay effective in the larger species assemblies (Tuda, 1996a). The experiment on the large species assembly in this section illustrated that it is not only the number of species itself but also the characteristics of species interactions that determine the persistence or stability of a host-parasitoid assembly.

V. POPULATION DYNAMICS IN A THREE-SPECIES SYSTEM: AT THE INTERFACE BETWEEN SIMPLICITY AND COMPLEXITY

Recent studies of insect and animal populations demonstrated complex nonlinear dynamics (including chaos) are ubiquitous in single – as well as paired – species populations (Nicholson, 1954; Hanski *et al.*, 1993; Costantino *et al.*, 1995, 1997; Ellner and Turchin, 1995; Falck *et al.*, 1995; Turchin, 1996; Turchin and Ellner, 2000). In nature, almost all populations are concurrently interacting with multiple species, as predator and prey or as host and parasite/parasitoid. Three-species systems are one of the simplest and most intriguing candidates to allow an understanding of irregular, complex population dynamics of multi-species assemblies, with emerging effects of indirect interactions (Holt, 1977; Sih *et al.*, 1985; Briggs, 1993). An example is a laboratory experimental system with a seed beetle host and two parasitoid wasps that persisted for a fairly long time with irregular population fluctuations of the three component species (Utida, 1957a).

Detection of mechanisms by parameterized models from population time-series has become a promising approach in ecology in recent years with the application of extensive statistical tools (e.g., Dennis *et al.*, 1995, 1997, 2001; Kendall *et al.*, 1999; Jost and Arditi, 2001; Kristoffersen *et al.*, 2001; Turchin and Hanski, 2001).

With parameterized semi-mechanistic models, this paper aims to: 1) characterize persistent population dynamics of the same three species as Utida (1957a); 2) detect any change in the dynamic property and effect induced by the addition of the third species, by comparing the three-species dynamics with the two-species dynamics; and 3) clarify which characteristic(s) of the three species induced alteration of dynamic behavior.

A. Experiment

All of the experiments were conducted at 30°C, 65% relative humidity and 16: 8-hour light: dark cycles. Our system was first designed with only the host, *C. chinensis*, in a square plastic case (15 cm × 15 cm × 4 cm) in which four round dishes (diameter 6 cm), each with 10 g of azuki beans (*Vigna angularis*) were set as resource patches. Live and dead adult insects were censused in each of three replicate systems at each 10-day interval, the oldest resource patch being replaced with uninfested azuki beans. The two parasitoid species were introduced sequentially into the system after the host population had converged to the equilibrium density: *A. calandrae* on days 130 and 140, then *H. prosopidis* on days 440 and 450. Population censuses were continued until the extinction of the host on day 1550. The number of adults that emerged during the preceding 10 days was calculated as: (number of the live adults at t) + (number of the dead adults at t) − (number of the live adults at $t - 1$) in each species.

In the host-only phase, *C. chinensis* reached the equilibrium population size of about 350 between day 50 and day 100 (Fig. 3). After adding the first parasitoid, *A. calandrae*, on day 130 the host-*A. calandrae* system showed stable population dynamics with small oscillations around the equilibrium population sizes of the host (70 to 90) and *A. calandrae* (150 to 180) in all three replicate systems. However, after introduction of the second parasitoid, *H. prosopidis*, on day 440, the stable dynamics of the host and *A. calandrae* were disrupted and changed to irregular fluctuations with a large amplitude (Fig. 3).

Persistence of the three-species system ranged from 100 days (replicate C, not shown) to longer than 800 days (replicate B, Fig. 3b). Replicate C collapsed after one peak of *H. prosopidis* population outbreak because the wasp parasitized all the hosts. In replicates A and B, however, the three-species assembly persisted for 450 and 860 days (Fig. 3). Since *H. prosopidis* went essentially extinct in replicate A on day 890 (all emergents were males), 4 males and 8 females of newly emerged *H. prosopidis* were introduced, the effect of which was small on the time-series of the host and *A. calandrae* in comparison with usual demographic stochasticity. Replicate A continued thereafter until day 1,230, when finally *H. prosopidis* went extinct again, which was followed by rapid convergence to the original host-*A. calandrae* equilibrium states (Fig. 3a). The coexistence of the host and *A. calandrae* persisted thereafter until day 1,550 when we ended the long-term experiment.

In replicate B, on the other hand, the three-species assembly persisted for longer than 800 days from day 440 to day 1,300, during which population fluctuations were much larger in the host and *A. calandrae* after introduction of *H. prosopidis* than before (Fig. 3b). Both parasitoids went extinct simultaneously on day 1,300 after very low population densities of the host. When the host population had recovered, we re-introduced *H. prosopidis* only into the system to observe population dynamics of the host-*H. prosopidis* assembly. After an outbreak of host and *H. prosopidis* populations, all of the hosts were parasitized and the system collapsed on day 1,450 (not shown).

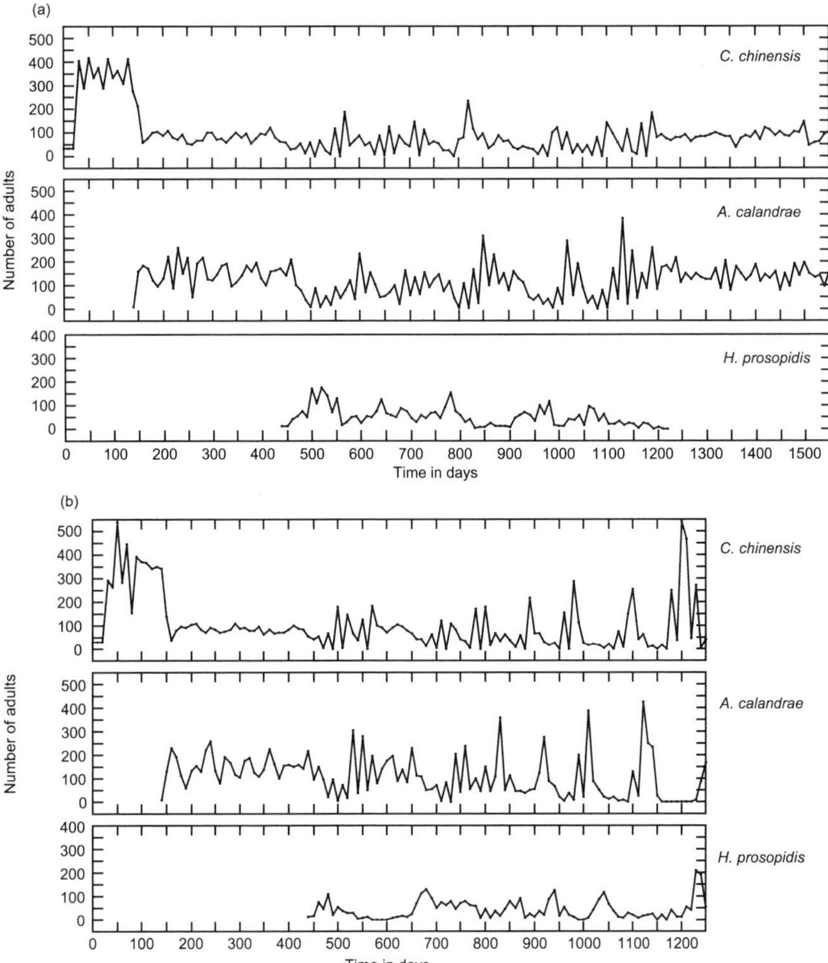

Figure 3 Observed population dynamics of *C. chinensis* (host), and its two parasitoids, *A. calandrae* and *H. prosopidis*. (a) replicate A and (b) replicate B. The numbers of adults are as explained in Figure 1. Redrawn from Kristoffersen *et al.* (2001).

Population fluctuation of the host became abruptly larger after a low-density phase of *A. calandrae* between day 1,180 and 1,210, so we omitted the time-series after day 1,200 of replicate B from later analyses in the host and *A. calandrae*.

B. Detection of Chaos by a Non-Mechanistic Model

The computer programs LENNS (Lyapunov Exponent of Noisy Nonlinear Systems) and RSM (Response Surface Methodology) were applied to estimate the dominant Lyapunov exponent (LE), used as a measure of chaotic dynamics (Ellner *et al.*, 1992; Turchin and Taylor, 1992). The 10 best fits by LENNS and the single best-fit estimates by RSM were selected to compute the LEs. Using LENNS, they were either positive (host in replicate B), zero (host in replicate A), or slightly negative (*A. calandrae* in both replicates) (Fig. 4a). Using RSM, the LEs in the three-species system were either positive (host population in both replicates and *A. calandrae* in replicate B) or slightly negative (*A. calandrae* in replicate A) (Fig. 4b). These estimates indicate that the one host-two parasitoid system can be chaotic but the result is not consistent among component species and replicates.

For comparison, the programs were also applied to the population dynamics of the two-species, or host-*A. calandrae* system (Fig. 4c, d). Both methods estimated large negative LEs for all four time-series except a positive exponent for the host in replicate A by LENNS.

By comparing the estimates between the two-species and three-species time series, the following characteristic differences were found: 1) the three-species system is less stable than the two-species system (Fig. 4); 2) the lag tends to be shorter in the three-species system (LENNS result in Table 1); and 3) the dimension is higher in the three-species system (Table 1).

C. Parameter Estimation and Reconstruction of Population Dynamics by a Semi-Mechanistic Model

1. Deterministic Model

The deterministic model is formulated with a Moran-Ricker equation for host density-dependent growth (Moran, 1950; Ricker, 1954) and a Nicholson-Bailey type equation for escape from parasitoid attack (Nicholson, 1933; Nicholson and Bailey, 1935). The selected model, based on the likelihood ratio test (Appendix 2), includes host reproduction followed by density-dependent survival of eggs according to adult density, density-dependent larval survival, and recruitment of adults from the previous time step in the host (Tuda, 2003, unpublished).

For generation time, two time units for the host and *A. calandrae* and one time unit for *H. prosopidis* were selected based on the time required for development from the egg to the reproducing female adult that is elongated by synergistic host feeding in *A. calandrae* (Appendix 1).

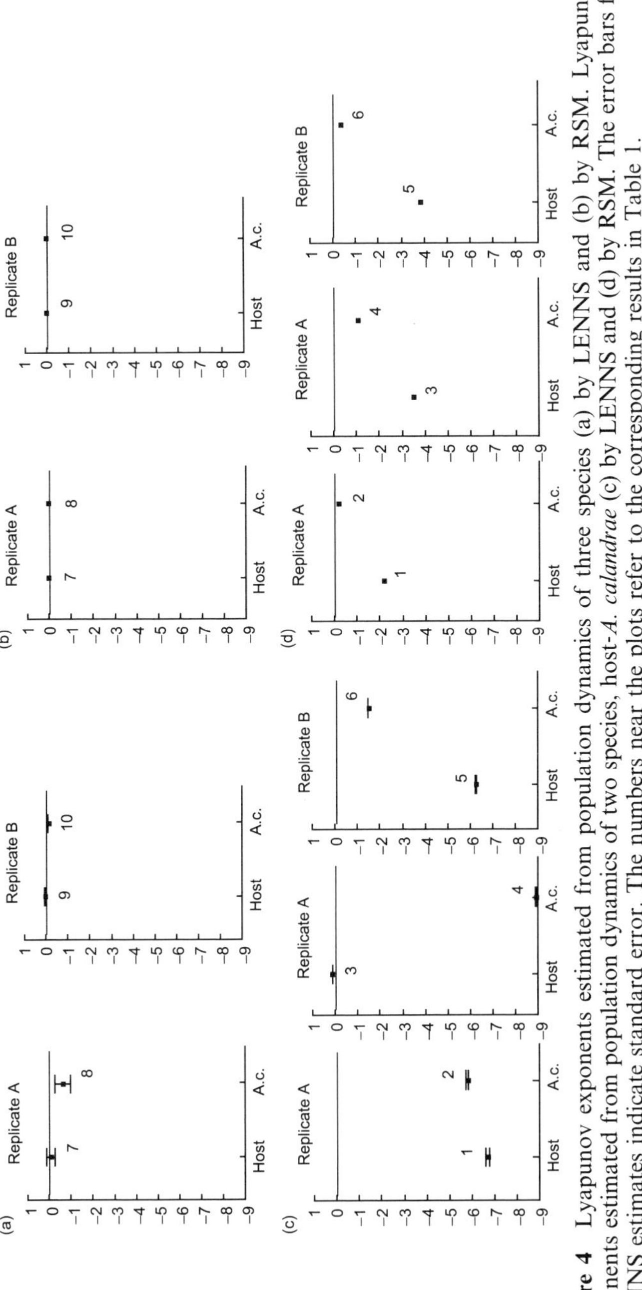

Figure 4 Lyapunov exponents estimated from population dynamics of three species (a) by LENNS and (b) by RSM. Lyapunov exponents estimated from population dynamics of two species, host-*A. calandrae* (c) by LENNS and (d) by RSM. The error bars for LENNS estimates indicate standard error. The numbers near the plots refer to the corresponding results in Table 1.

Table 1 Best fit parameters and the coefficient of determination (r^2) in LENNS and RSM. LENNS-estimated Lyapunov exponents, Λ's, are the means of the 10 best fits and l (the lag), d (the number of lags or embedding dimensions), and k (the number of nodes in the hidden layer), based on GCV (generalized cross validation) criterion (Ellner and Turchin, 1995). Numbers in parentheses indicate the second frequent l, d, or k in the 10 best fits. The RSM-estimated Λ's are, on the other hand, the best fit at the best l (the time lag = 1), d (number of lags or dimensions), and q (the order of a polynomial), based on cross-validation criterion (Turchin and Taylor, 1992). The number at the head of each time series corresponds to the number in Figure 4. The range of days shows the time series that was used for calculation

	LENNS				RSM			
	l	d	k	lnGCV	l	d	q	r^2
Host-*A. calandrae*								
Replicate A (day 140–440)								
[1]Host	2	3	1	0.996	1	1	1	0.433
[2]*A. calandrae*	4 (1, 3, 2)	1	1	0.935	1	3	1	0.801
Replicate A (day 1,220–1,550)								
[3]Host	1 (2, 3)	1	1	0.687	1	1	1	0.468
[4]*A. calandrae*	1 (3)	1	1	0.500	1	1	1	0.747
Replicate B (day 140–440)								
[5]Host	2 (4, 1)	1	1	1.07	1	1	1	0.624
[6]*A. calandrae*	1	2	1	0.427	1	2	1	0.696
Host-2 parasitoids								
Replicate A (day 440–1,230)								
[7]Host	1	3 (4, 2)	2	−0.319	1	2	2	0.818
[8]*A. calandrae*	1	4 (2)	2	−0.446	1	3	2	0.788
Replicate B (day 440–1,190)								
[9]Host	1	4 (3)	2	−0.486	1	2	2	0.807
[10]*A. calandrae*	1	4	1	−0.322	1	3	2	0.689

Multiparasitized (i.e., co-parasitized) hosts were assumed to produce each parasitoid with an equal chance (Appendix 1), thus the probability of producing *A. calandrae* is the sum of the probability of being attacked at least once by *A. calandrae* but not by *H. prosopidis*, which is $\exp(-a'_Z Z'_{t-1})\{1 - \exp(-a'_Y Y'_{t-1})\}$ and the probability of being attacked by both parasitoids and only *A. calandrae* surviving, that is $\{1 - \exp(-a'_Z Z'_{t-1})\}\{1 - \exp(-a'_Y Y'_{t-1})\}/2$. The sum is reduced to the form: $\{1 + \exp(-a'_Z Z'_{t-1})\}\{1 - \exp(-a'_Y Y'_{t-1})\}/2$. The same can be done for *H. prosopidis*. Population sizes of each developmental stage of the host

Table 2 Definitions of variables and parameters

Variables/Parameters	Definition
$X_{l,t}$	number of host larvae
$X_{a,t}$	number of host adults
$Y_{l,t}$	number of larvae of parasitoid *A. calandrae*
$Y_{a,t}$	number of adults of parasitoid *A. calandrae*
Z_t	number of adults of parasitoid *H. prosopidis*
λ	number of eggs deposited by a female host × 0.5
$b0$	density dependence in egg mortality in host
$b1$	density dependence in larval mortality in host
s	survival rate of host adults from t-1 to t
a_Y	searching efficiency of parasitoid *A. calandrae*
f	coefficient for host-feeding by *A. calandrae* that contributes to egg production
m	mutual interference in *A. calandrae*
a_Z	searching efficiency of parasitoid *H. prosopidis*
c_Z	handling time (or inverse of number of eggs) in parasitoid *H. prosopidis*

(X_l, X_a) and parasitoids, *A. calandrae* (Y_l, Y_a) and *H. prosopidis* (Z), at time t are thus described as:

$$X_{l,t} = \lambda X_{a,t-1}\exp\{-b_0 X_{a,t-1} - b_1 \lambda X_{a,t-1}\exp(-b_0 X_{a,t-1})\} \tag{1}$$

$$X_{a,\mathrm{t}=X_{l,t-}\ \mathrm{fff}1\exp(-a^{4\prime}_{Y}Y'_{t-1}-a'_{Z}Z'_{t-1})+sX_{a,t-1}} \tag{2}$$

$$Y_{l,t} = X_{l,t-1}\{1+\exp(-a'_Z Z'_{t-1})\}\{1-\exp(-a'_Y Y'_{t-1})\}/2 \tag{3}$$

$$Y_{a,t} = Y_{l,t-1} \tag{4}$$

$$Z_t = X_{l,t-1}\{1+\exp(-a'_Y Y'_{t-1})\}\{1-\exp(-a'_Z Z'_{t-1})\}/2. \tag{5}$$

The variables and parameters are defined in Table 2. The gender ratios of parasitoid was approximated by 0.5, so that the numbers of females are described as $Y'_t = Y_{a,t}/2$ and $Z'_t = Z_t/2$ for each respective parasitoid. Female *A. calandrae* feeds on hosts and this increases the number of her mature eggs. Therefore, host-feeding by adult females that contributes exponentially to the attack rate was incorporated. It is observed that the female *A. calandrae* interrupt with parasitizing conspecifics on encounter by jumping upon them (Tuda, 2003, personal observation), while female *H. prosopidis* do not show interference behavior, although pseudo-interference appears at extremely high density (Fig. 4 in Shimada, 1999). For *A. calandrae*, mutual interference (Hassell and Varley, 1969) was incorporated that reduces the attack rate as adult densities of *A. calandrae* increase (Shimada, 1999). *H. prosopidis* searches and parasitizes hosts efficiently but soon depletes its eggs by doing so, as in a typical type II functional response (Holling, 1959; Royama, 1971; Rogers, 1972). And it does not host-feed. Thus, the attack rates of the parasitoids are formulated as:

$$a'_Y = a_Y \exp(fX_{l,t-1})\, Y'^{-m}_{t-1} \tag{6}$$

$$a'_Z = a_Z/(1 + a_Z c_Z X_{l,t-1}). \tag{7}$$

Not all of the developmental stages in equations 1 to 5, however, can be observed. Larval stages of the host and parasitoid are invisible from outside the beans. The equations for the larval stages (equations 1 and 3) are embedded into equations 2 and 4, respectively. Thus, the deterministic model is represented by three state variables:

$$X_{a,t} = H_{t-2}\exp(-a'_{Y,t-1}Y'_{t-1} - a'_{Zt-1}Z'_{t-1}) + sX_{a,t-1} \tag{8}$$

$$Y_{a,t} = H_{t-3}\{1 + \exp(-a'_{Zt-2}Z'_{t-2})\}\{1 - \exp(-a'_{Y,t-2}Y'_{t-2})\}/2 \tag{9}$$

$$Z_t = H_{t-2}\{1 + \exp(-a'_{Y,t-1}Y'_{t-1})\}\{1 - \exp(-a'_{Zt-1}Z'_{t-1})\}/2, \tag{10}$$

where

$$a'_{Y,t-i} = a_Y \exp(fH_{t-i-1})\, Y'^{-m}_{t-i} \tag{11}$$

$$a_{Z,t-i} = a_Z/(1 + a_Z c_Z H_{t-i-1}) \tag{12}$$

$$H_{t-i} = \lambda X_{a,t-i}\exp\{-b_0 X_{a,t-i} - b_1 \lambda X_{a,t-i}\exp(-b_0 X_{a,t-i})\}. \tag{13}$$

2. *Stochastic Model and Fitting to Time-Series Data*

In the observed time series, noise is added to the deterministic dynamics. Demographic noise presumably dominates, relative to environmental noise in our controlled growth-chamber environment. Host reproduction is assumed to follow a Poisson distribution with mean $\lambda X_{a,t-1}$. Survival through density-dependent competition and parasitism can be described as a binomial distribution. A binomially distributed variable with Poisson-distributed mean n and probability p is known to follow Poisson distribution with mean μp, where μ is the mean of the Poisson distribution of n (Boswell *et al.*, 1979). Therefore, all the variables on the left hand of equations 1 to 5 can be approximated by a Poisson distribution. The numbers in the present data set are square-root transformed to normalize the Poisson distribution (Dennis *et al.*, 2001). Stochastic realization of the model with demographic noise is as follows:

$$x_{a,t}^{1/2} = \{h_{t-2}\exp(-a'_Y y'_{t-1} - a'_Z z'_{t-1}) + s x_{a,t-1}\}^{1/2} + E_{x,t} \quad (14)$$

$$y_{a,t}^{1/2} = [h_{t-3}\{1 + \exp(-a'_Z z'_{t-2})\}\{1 - \exp(-a'_Y y'_{t-2})\}/2]^{1/2} + E_{y,t} \quad (15)$$

$$z_t^{1/2} = [h_{t-2}\{1 + \exp(-a'_Y y'_{t-1})\}\{1 - \exp(-a'_Z z'_{t-1})\}/2]^{1/2} + E_{z,t}. \quad (16)$$

For the assumptions on the noise and parameter estimation, see Appendix 3.

Numbers of adults of each species from day 430 to 870 in replicate A and day 430 to 1270 in replicate B were used as time series to fit the model.

3. *Two-Species Dynamics*

The host-*A. calandrae* dynamics before the introduction of *H. prosopidis* were fit by a subset of the three-species model to see if any factor was modified by the presence of *H. prosopidis*. The deterministic model for the host (X_a) and *A. calandrae* (Y_a) was a modified version of equations 8 and 9:

$$X_{a,t} = H_{t-2}\exp(-a'_{Y,t-1} Y'_{t-1}) + sX_{a,t-1} \quad (17)$$

$$Y_{a,t} = H_{t-3}\{1 - \exp(-a'_{Y,t-2} Y'_{t-2})\}. \quad (18)$$

A stochastic model for the two species with normally-distributed demographic noise was modified from equations 14 and 15:

$$x_{a,t}^{1/2} = \{h_{t-2}\exp(-a'_{Y,t-1} y'_{t-1}) + s x_{a,t-1}\}^{1/2} + E_{x,t} \quad (19)$$

$$y_{a,t}^{1/2} = [h_t - 3\{1 - \exp(-a'_{Y,t-2} y'_{t-2})\}]^{1/2} + E_{y,t}. \quad (20)$$

Observed population sizes of each replicate were used for the initial values in the simulations: replicate A; $X_1 = 80$, $X_2 = 61$, $X_3 = 57$, $Y_2 = 169$, $Y_3 = 142$, $Z_2 = 12$, and $Z_3 = 13$; and replicate B; $X_1 = 85$, $X_2 = 55$, $X_3 = 42$, $Y_2 = 218$, $Y_3 = 96$, $Z_2 = 12$, and $Z_3 = 17$ for the three-species system (day 430 to 450). Replicate A; $X_1 = 410$, $X_2 = 274$, $X_3 = 212$, $Y_2 = 8$, and $Y_3 = 158$ and replicate B; $X_1 = 351$, $X_2 = 343$, $X_3 = 140$, $Y_2 = 8$, and $Y_3 = 132$ were used for the two-species system (day 130 to 150).

4. *Estimation of Dominant LE*

The dominant LE was estimated from a time-series of time unit (t) 1,001–2,000 (1 time unit = 10 days), omitting the first 1,000 points to remove any transient phase. The Jacobian matrix of partial derivatives of equations

1 to 5 was used to estimate LE by applying the chain rule and then averaging the exponential divergence rates of initial perturbation over time. Bootstrapped datasets were constructed to estimate the confidence intervals of LEs, following Dennis *et al.* (2001). The error terms of the three populations as a set were resampled 1,000 times with replacement after standardization. Except for the first three data points, the bootstrapped errors were added to (square-rooted) population densities generated by deterministic simulation of the models with maximum likelihood (ML) estimates, and squared. When the population densities with added noise were <0, they were set to 0.5 before squared. The resultant data with the first three observed data in each replicate were fitted by the model to estimate LE. The 2.5th and 97.5th percentiles of the LE distribution were considered as 95% confidence limits of the LE.

D. Results and Discussion

1. *Three-Species Population Dynamics*

The results of the fit by the model and stochastic simulation are shown in Fig. 5. The ML estimates and 95% confidence intervals (CI) were as shown in Table 3.

The dominant LE estimated from a deterministic simulation of the model with the ML estimates was slightly negative, −0.0510 (bootstrap 95% CI, −0.684, 0.0014 for lower and upper confidence limits (CL) Fig. 6a). Simulated population dynamics with ML estimates had transients to stable equilibrium (Table 3) reached at time unit 472 (bootstrap 95% CI for duration of transient dynamics, 46, 2000 <). The equilibrium may not be reached in the experiments that persisted for 10, 45, and 86 time units. Among estimates for all the bootstrapped data, 9.3% exhibited complex dynamics (Fig. 6b) (about 8% were either chaotic or quasiperiodic and about 1% were long chaotic transient dynamics). The LEs were also estimated with the 95% CL of the parameters and were either positive or 0 at the lower limits of parameters a_Y and m for *A. calandrae* (equation 11), whereas at the other limits the LEs were all negative but with small absolute values (Table 3; Fig. 7 for one-dimensional bifurcation diagrams). The LE can be used as an index of resilience (i.e., ability to recover from a small perturbation; Gunderson, 2000) because the larger absolute values of negative LE, the more rapid convergence to an equilibrium (a strong positive correlation between inverse of the duration of transients and absolute values of negative LE, $r = 0.992$, $n = 128$, $p < 0.0001$). In multi-dimensional bifurcation, where all parameters were estimated for each value of a parameter changed within its confidence interval, LEs monotonically increased with decreasing a_Y and

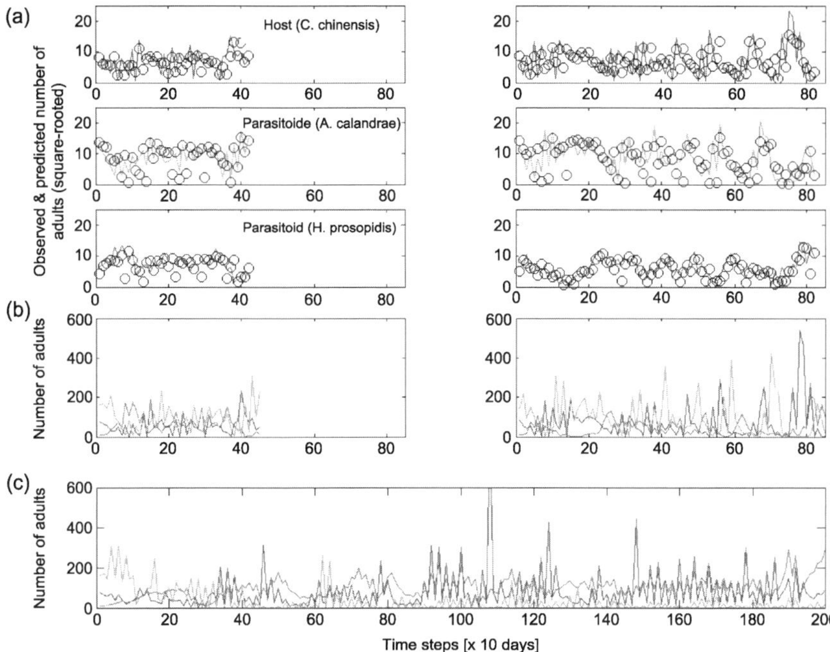

Figure 5 (a) One-step-ahead predictions by the model (+), superimposed on population dynamics of adult numbers of three species (square-root transformed) (connected by lines). (b) Observed population dynamics of the three species (not transformed). Upper panel, solid lines indicate parasitoid *A. calandrae*. Lower panel, parasitoid *H. prosopidis*. (c) Stochastic simulation of the three-species model with maximum likelihood estimates. Demographic noise and environmental noise (standard deviation of the environmental noise = 0.25) are incorporated. Upper panel, solid lines indicate host adults ($X_{a,t}$), dotted lines indicate adults of parasitoid *A. calandrae* ($Y_{a,t}$). Lower panels, adults of parasitoid *H. prosopidis* (Z_t).

m that can induce complex population dynamics (i.e., chaos and quasiperiodicity).

Within the confidence interval of LE, these chaotic, quasiperiodic, or chaotic transient behaviors resembled cycles of periods that varied from 3–4 to 18–19 time steps (typically 4 and 11 time steps, that is, 2 and 5.5 host generations). When the attractor is a point equilibrium, long transients of quasi-period of 10–25 (about 5–12.5 host generations) were seen before reaching the equilibrium state. During low-density phases of such periods, a population is vulnerable to extinction especially with demographic noise. Extinctions that occurred at time units 10 (replicate C), 45 (replicate A), and 86 (replicate B) in the three experimental replicates may correspond to such low densities in quasi-cycles. Incorporation of demographic stochasticity

Table 3 Estimated values and their 95% confidence intervals of parameters, and property of attractors at the confidence limits of the three-species system, *C. chinensis* (host) and two parasitoids, *A. calandrae* and *H. prosopidis*. An asterisk indicates a significant difference from the parameter value estimated from the two-species system. The Lyapunov exponent for ML estimate was –0.0510 (bootstrap 95% CI, –0.684, 0.0014) and simulated dynamics showed chaotic transients to stable equilibrium (X_a, Y_a, Z) = (30.33, 42.83, 87.92). The absolute of eigen values at the equilibrium were all <1. Estimated elements in variance-covariance matrix were $\sigma_{11} = 10.38$, $\sigma_{22} = 12.40$, $\sigma_{33} = 6.899$, $\sigma_{12} = \sigma_{21} = 0.1411$, $\sigma_{13} = \sigma_{31} = 1.292$, $\sigma_{23} = \sigma_{32} = -1.719$

Parameters	ML Estimate (95% CI)	Lyapunov Exponent With 95% CL	Attractor With 95% CL
λ	7.580 (5.726, 11.52)	(–0.0219, –0.143)	(equilibrium, equilibrium)
b_0	0.004658 (0.002584, 0.006560)	(–0.0325, –0.124)	(equilibrium, equilibrium)
b_1	0.001168 (0.0006660, 0.001529)	(–0.0192, –0.141)	(equilibrium, equilibrium)
s	0.09774(0.03996, 0.1723)	(–0.0502, –0.0655)	(equilibrium, equilibrium)
a_Y	0.07168(0.02517, 0.1796)	(0.0003, –0.0780)	(chaos, equilibrium)
f	0.002082 (0, 0.005473)	(–0.0704, –0.0370)	(equilibrium, equilibrium)
m	0.4480 *(0.2684, 0.6024)	(–0.0003, –0.0903)	(quasiperiod, equilibrium)
a_Z	0.1580 (0.04885, 1.0)	(–0.0546, –0.0223)	(equilibrium, equilibrium)
c_Z	0.1891 (0.1156, 0.2545)	(–0.0257, –0.152)	(equilibrium, equilibrium)

was likely to terminate the system. The dynamics of the three-species assembly is concluded to be long cyclic/chaotic transients towards an attractor that is either point equilibrium, cycles, quasi-cycles, or chaos, judging on a realistic time scale. In summary, complex population dynamics in the three-species assembly are likely either when the searching efficiency of *H. prosopidis* is high (Fig. 7a), when the searching efficiency of *A. calandrae* is low (Fig. 7b) that it associates with the high rate of synergistic host feeding, or when the mutual interference of *A. calandrae* is low (Fig. 7c). Both synergistic host feeding and low mutual interference are generally destabilizing (May and Hassell, 1973, 1981; Briggs *et al.*, 1995).

The model that assumes different host developmental stages are vulnerable to the two parasitoids (with *H. prosopidis* attacking the host first), had a lower ML (-948.972) and, therefore, was declined despite its simplicity. The

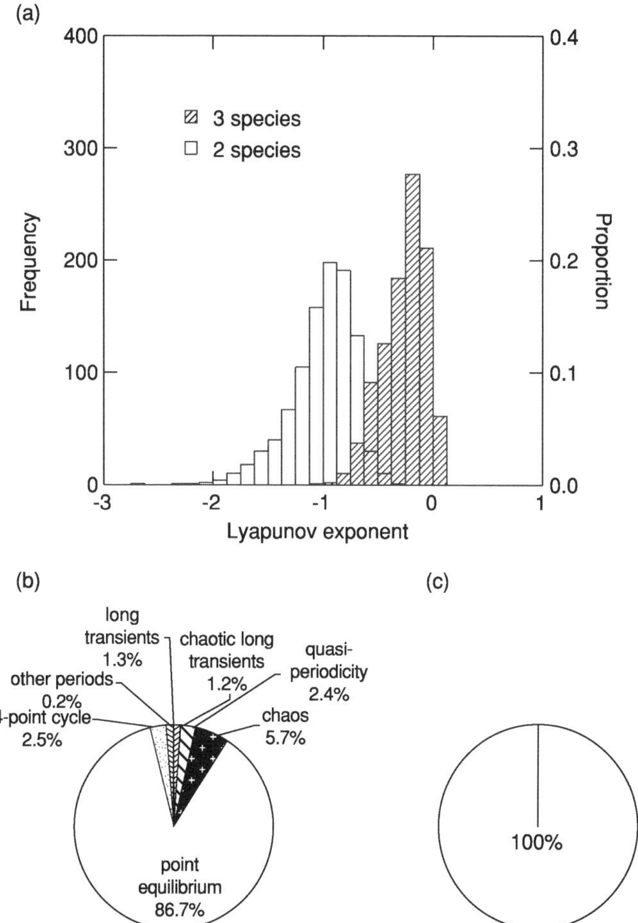

Figure 6 Estimated property of bootstrapped data sets. (a) Frequency distribution of Lyapunov exponents of deterministic three- and two-species dynamics. Population dynamic properties of attractors for (b) three-species system and (c) two-species system.

model with a generation time of *A. calandrae* of 1 time unit instead of 2 had a lower ML (-992.770), which was also declined. Our semi-mechanistic model is different from the non-parametric (or non-mechanistic) model proposed by Kristoffersen *et al.* (2001), in that multiparasitism (i.e., co-parasitism) is incorporated and mutual interference is present in *A. calandrae*. In their study, multiparasitism was assumed negligible based on experimental results by Shimada (1985). Our explicit statistical comparison of models with and without multiparasitism showed that multiparasitism by the two parasitoids is not negligible. We interpret the experimental result on the niche modification by the two coexisting parasitoid species (Shimada and Fujii, 1985) as follows: The separation of the host stages that the two parasitoids utilize was only partial, which probably enabled the coexistence of the parasitoids for more than 100 days (7–9 generations) even when the single stage of hosts was provided (Shimada, 1985). Previous experiments showed that mutual interference is present in *A. calandrae* (Shimada, 1999) which is consistent with our result. In *H. prosopidis*, on the other hand, the effect of mutual interference becomes manifest only when its density is extremely high (Shimada, 1999), which may be the cause of exclusion of mutual interference by the present analysis. The time series used for analysis in Kristoffersen *et al.* (2001) was different from the one used here (i.e., the former excluded the initial part of both replicates and the last part of replicate B and included the last part of replicate A), which might explain the different density-dependence structures. Furthermore, mutual interference that remains constant along its density (i.e., a constant negative slope along logged densities), as in *A. calandrae* (Shimada, 1999), is more likely to be detected in functional response, embedded in Nicholson-Bailey type model (our model), than in total density (Kristoffersen *et al.* [2001] model).

The estimated parameter set at the confidence intervals indicated some correlations between parameters; positive correlations among λ, b_0, and b_1 (between λ and b_0, $r = 0.823$, $n = 19$, $p < 0.001$; between λ and b_1, $r = 0.863$, $p < 0.001$) and between a_Z and c_Z ($r = 0.690$, $p < 0.05$), and a negative correlation between a_Y and f ($r = -0.743$, $p < 0.01$).

2. Two-Species Population Dynamics

The host-*A. calandrae* dynamics before introduction of *H. prosopidis* were fit by a subset of the three-species model (Equations 19 and 20) to determine if any factor was modified by the presence of *H. prosopidis* (Fig. 8). The ML estimates and 95% confidence intervals are given in Table 4. The LE with ML estimates was negative, –0.889 (bootstrap 95% CI, –1.692, –0.585). The bootstrapped LE for the two-species system was significantly smaller than that for the three-species (Fig. 6a) (Bartlett's test for homogeneity of variance, $F = 141.5$, d.f. = 1, $p < 0.0001$; Welch's ANOVA, $F = 4{,}533.2$, numerator d.f. = 1, denominator d.f. = 1,764.8, $p < 0.0001$) and the simulated dynamics for all the bootstrapped data quickly converged to a stable equilibrium at the time unit, 35 (bootstrap 95% CI, 19, 49 time units) (Fig. 6c, Table 4). Among the estimated parameters, m for mutual

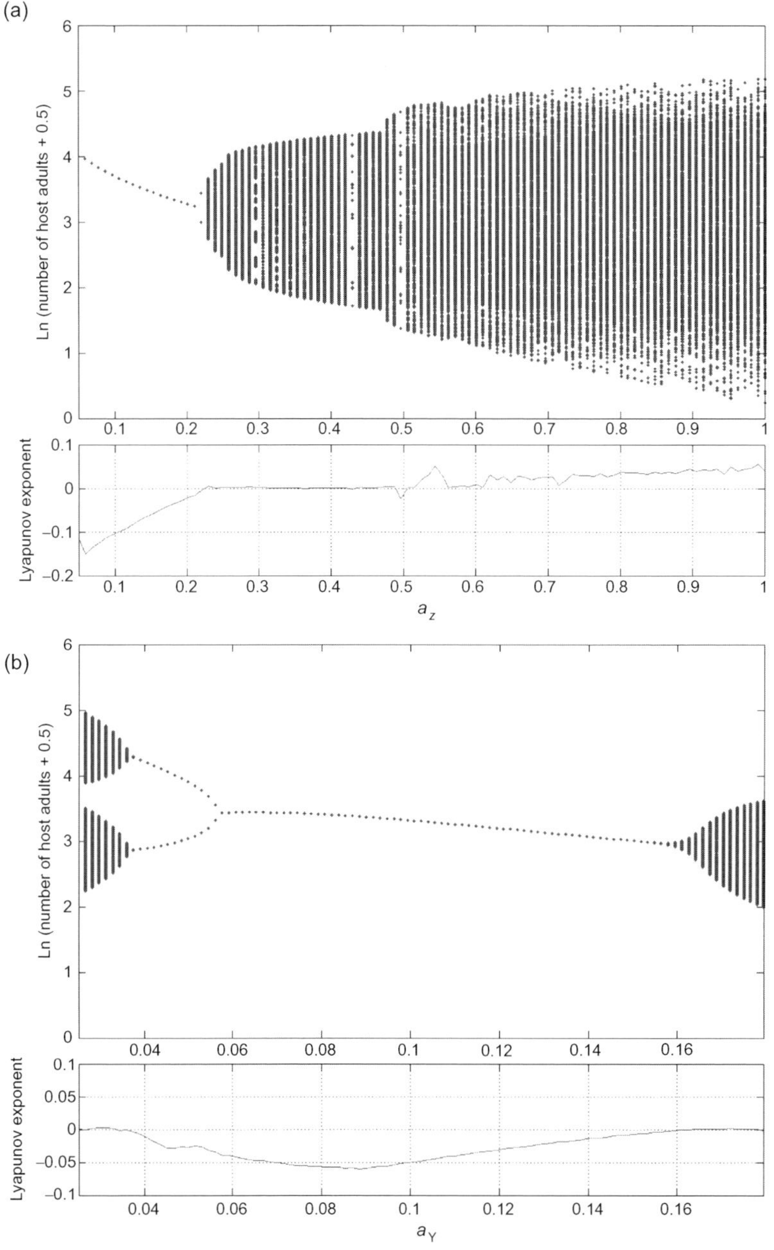

Figure 7 (*continued*)

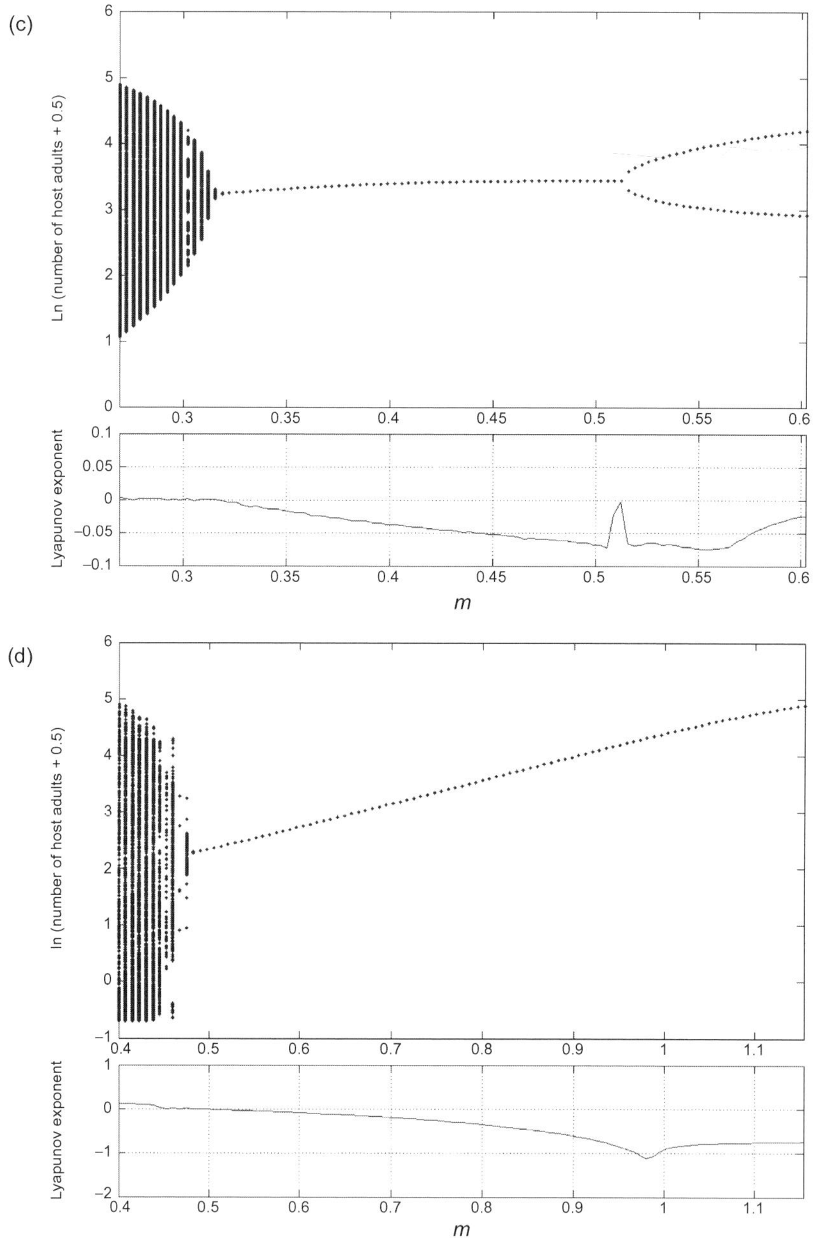

Figure 7 Bifurcation diagrams and corresponding dominant Lyapunov exponents for (a) a_Z, (b) a_Y, and (c) m in three-species system within 95% confidence interval of respective parameter and (d) m in two-species system with extended lower limit of its 95% confidence interval to compare to that of three species.

Table 4 Estimated parameter values and their 95% confidence intervals for the two-species system, *C. chinensis* (host) and *A. calandrae*. An asterisk indicates a significant difference from the parameter value estimated from the three-species system. The Lyapunov exponent with ML estimate was −0.889 (bootstrap 95% CI, −1.692, −0.585) and simulated dynamics showed quick convergence to stable equilibrium ($X_a = 81.56$, $Y_a = 147.32$). Estimated elements in variance-covariance matrix were $\sigma_{11} = 0.8513$, $\sigma_{22} = 2.965$, $\sigma_{12} = \sigma_{21} = -0.2112$

Parameters	ML Estimate (95% CI)	Lyapunov Exponent With 95% CL	Attractor With 95% CL
λ	6.776 (5.157, 9.529)	(−1.06, −0.843)	(equilibrium, equilibrium)
b_0	0.003654 (0.001540, 0.005352)	(−0.986, −0.817)	(equilibrium, equilibrium)
b_1	0.001398 (0.001013, 0.001585)	(−1.15, −0.708)	(equilibrium, equilibrium)
s	0 (0, 0.1506)	(−0.888, −0.698)	(equilibrium, equilibrium)
a_Y	0.2473 (0.08347, 0.6547)	(−0.670, −0.741)	(equilibrium, equilibrium)
f	0.006286 (0.003209, 0.01107)	(−1.01, −0.546)	(equilibrium, equilibrium)
m	1.005 * (0.8636, 1.154)	(−0.638, −0.470)	(equilibrium, equilibrium)

interference that can stabilize the system was significantly larger in the two-species system than in the three-species system with *H. prosopidis* (Welch's $t = -4.93, p < 0.001$) (Tables 3 and 4). A bifurcation diagram showed that a reduction in m results in chaotic dynamics (Fig. 7d).

The larger values of m in the two-species system can be explained by the higher density of *A. calandrae* relative to the host density, in the absence of *H. prosopidis*. With the addition of the third species (or the second parasitoid), hosts, on average, became less available and consequently *A. calandrae* did not increase to the population level that triggers mutual interference behavior because of limited host-feeding on a less abundant host. However, during chaotic transients the host can become temporally super-abundant following low density of parasitoids, which may be the cause of the reduced self-regulation effect of mutual interference among *A. calandrae*.

On the other hand, the *C. chinensis-H. prosopidis* system was diagnosed as a long chaotic transient, and there were no significant changes in the parameter values from those in the three-species system, indicating asymmetric interspecific interactions between the two parasitoids (Tuda, 2004, unpublished).

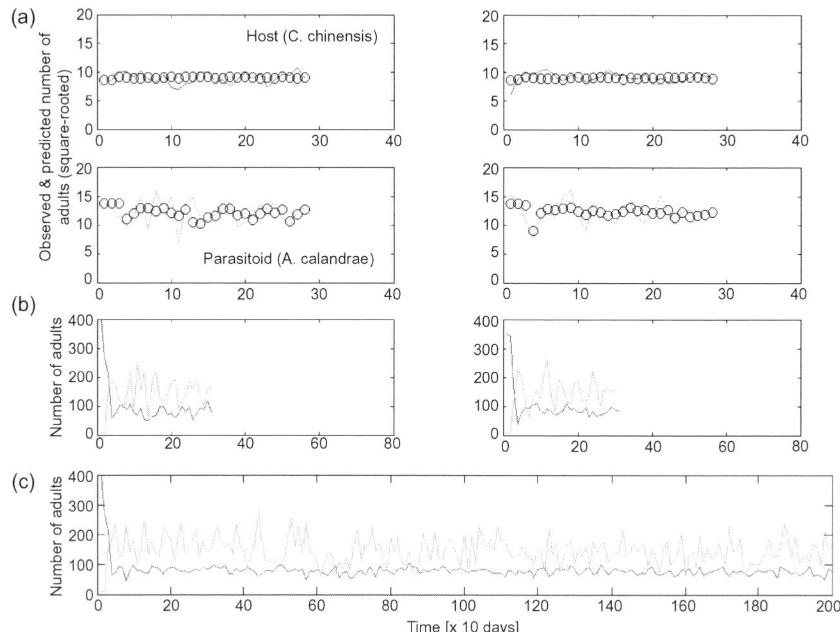

Figure 8 (a) One-step-ahead predictions by the model (+), superimposed on population dynamics of adult numbers of two species (square-root transformed) (connected by lines). (b) Observed population dynamics of the two species (not transformed). (c) Stochastic simulation of the three-species model with maximum likelihood estimates. Demographic noise and environmental noise (standard deviation of the environmental noise = 0.15) are incorporated. The lines and symbols are as indicated in Figure 5.

3. Changes in Dynamical Behavior From Two- to Three-Species Assemblies

Complex host-parasitoid dynamics in the three-species system were generated when self regulations in the host and the parasitoid *A. calandrae* were low and when the searching efficiency of the non–self-regulating parasitoid *H. prosopidis* was high. This is induced by the introduction of a non–self-regulating parasitoid that also may have reduced the self-regulation in the other parasitoid indirectly through reduced host density. It is also important to realize that the observed population dynamics correspond to initial transients towards such an attractor. A long transient phase is known for high-dimensional physical and biological models (Crutchfield and Kaneko, 1988; Kaneko, 1990; Hastings and Higgins, 1994; Chen and Cohen, 2001a). The asymmetric stronger effect of a destabilizing species (*H. prosopidis*) on a stabilizing species (*A. calandrae*) is the key to generating complex transient population behavior in our three-species system. *H. prosopidis* has a higher searching efficiency and a priority effect, both of them contribute to competitive superiority of *H. prosopidis* to *A. calandrae*, which explains the larger influence of *H. prosopidis* to *A. calandrae* than vice versa.

Another three-species system (Begon *et al.*, 1996c, 1997) showed multiple-generation cycles, compared to the generation cycles with smaller magnitudes of either single-species (host only) or any combination of two component species. Three-, or more, species systems are likely to have indirect interactions besides direct interactions, which can generate irregular, unpredictable dynamics (Holt, 1997). Several prey and predator species may be involved in the 10-year population cycles of the hare-*Lynx* system (Keith, 1983). Many natural systems contain large numbers of interactions, some destabilizing, some stabilizing, and their combined effects almost always include an indirect interaction that could not be predicted from independent pairwise interactions (Hassell and Anderson, 1989; Begon *et al.*, 1996a,b).

The dynamics of interacting populations that we observe may often display transient behavior because it is likely that they have been driven away from intrinsic attractors by recent natural as well as human-induced changes. The characteristics of attractors themselves may be altered by evolution of component species in the assemblies (e.g., Tuda and Iwasa, 1998). Understanding non-equilibrium (i.e., transient and/or evolutionary) dynamics is crucial for the prediction on dynamics of biological assembly under changing environments. Laboratory experimental host-parasitoid systems will provide us with tools for further testing of ecological and evolutionary topics.

VI. HOST-PARASITOID INTERACTION AND BEYOND

In brief, ecological rules on population dynamics of interacting species discovered from host-parasitoid experimental systems with bruchid beetles as the host are as follows:

Transient dynamics: The attractor of the three-species system has both stable and chaotic features. In the chaotic region of the attractor, quasi-cycles of a period of multiple (and, in some cases, partial) generations of host and parasitoids appeared in the simulation. Even when the attractor is a point equilibrium, a long transient phase of quasi-cycles of multiple generations preceded, and the time required to reach equilibrium states is beyond the time of observed coexistence. The population dynamics of three-species host-parasitoid dynamics, therefore, is ascribed to a chaotic transient to a stable/chaotic attractor. Such intrinsically long transients can easily be elongated by demographic and environmental noise that constantly excites the population dynamics away from an equilibrium state (Bauch and Earn, 2003).

Bottom-up cascade control: Species of bean resources (plant) can control persistence of interactions of herbivores and parasitoids. In the simple *Callosobruchus*-parasitoid systems, the window of vulnerability was elongated by host-plant shift, which resulted in termination of the *Callosobruchus*-parasitoid interaction because of increased vulnerability of the host *Callosobruchus* to parasitism.

Evolutionary cascade effect: Evolutionary change occurred in the *Callosobruchus* beetle within a relatively short time span, as it shifted to a new bean resource that differs in size from its original bean. This change not only affected its own population dynamics but also modified host-parasitoid dynamics. An implication of the result for agricultural selection of larger and nutrient-rich crops is that pests can become more scramble-type competitors that allow more serious outbreaks of a pest population even under a control of parasitoid. It should be noted that the biology of evolution of contest/scramble competition can be more profound than shown in the present paper; the selection background of different geographical populations constrains artificial selection in the laboratory (Takano *et al.*, 2001; Kawecki and Mery, 2003; Tuda, 2005, unpublished).

In the future, how evolutionary changes modify behavior of interacting species and eventually stability/persistence of overall assembly of hosts and parasitoids will be one of the next themes to pursue. As described in the section on bottom-up control, experimental tests on the possibility of evolution 'against coexistence' are intriguing future challenges. Does evolutionary capacity of species tend to vary between trophic levels? Does it depend on the number of interacting species (generalist vs. specialist, herbivore vs. carnivore, intraguild predator vs. non-intraguild predator, and so on)? These are some of the questions that do not yet have concrete empirical answers, however, they are crucial for the prediction of possible outcomes on the whole assembly when a species either is introduced or goes extinct. Evolutionary scheme of ecological stability/persistence of biological interactions may also be extended to include a lower trophic level (i.e., producer or plant) which could be much more complex. What is known from the investigation on insect fauna associated with leguminous plants, however, suggests that it may not complicate our understandings but rather simplify them: ecological characteristics of legume species such as their distributional range (similar to island-size effect), morphology (tree or herb), and historical background (introduced, native or endemic) are found to be excellent predictors of species richness of their seed predators, including bruchid beetles, and consequently their parasitoids (Tuda *et al.*, 1998, unpublished). The process of accumulation of insect herbivores, followed by parasitoids on introduced plants or that of parasitoid on invading insect herbivores is not well understood (Tuda *et al.*, 2001); this should include learning and evolutionary processes in insects. Patterns of species richness in nature such as this will be readily tested for their ecological and evolutionary processes, using laboratory experimental systems of bruchid beetles and their parasitoid (for a list of new candidates of *Callosobruchus* species for future laboratory experiments on species richness, see Tuda *et al.*, 2005, and Tuda, 2003).

ACKNOWLEDGMENTS

We are grateful to R.F. Costantino, J.M. Cushing, B. Dennis and R.A. Desharnais for the opportunity to present this paper. Thanks are also extended to J.E. Cohen, B. Dennis, H. Hakoyama, T. Ikegami, Y. Iwasa, G.J. Kenicer, T. Yanagawa, and two anonymous referees for their valuable advice. Discussions at the Complex Population Dynamics working group at the National Center for Ecological Analysis and Synthesis, Santa Barbara, and at the Centre for Population Biology, Silwood Park, stimulated this work. This study was supported in part by Grant-in-Aid for Scientific Research (08640798 and 09640747) and for International Scientific Research (07044180 and 09044202) from MESC and by JSPS fellowship (083895).

APPENDICES

Appendix 1. Basic Ecology of the Experimental Organisms

Beetles of the genus *Callosobruchus* (Chrysomeloidea: Bruchidae) utilize beans or seeds of leguminous plants as the resource for their larvae. Adults deposit eggs and hatching larvae bore into not only immature, soft seeds but also mature, dried ones. Consequently, the beetles are stored bean pests. Since *Callosobruchus* larvae, except those at the late fourth (last) stage, cannot survive once they are removed from beans, the beans they bore into are the only resource they consume. During development, therefore, the larvae can suffer from competition with other individuals in the beans when the density of the beetles is high relative to the resource requirement. There are two parasitoids of *Callosobruchus* that are well studied for host-parasitoid population dynamics. They are *Anisopteromalus calandrae* (Hymenoptera: Pteromalidae) and *Heterospilus prosopidis* (Hymenoptera: Braconidae), which are solitary ectoparasitoids of late stages of larvae and pre/pupae of Bruchidae. The former widely utilizes coleopterous pests of stored products such as grain weevils (Curculionoidea: Rhynchophoridae) and biscuit beetles (Bostrychoidea: Anobiidae) as its hosts (Ghani and Sweetman, 1955). These parasitoids deposit a single egg on the host by inserting their ovipositor into beans. When they superparasitize (i.e., deposit multiple eggs) a host, only a single wasp individual can survive and emerge as an adult.

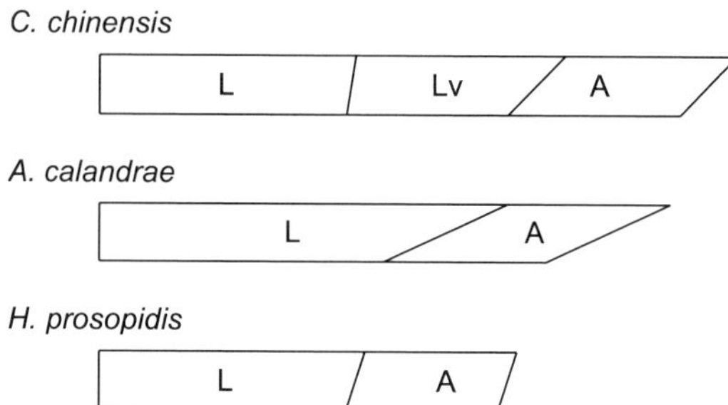

Figure 9 Life cycles of the host beetle *C. chinensis*, and the two parasitoid *A. calandrae* and *H. prosopidis*. L: larval stage, Lv: larval vulnerable stage of host, A: adult stage.

Developmental times from the egg deposition to adult emergence of females are 22 days (range 20 to 24) for the host and about 12 days (range 12 to 13) and 15 days (range 13 to 20) for *H. prosopidis* and *A. calandrae*, respectively, under our growth-chamber conditions (Utida, 1941, 1944a,b, 1948a; Tuda and Shimada, 1995; Tuda, 1999, unpublished data) (Fig. 9). Durations of the reproductive adult stage are approximately one week in all three species (Utida, 1944a,b, 1948; Tuda and Shimada, 1995), which can be extended under high population density (Shimada and Tuda, 1996). Fecundity of a female *C. chinensis* is about 80 (Utida, 1941; Tuda and Shimada, 1995). Fecundity of *H. prosopidis* is about 35 (Utida, 1948a) and of *A. calandrae* it can increase up to 260 (Utida, 1943b) by synergistic host feeding at higher host density. Host-feeding in the latter species also elongates longevity (Utida and Nagasawa, 1949; Ghani and Sweetman, 1955). Each parasitoid has an equal chance of survival on multiparasitized (i.e., co-parasitized) hosts (Wai, 1990).

The host population undergoes two density-dependent processes: first, the eggs suffer from mortality caused by mechanical injury by adult trampling; second, density-dependent mortality is induced by competition among larvae over a limited (bean) resource. The integration of these two processes exhibits weak overcompensation, so a single-species system converges to an equilibrium with short-damped oscillations (Fujii, 1968; Shimada, 1989; Shimada and Tuda, 1996).

Appendix 2. Selection of Model

A mechanistic model was first designed incorporating functions that are maximally complex, regarding the real biology we know about the component species. Then parameters in the model were removed one by one, according to the likelihood ratio statistic:

$$G^2 = -2(\ln L_N - \ln L_A) \sim \chi^2, \qquad (a2.1)$$

where L_N and L_A are the likelihoods of null and alternative models, respectively. This statistic approximately follows a chi-square distribution. A parameter was removed when the likelihood ratio is less than $\chi^2_{0.05}$ with the degree of freedom set at 1, which is a difference in the numbers of parameters between the two models that are compared, where $\chi^2_{0.05}$ is the 95th percentile of a chi-square distribution (Dennis *et al.*, 1995).

For *A. calandrae,* we incorporated mutual interference but not handling time, based on the result of the likelihood ratio test. A host-feeding parameter of the parasitoid was incorporated despite its insignificance in the three-species model but because of its statistical significance in the two-species model (Tuda, 2003, unpublished).

For *H. prosopidis*, a simple type II functional response was applied since the likelihood ratio test indicated the model with mutual interference that is either constant or increasing with its density was not significantly different from that without it (Tuda, 2003, unpublished).

Appendix 3. Assumption of Noise and Parameter Estimation

The noise is assumed to follow a multivariate normal distribution with multiple parameters in the model and elements in variance-covariance matrix Σ in maximum likelihood function to be estimated (Dennis *et al.*, 1995, 2001):

$$L(\boldsymbol{\theta}, \boldsymbol{\Sigma}) = \prod_{t=i+1}^{q} p(\mathbf{w}_t|\mathbf{w}_{t-i}), \qquad (a3.1)$$

where $p\ (\mathbf{w}_t|\mathbf{w}_{t-i})$ is the joint transition probability density function for $\mathbf{W}_\mathrm{t}$ conditional on $\mathbf{W}_{t-i} = \mathbf{w}_{t-i}$, q is the length of time series and $\boldsymbol{\theta}$ is the vector of unknown parameters of functions:

$$\mathrm{g}(\mathbf{w}_{t-i}) = \begin{bmatrix} [h_{t-2}\exp(-a_Y y'_{t-1} - a_Z z'_{t-1}) + s x_{a,t-1}]^{1/2} \\ [h_{t-3}\{1 + \exp(-a_Z z'_{t-2})\}\{1 - \exp(-a_Y y'_{t-2})\}/2]^{1/2} \\ [h_{t-2}\{1 + \exp(-a_Y y'_{t-1})\}\{1 - \exp(-a_Z z'_{t-1})\}/2]^{1/2} \end{bmatrix} \qquad (a3.2)$$

and

$$p(\mathbf{w}_t|\mathbf{w}_{t-i}) = (2\pi)^{d/2}|\boldsymbol{\Sigma}|^{1/2}\exp(-\mathbf{e}_t^{\tau}\boldsymbol{\Sigma}^{-1}\mathbf{e}_t/2), \qquad (a3.3)$$

where $\mathbf{e}_t = \mathbf{w}_t - \mathrm{h}\mathbf{w}_{t-i})$, $\mathbf{e}_t^{\tau}$ is the transpose of the vector $\mathbf{e}_t$ and d is the number of dimensions or state variables. The distribution of the noise is assumed uncorrelated through time but correlated with each other between species at time t. The latter assumption requires estimation of three covariance elements in addition to three variance elements in the variance-covariance matrix. Log-likelihood was maximized by Nelder-Mead's simplex method, implemented in MATLAB 6.5.1 for Windows. The 95% confidence intervals of parameters were calculated based on profile likelihood that uses a likelihood ratio test (Venzon and Moolgavkar, 1988; McCullagh and Nelder, 1989; Lebreton *et al.*, 1992; Dennis *et al.*, 1995).

REFERENCES

Abrams, P.A. (2000) The evolution of predator-prey interactions: Theory and evidence. *Annu. Rev. Ecol. Syst.* **31**, 79–105.

Agrawal, A.A. (2001) Ecology-phenotypic plasticity in the interactions and evolution of species. *Science* **294**, 321–326.

Bauch, C.T. and Earn, D.J.D. (2003) Transients and attractors in epidemics. *Proc. Roy. Soc. Lond. B* **270**, 1573–1578.

Begon, M., Bowers, R.G., Sait, S.M. and Thompson, D.J. (1996a) Population dynamics beyond two species: Hosts, parasitoids and pathogens. In: *Frontiers of Population Ecology* (Ed. by R.B. Floyd, A.W. Sheppard and P.J. De Barro), pp. 115–126. CSIRO Publ., Melbourne.

Begon, M., Harper, J.L. and Townsend, C.R. (1996b) *Ecology: Individuals, Populations and Communities*, 3rd ed. Blackwell Sci., Oxford.

Begon, M., Sait, S.M. and Thompson, D.J. (1995) Persistence of a predator-prey system: refuges and generation cycles? *Proc. Roy. Soc. Lond. B* **260**, 131–137.

Begon, M., Sait, S.M. and Thompson, D.J. (1996c) Predator-prey cycles with period shifts between two- and three-species systems. *Nature* **381**, 311–315.

Begon, M., Sait, S.M. and Thompson, D.J. (1997) Two's company, three's a crowd: host-pathogen-parasitoid dynamics. In: *Multitrophic Interactions in Terrestrial Ecosystems* (Ed. by A.C. Gange and V.K. Brown), pp. 307–332. Blackwell, Oxford.

Benrey, B. and Denno, R.F. (1997) The slow-growth-high-mortality hypothesis: A test using the cabbage butterfly. *Ecology* **78**, 987–999.

Boswell, M.T., Ord, J.K. and Patil, G.P. (1979) Chance mechanisms underlying univariate distributions. In: *Statistical Distributions in Ecological Work* (Ed. by J.K. Ord, G.P. Patil and C. Taillie). International Cooperative Publishing House, Fairland, Maryland.

Briggs, C.J. (1993) Competition among parasitoid species on a stage-structured host and its effect on host suppression. *Am. Nat.* **141**, 372–397.

Briggs, C.J., Nisbet, R.M., Murdoch, W.W., Collier, T.R. and Metz, J.A.J. (1995) Dynamical effects of host-feeding in parasitoids. J. *Anim. Ecol.* **64**, 403–416.

Chen, X. and Cohen, J.E. (2001a) Transient dynamics and food-web complexity in the Lotka-Volterra cascade model. *Proc. R. Soc. Lond. B* **268**, 869–877.

Chen, X. and Cohen, J.E. (2001b) Global stability, local stability and permanence in model food webs. J. *Theor. Biol.* **212**, 223–235.

Clancy, K.M. and Price, P.W. (1987) Rapid herbivore growth enhances enemy attack, sublethal plant defenses remain a paradox. *Ecology* **68**, 736–738.

Costantino, R.F., Cushing, J.M., Dennis, B. and Desharnais, R.A. (1995) Experimentally-induced transitions in the dynamic behavior of insect populations. *Nature* **375**, 227–230.

Costantino, R.F., Desharnais, R.A., Cushing, J.M. and Dennis, B. (1997) Chaotic dynamics in an insect population. *Science* **275**, 389–391.

Crutchfield, J.P. and Kaneko, K. (1988) Are attractors relevant to turbulence? *Phys. Rev. Lett.* **60**, 2715–2718.

Dennis, B., Desharnais, R.A., Cushing, J.M. and Costantino, R.F. (1995) Nonlinear demographic dynamics-mathematical models, statistical methods, and biological experiments. *Ecol. Monogr.* **65**, 261–281.

Dennis, B., Desharnais, R.A., Cushing, J.M. and Constantino, R.F. (1997) Transitions in population dynamics: Equilibria to periodic cycles to aperiodic cycles. *J. Anim. Ecol.* **66**, 704–729.

Dennis, B., Desharnais, R.A., Cushing, J.M., Henson, S.M. and Costantino, R.F. (2001) Estimating chaos and complex dynamics in an insect population. *Ecol. Monogr.* **71**, 277–303.

Ellner, S., Nychka, D.W. and Gallant, A.R. (1992) *LENNS, a program to estimate the dominant Lyapunov exponent of noisy nonlinear systems from time series data.* Institute of Statistics Mimeo Series #2235 (BMA Series #39), Stat. Dept., North Carolina State Univ., Raleigh, NC.

Ellner, S. and Turchin, P. (1995) Chaos in a 'noisy' world: New methods and evidence from time series analysis. *Am. Nat.* **145**, 343–375.

Elton, C.S. (1927) *Animal Ecology*. Sidwick and Jackson, London.

Falck, W., Bjornstad, O.N. and Stenseth, N.J. (1995) Voles and lemmings: Chaos and uncertainty in fluctuating populations. *Proc. R. Soc. Lond. B* **262**, 363–370.

Fujii, K. (1968) Studies on interspecies competition between the azuki bean weevil and the southern cowpea weevil. III. Some characteristics of strains of two species. *Res. Popul. Ecol.* **10**, 87–98.

Fujii, K. (1981) Interspecific interactions and community structure. In: *Recent Advances in Entomology* (Ed. by S. Ishii), pp. 97–181. Tokyo Daigaku Shuppankai, Tokyo. In Japanese.

Fujii, K. (1983) Resource dependent stability in an experimental laboratory resource-herbivore-carnivore system. *Res. Popul. Ecol.* **3**(Suppl.), 155–165.

Fujii, K. (1994) Bruchids in a Petri dish. In: *The Earth in a Petri Dish* (Ed. by K. Fujii, M. Shimada and Z. Kawabata), pp. 97–181. Heibonsha, Tokyo. In Japanese.

Ghani, M.A. and Sweetman, H.L. (1955) Ecological studies on the granary weevil parasite, *Aplastomorpha calandrae* (Howard). *Biologia* **1**, 115–139.

Gunderson, L.H. (2000) Ecological resilience in theory and application. *Annu. Rev. Ecol. Syst.* **31**, 425–439.

Hairston, N.G., Allan, J.D., Colwell, R.K., Futuyma, D.J., Howell, J., Lubin, M.D., Mathias, J. and Vandermeer, J.H. (1968) Relationship between species diversity and stability: an experimental approach with protozoa and bacteria. *Ecology* **49**, 1091–1101.

Hanski, I., Turchin, P., Korpimaki, E. and Henttonen, H. (1993) Population oscillations of boreal rodents: regulation by mustelid predators leads to chaos. *Nature* **364**, 232–235.

Hassell, M.P. (1978) *The Dynamics of Arthropod Predator-Prey Systems*. Princeton Univ. Press, New Jersey.

Hassell, M.P. and Anderson, R.M. (1989) Predator-prey and host-pathogen interactions. In: *Ecological Concepts* (Ed. by J.M. Cherrett), pp. 147–196. Blackwell, Oxford.

Hassell, M.P. and May, R.M. (1973) Stability in insect host-parasite models. *J. Anim. Ecol.* **42**, 693–726.

Hassell, M.P. and Varley, G.C. (1969) New inductive population model for insect parasites and its bearing on biological control. *Nature* **223**, 1133–1136.

Hastings, A. and Higgins, K. (1994) Persistence of transients in spatially structured ecological models. *Science* **263**, 1133–1136.

Hastings, A., Hom, C.L., Ellner, S., Turchin, P. and Godfray, H.C.J. (1993) Chaos in ecology: Is mother nature a strange attractor? *Ann. Rev. Ecol. Syst.* **24**, 1–33.

Holling, C.S. (1959) Some characteristics of simple types of predation and parasitism. *Can. Entomol.* **91**, 385–398.

Holt, R.D. (1977) Predation, apparent competition, and structure of prey communities. *Theor. Pop. Biol.* **12**, 197–229.

Holt, R.D. (1997) Community modules. In: *Multitrophic Interactions in Terrestrial Ecosystems* (Ed. by A.C. Gange and V.K. Brown), pp. 333–350. Blackwell, Oxford.

Imura, D., Toquenaga, Y. and Fujii, K. (2003) Genetic variation can promote system persistence in an experimental host-parasitoid system. *Popul. Ecol.* **45**, 205–212.

Johnson, M. T. and Gould, F. (2002) Interaction of genetically engineered host plant resistance and natural enemies of *Heliothis virescens* (Lepidoptera, Noctuidae) in tobacco. *Environ. Entomol.* **21**, 586–597.

Jost, C. and Arditi, R. (2001) From pattern to process: Identifying predator-prey models from time-series data. *Popul. Ecol.* **43**, 229–243.

Kaneko, K. (1990) Supertransients, spatiotemporal intermittency, and stability of fully developed spatiotemporal chaos. *Phys. Lett.* **149 A**, 105–112.

Kawecki, T.J. and Mery, F. (2003) Evolutionary conservatism of geographic variation in host preference in *Callosobruchus maculatus*. *Ecol. Entomol.* **28**, 449–456.

Keith, L.B. (1983) Role of food in hare population cycles. *Oikos* **40**, 385–395.

Kendall, B.E., Briggs, C.J., Murdoch, W.W., Turchin, P., Ellner, S.P., McCauley, E., Nisbet, R.M. and Wood, S.N. (1999) Why do populations cycle? A synthesis of statistical and mechanistic modeling approaches. *Ecology* **80**, 1789–1805.

Kondoh, M. (2003) Foraging adaptation and the relationship between food-web complexity and stability. *Science* **299**, 1388–1391.

Kristoffersen, A.B., Lingjaerde, O.C., Stenseth, N.Chr. and Shimada, M. (2001) Non-parametric modelling of non-linear density dependence: A three-species host-parasitoid system. *J. Anim. Ecol.* **70**, 808–819.

Lawler, S.P. (1993) Species richness, species composition and population dynamics of protests in experimental microcosms. *J. Anim. Ecol* **62**, 711–719.

Lawler, S.P. and Morin, P.J. (1993) Food-web architecture and population dynamics in laboratory microcosms of protists. *Am. Nat.* **141**, 675–686.

Lebreton, J.D., Burnham, K.P., Clobert, J. and Anderson, D.R. (1992) Modeling survival and testing biological hypotheses using marked animals: A unified approach with case studies. *Ecol. Monogr.* **62**, 67–118.

May, R.M. (1972) Will a large complex system be stable? *Nature* **238**, 413–414.

May, R.M. and Hassell, M.P. (1981) The dynamics of multiparasitoid-host interactions. *Am. Nat.* **117**, 234–261.

McCullagh, P. and Nelder, J.A. (1989) *Generalized Linear Models*, 2nd ed. Chapman & Hall, London.

Mitsunaga, T. and Fujii, K. (1997) The effects of spatial and temporal environmental heterogeneities on persistence in a laboratory experimental community. *Res. Popul. Ecol.* **39**, 249–260.

Mitsunaga, T. and Fujii, K. (1999) An experimental analysis on the relationship between species combination and community persistence. *Res. Popul. Ecol.* **41**, 127–134.

Moran, P.A.P. (1950) Some remarks on animal population dynamics. *Biometrics* **6**, 250–258.

Nicholson, A.J. (1933) The balance of animal populations. *J. Anim. Ecol.* **2**, 131–178.

Nicholson, A.J. (1954) An outline of the dynamics of animal populations. *Aust. J. Zool.* **2**, 9–65.

Nicholson, A.J. (1957) The self-adjustment of populations to change. *Cold Spring Harbor Symp. Quant. Biol.* **22**, 153–173.

Nicholson, A.J. and Bailey, V.A. (1935) The balance of animal populations. *Proc. Zool. Soc. Lond.* **3**, 551–598.

Ohdate, K. (1980) Relationship between diversity and stability in bean-bruchid-parasitoid experimental systems. Thesis for M. S., University of Tsukuba. In Japanese.

Pearl, R. (1927) The growth of populations. *Quart. Rev. Biol.* **2**, 532–548.

Pimentel, D. (1968) Population regulation and genetic feedback. *Science* **159**, 1432–1437.

Pimentel, D. and Stone, F.A. (1968) Evolution and population ecology of parasite-host systems. *Can. Entomol.* **100**, 655–662.

Ricker, W.E. (1954) Stock and recruitment. *J. Fish. Res. Bd. Can.* **11**, 559–623.

Rogers, D.J. (1972) Random search and insect population models. *J. Anim. Ecol.* **41**, 369–383.

Rosenzweig, M.L. (1971) Paradox of enrichment: Destabilization of exploitation ecosystems in ecological time. *Science* **171**, 385–387.

Royama, T. (1971) A comparative study of models for predation and parasitism. *Res. Popul. Ecol.* (Suppl. 1), 1–91.

Royama, T. (1992) *Analytical Population Dynamics*. Chapman & Hall, London.

Shimada, M. (1985) Niche modification and stability of competitive systems. II. Persistence of interspecific competitive systems with parasitoid wasps. *Res. Popul. Ecol.* **27**, 203–216.

Shimada, M. (1989) Systems analysis of density-dependent population processes in the azuki bean weevil, *Callosobruchus chinensis*. *Ecol. Res.* **4**, 145–156.

Shimada, M. (1999) Population fluctuation and persistence of one-host-two-parasitoid systems depending on resource distribution: From parasitizing behavior to population dynamics. *Res. Popul. Ecol.* **41**, 69–79.

Shimada, M. and Fujii, K. (1985) Niche modification and stability of competitive systems. I. Niche modification process. *Res. Popul. Ecol.* **27**, 185–201.

Shimada, M. and Tuda, M. (1996) Delayed density dependence and oscillatory population dynamics in overlapping-generation systems of a seed beetle *Callosobruchus chinensis*: Matrix population model. *Oecologia* **105**, 116–125.

Sih, A., Crowley, P., McPeek, M., Petranka, J. and Strohmeier, K. (1985) Predation, competition and prey communities: A review of field experiments. *Annu. Rev. Ecol. Syst.* **16**, 269–311.

Stokes, T.K., Gurney, W.S.C., Nisbet, R.M. and Blythe, S.P. (1988) Parameter evolution in a laboratory insect population. *Theor. Popul. Biol.* **34**, 248–265.

Takano, M., Toquenaga, Y. and Fujii, K. (2001) Polymorphism of competition type and its genetics in *Callosobruchus maculatus* (Coleoptera:Bruchidae). *Popul. Ecol.* **43**, 265–273.

Thompson, J.N. (1998) Rapid evolution as an ecological process. *Trends Ecol. Evol.* **13**, 329–332.

Toquenaga, Y. and Fujii, K. (1991) Contest and scramble competition in two bruchid species, *Callosobruchus analis* and *C. phaseoli* (Coleoptera:Bruchidae). III. Multiple-generation competition experiment. *Res. Popul. Ecol.* **33**, 187–197.

Toquenaga, Y., Ichinose, M., Hoshino, T. and Fujii, K. (1994) Contest and scramble competitions in an artificial world: Genetic analysis with GA. In: *Artificial Life III* (Ed. by C.G. Langton), pp. 177–199. Addison-Wesley.

Tuda, M. (1996a) Mechanism for coexistence in the laboratory community of bean-bean weevil-parasitic wasp. *Jap. J. Ecol.* **46**, 313–320. In Japanese.

Tuda, M. (1996b) Temporal/spatial structure and the dynamical property of laboratory host-parasitoid systems. *Res. Popul. Ecol.* **38**, 133–140.

Tuda, M. (1998) Evolutionary character changes and population responses in an insect host-parasitoid experimental system. *Res. Popul. Ecol.* **40**, 293–299.

Tuda, M. (2003) A new species of *Callosobruchus* (Coleoptera:Bruchidae) feeding on seeds of *Dunbaria* (Fabaceae), a closely related species to a stored-bean pest, *C. chinensis*. *Appl. Entomol. Zool.* **38**, 197–201.

Tuda, M. and Bonsall, M.B. (1999) Evolutionary and population dynamics of host-parasitoid interactions. *Res. Popul. Ecol.* **41**, 81–91.

Tuda, M., Chou, L.-Y., Niyomdham, C., Buranapanichpan, S. and Tateishi, Y. (2005) Ecological factors associated with pest status in *Callosobruchus* (Coleoptera:Bruchidae): High host specificity of non-pests to Cajaninae (Fabaceae). *J. Stor. Prod. Res.* **41**, 31–45.

Tuda, M. and Iwasa, Y. (1998) Evolution of contest competition and its effect on host-parasitoid dynamics. *Evol. Ecol.* **12**, 855–870.

Tuda, M., Shima, K., Johnson, C.D. and Morimoto, K. (2001) Establishment of *Acanthoscelides pallidipennis* (Coleoptera:Bruchidae) feeding in seeds of the introduced legume *Amorpha fruticosa*, with a new record of its *Eupelmus* parasitoid in Japan. *Appl. Entomol. Zool.* **36**, 269–276.

Tuda, M. and Shimada, M. (1995) Developmental schedules and persistence of experimental host-parasitoid systems at two different temperatures. *Oecologia* **103**, 283–291.

Turchin, P. (1996) Nonlinear time-series modeling of vole Population fluctuations. *Res. Popul. Ecol.* **38**, 121–132.

Turchin, P. and Ellner, S.P. (2000) Living on the edge of chaos: population dynamics of Fennoscandian voles. *Ecology* **81**, 3099 3116.

Turchin, P. and Hanski, I. (2001) Contrasting alternative hypotheses about rodent cycles by translating them into parameterized models. *Ecol. Lett.* **4**, 267–276.

Turchin, P. and Taylor, A.D. (1992) Complex dynamics in ecological time-series. *Ecology* **73**, 289–305.

Utida, S. (1941) Studies on experimental population of the azuki bean weevil, *Callosobruchus chinensis* (L.). II. The effect of population density. *Mem. Coll. Agr. Kyoto Imp. Univ.* **49**, 1–20.

Utida, S. (1943a) Studies on experimental population of the azuki bean weevil, *Callosobruchus chinensis* (L.). VIII–IX. *Mem. Coll. Agr. Kyoto Imp. Univ.* **54**, 1–40.

Utida, S. (1943b) Host-parasite interaction in the experimental populations of the azuki bean weevil, *Callosobruchus chinensis* (L.). III. The effect of density of host population on the growth of the parasite population. *Seitaigakukenkyu* **9**, 40–53.

Utida, S. (1944a) Host-parasite interaction in the experimental populations of the azuki bean weevil, *Callosobruchus chinensis* (L.). I. The effect of density of parasite population on the growth of host population and also of the parasite population. *Oyo-Kontyu* **4**, 117–128. In Japanese with English summary.

Utida, S. (1944b) Host-parasite interaction in the experimental populations of the azuki bean weevil, *Callosobruchus chinensis* (L.). II. The effect of density of parasite population on the growth of host population and also of the parasite population II. *Oyo Dobutsugaku Zasshi* **15**, 1–18. In Japanese.

Utida, S. (1948a) The effect of host density on the growth of host and parasite populations II. *Oyo-Kontyu* **4**, 164–174.

Utida, S. (1948b) Population fluctuations caused by host-parasite interaction. *Physiol. Ecol.* **2**, 1–11. In Japanese with English summary.

Utida, S. (1957a) Population fluctuation, an experimental and theoretical approach. *Cold Spring Harbor Symp. Quant. Biol.* **22**, 139–151.

Utida, S. (1957b) Cyclic fluctuation of population density intrinsic to the host-parasite system. *Ecology* **38**, 442–449.

Utida, S. and Nagasawa, S. (1949) On the developmental period and that of adult life of *Neocatolaccus mamezophagus*, a pteromalid parasite of the azuki bean weevil. *Kontyu* **17**, 7–21. In Japanese.

Venzon, D.J. and Moolgavkar, S.H. (1988) A method for computing profile-likelihood-based confidence intervals. *Appl. Stat.* **37**, 87–94.

Wai, K. M. (1990) Intra- and interspecific larval competition among wasps parasitic to bean weevil larvae. Thesis–University of Tsukuba, D.Sc. (A), no. 714.

Population Dynamics, Life History, and Demography: Lessons From *Drosophila*

LAURENCE D. MUELLER, CASANDRA L. RAUSER AND MICHAEL R. ROSE

I. INTRODUCTION

The many scientific advances over the last 200 years have clearly been one of the greatest achievements of human civilization. Much of this scientific progress has been accomplished by employing the paradigm we call 'the scientific method.' Hypotheses are proposed, experiments are designed to test these hypotheses, then ideas are revised based on the outcome of these experiments. For many reasons this paradigm has not always been embraced by ecologists. Often scientific hypotheses must be simple and address rudimentary aspects of a problem that have not been studied previously. Many ecologists do not accept the simplicity of explicitly stated hypotheses. For example, there is the early view of Thompson (1948): "The tremendous multiplicity of factors acting on the real world has not merely the complexity of an elaborate mathematical equation, which is theoretically but not practically manageable, but implies a genuine unpredictability because the actual combination of factors has never been observed to operate and until it has, we cannot really be sure what its effect will be. Much less can we see this effect in its causes."

ADVANCES IN ECOLOGICAL RESEARCH VOL. 37 0065-2504/05 $35.00

DOI: 10.1016/S0065-2504(04)37003-0

Thompson's lament concerns the great complexity of nature as well as its unpredictability. Implicit in Thompson's comments is the assumption that scientific theory needs to explain nature in all its details. Think of where genetics would be, had the early geneticists required that Mendel's laws also explain the distribution of progeny height and weight from known crosses. While it would be magnificent to be able to develop ecological theories that explain all aspects of the number and distribution of organisms in their natural environment, perhaps there is some benefit to starting with more modest goals.

This compromise has been the path chosen by many theoreticians. They have focused on just a few environmental variables, e.g., population density or predators. A theoretician might be satisfied with merely understanding the logical consequences of his simple theory. However, the empiricist will want some validation that, when the model assumptions are met, biological systems will really do what the model predicts. Here is where we feel there is the greatest disagreement about what constitutes a strong empirical test. Experimental ecologists, like many of those contributing to this volume, would suggest developing an experimental system where the scientists can ensure all the assumptions of the model are met. A critical test of the theory is thus guaranteed. Other ecologists feel that the theory must be tested in a natural ecosystem, albeit one they think is congruent with as many of the model assumptions as possible. The justification for this latter approach often appeals to a visceral notion that there are some ephemeral qualities of nature that could never be reproduced in the laboratory, which makes results from the laboratory suspect.

We not only reject this point of view, we feel that many field tests hinder scientific progress. Our view follows from the ability of the field ecologist to take the results of a field study that have falsified a theory and argue that the theory is still valid, because the contrary results were simply a consequence of uncontrolled factors playing havoc with the critical observations. We believe that testing ecological theory in the laboratory will instead allow us to build an understanding of ecological phenomena that will ultimately let us understand the complexity of nature. With this perspective in mind, we review some of the important lessons in ecology and life history evolution that have been learned from studies of laboratory populations of *Drosophila*.

Having made this argument for experimental ecological research we note that there are oftentimes experiments in the field that can be particularly useful. One notably good example of this is the work of Reznick and his colleagues on the evolution of life-histories in the guppy *Peocilia reticulate*. These studies have used both observations in natural populations (Reznick and Endler, 1982; Reznick, 1989; Reznick and Bryga, 1996; Reznick *et al*, 1996) and replicated introduction experiments in natural populations

(Reznick and Bryga, 1987; Reznick *et al.*, 1990, 1997). Together these studies have developed a solid understanding of the role of predator-mediated mortality on the evolution of guppy life history. These results are not only of general theoretical interest but they can be used to understand some of the variation in life history patterns in natural populations of guppies.

II. POPULATIONS WITHOUT AGE STRUCTURE

In this section, we review topics that are motivated by theories of density-dependent population growth and selection in populations without age structure. These topics include the evolution of density-dependent rates of population growth and the evolution of population stability. In the next section, we cover material that explicitly accounts for age structure. The topics to be reviewed in that section will include the evolution of senescence, and the evolution of age-specific mortality and fecundity patterns.

In each of these sections, we first review the important theories that motivate the experimental research. An important take-home message from this review will be the manner in which experiments are designed to test the critical concepts of these theories. Potential artifacts and confounding factors are often avoided by the experimental design or can be independently investigated so their contribution to certain experimental results can be evaluated. Consequently, the results from these experimental studies will often have clear interpretations.

A. Evolution of Density-Dependent Rates of Population Growth

1. Theory

The most important event in the development of the theory of density-dependent natural selection was the book, *The Theory of Island Biogeography* by MacArthur and Wilson (1967). They called their theory r- and K-selection and many of their ideas were presented as verbal models. However, these ideas were subsequently made more rigorous, as exemplified by the work of Roughgarden (1971).

The history of r- and K-selection has been reviewed numerous times (Stearns, 1976, 1977; Boyce, 1984; Mueller, 1997; Reznick *et al.*, 2002). In many respects this particular field can be viewed as a case study for the advantages of mathematical theories versus verbal theory and the strength of controlled laboratory experiments versus field studies. A great weakness of verbal theories concerning density-dependent selection was the attempt to extend the reasoning of MacArthur and Wilson, which was framed within the context of logistic population growth, to populations with age structure. Meanwhile, empirical studies used wild populations where only very crude inferences about past densities could be made, and there was no ability to control for factors that might affect the evolution of life history other than density (Gadgil and Solbrig, 1972; McNaughton, 1975).

The theory developed by Roughgarden assumed that populations harbor genetic variation for density-dependent rates of population growth. Under this theory the effects of density are felt adversely by different genotypes. Using the simple structure of single-locus genetics and the logistic equation, then the per-capita growth rate or fitness (W_{ij}) for genotype A_iA_j at a population density N is:

$$W_{ij} = 1 + r_{ij} - r_{ij}NK_{ij}^{-1}. \qquad (1)$$

The most interesting prediction from this theory requires two assumptions: (1) W_{ij} adequately summarizes fitness, and (2) genotypes show trade-offs. Trade-offs mean that genotypes with high values of r, have relatively low values of K and vice versa. With these assumptions granted, we expect populations evolving at very high and very low population densities to become phenotypically distinct, at least with respect to their per-capita growth rates. Particular aspects of life history, like competitive ability or adult size, may have to change to accomplish these changes in per-capita growth rates but this theory says nothing about what those changes might be.

2. *Laboratory Experiments*

Given the theory above, it is apparent that a suitable test would be to create different populations that differ only with respect to the level of crowding they experience. To test whether such populations achieve the predicted phenotypic differentiation would then require measuring the density-dependent per-capita growth rates in each population. There are no other surrogate phenotypes that can be used to test this theory. If one were to measure some other phenotype, then failure of the theory could always be argued to be a consequence of measuring an inappropriate phenotype. The power of strong inference would have been effectively thwarted.

Mueller and Ayala (1981a) were the first to test this theory by directly measuring density-dependent rates of population growth in laboratory populations that had evolved at very high and low densities. These experiments revealed the trade-off assumed by the Roughgarden theory. The laboratory populations used by Mueller and Ayala differed in adult density and in effective population size during their evolution. This left open the possibility that the observed differences might be due to inbreeding or some other aspect of the effective population size differences. A separate set of independent experiments (Mueller *et al.*, 1991) controlled for the effects of inbreeding and were able to replicate the earlier results.

While these laboratory populations appear highly simplified, there have been results that connect these studies to more complicated field environments. For instance, Borash *et al.* (1998) studied simple cultures with high larval densities. Over the course of the two-week developmental period of these larvae the levels of ammonia increased exponentially, while the levels of food and ethanol declined substantially. This environmental deterioration has important consequences for evolution in these crowded environments. Borash *et al.* (1998) document a genetic polymorphism that appears to be stably maintained in these environments. One early-developing genotype is characterized by rapid development and high feeding rates, but by low viability in ammonia-laced food. A second slow-developing genotype has longer development time and slower feeding rates, but has high survival in ammonia-laced food. Additionally, the type of within-generation temporal variation is similar to variability in many naturally occurring ephemeral habitats. Some examples include excrement from large mammals or fresh fruit that falls off trees.

There have been a number of other laboratory studies of density-dependent natural selection using *Drosophila* (Taylor and Condra, 1980; Barclay and Gregory, 1981, 1982; Sokolowski *et al.*, 1997). None of these studies measured rates of population growth directly and some of the early studies had methodological flaws that have been previously reviewed (Mueller, 1985). Sokolowski *et al.* (1997) created a set of populations similar to the *r*- and *K*-lines created by Mueller and Ayala (1981a). Sokolowski *et al.* (1997) showed that their high-density lines evolved increased foraging path lengths as did the high-density *K*-populations created by Mueller and Ayala. Consequently, we have independent corroboration of the type of phenotypic evolution that results from evolution in crowded environments.

The tension between the controlled laboratory environment and the desire to create natural conditions is reflected in the experiment of Barclay and Gregory (1982). In this study, the goal was to create experimental treatments that differed in levels of adult mortality. In most theories that permit adult mortality rates to vary, it is relatively unimportant what causes death since the event itself is fairly unambiguous. Barclay and Gregory decided to put frogs in their fly cages as a means of increasing adult mortality rather than simply removing the adults manually. In this example, we think that whatever gain in reality is achieved by having frogs kill flies does not compensate for the precision that manual control of mortality affords the experimentalist. Thus, like the decision to perform laboratory experiments versus field work, other decisions to make conditions more natural must be evaluated with respect to their ability to permit cogent interpretation of the experimental results.

B. Evolution of Population Stability

1. Theory

Models of density-dependent population growth typically assume that per-capita growth rates decline with increasing population density. As a result there will typically be an equilibrium density at which the population will show no change in size from one generation to the next. An important property of population dynamic models is whether the population will return to an equilibrium after a small perturbation away from it. The behavior of the dynamic system around an equilibrium is called the 'stability of the equilibrium.'

It might seem as if predictions concerning population stability should follow easily from the evolution of density-dependent population growth rates (see Mueller and Joshi, 2000, Chapter 2 for a review of this theory). In the simple discrete time logistic, stability is determined by the Malthusian parameter r. Thus, once we know how r evolves we ought to understand how stability evolves. However, the evolution of stability is not that straightforward. A disconcerting complication with these simple models is that the outcome of evolution often depends on small model details. Using the discrete-time logistic, exponential and hyperbolic models, Turelli and Petry (1980) found that in variable environments the evolution of r could result in increased, decreased, or no change in stability. Turelli and Petry obtained more consistent results when they used the model,

$$N_{t+1} = N_t G[(N_t/K)^{\theta}]. \tag{2}$$

and let the parameter θ evolve. Even as $G[.]$ changed to one of the three functional forms (logistic, exponential, and hyperbolic) evolution typically favored increased stability. However, this result depends on genetic variation affecting only θ, and not r and K. This is an empirical problem, which requires an experimental resolution.

The theory developed by Turelli and Petry started with populations at a deterministically stable equilibrium. In an early study, Mueller and Ayala (1981b) examined the evolution of stability in a population undergoing a two-point cycle in population density. The results suggested that increased stability could evolve if there was a trade-off between density-dependent survival and fertility. Subsequent theoretical work has also confirmed the need for trade-offs to have stability evolve (Stokes *et al.*, 1988; Gatto, 1993). As with the theory of density-dependent natural selection, there is no theoretical way to determine if the requisite trade-offs exist. This is entirely an empirical question.

2. *Laboratory Experiments*

To test the theory that natural selection may affect the stability of populations required, at a minimum, populations that differed in their stability characteristics but were otherwise similar. For *Drosophila* we determined that a crucial determinant of stability was the level of resources provided to adults and larvae (Mueller and Huynh, 1994). High levels of food for the larval stages but low levels for adults generally enhanced population stability. The opposite conditions, high adult food and low larval food levels, typically produced stable cycles or chaos (Sheeba and Joshi, 1998). This information was then used to design experimental populations to monitor the evolution of stability.

Twenty populations were created. Ten were created from populations called CU that had evolved at high larval and low adult densities prior to this experiment. Ten additional populations were created from populations called UU that had been kept at low larval and low adult density prior to the start of the experiment. The 10 CU populations were further divided into five populations maintained under high larval food and low adult food conditions (called HL). These are the conditions most conducive to stability. The other five CU populations were maintained under low larval food and high adult food conditions (called LH). These are the conditions that are the least conducive to stability. The 10 UU populations were divided into two groups of five in a similar fashion.

Visual inspection of the cultures suggested that the larval densities were typically much higher in the LH treatments. These impressions are confirmed by the fact that on average the adults that emerge from the LH treatments were much smaller than the adults in the HL treatments (Mueller *et al.*, 2000; Fig. 1A). The actual population size fluctuations are quite dramatic in both types of populations (Fig. 1B). However, the HL treatments tend to be more stable (Mueller *et al.*, 2000). Formally, the stability of an equilibrium is assessed by determining the rate of exponential growth either away from or towards the equilibrium in a small neighborhood around the equilibrium. The parameter describing this rate of exponential growth is often called the 'eigenvalue'. Since the eigenvalue may be an imaginary number it is common to evaluate the modulus of the eigenvalue. For real numbers the modulus is simply its absolute value and for imaginary numbers it is the square root of the sum of the squares of the real and imaginary parts of the eigenvalue. For discrete time models the modulus must be <1 to ensure stability.

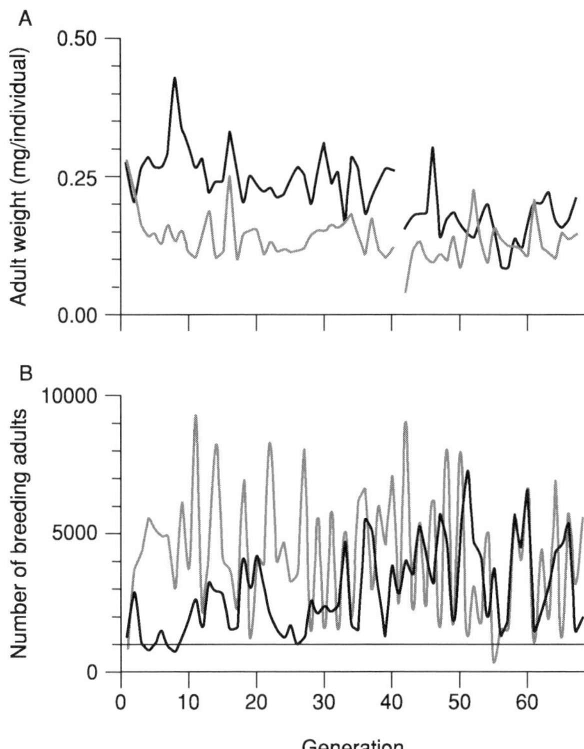

Figure 1 Adult size (A) and numbers of adults (B) for 68 generations in the experimental populations derived from the CU_5 ancestral population. The solid black lines are the HL populations and the dashed (gray) lines are the LH populations.

To assess whether the dynamics of the populations in the less stable LH treatments changed over 68 generations of lab evolution, we determined the modulus of the leading eigenvalue during the first 15 generations of the experiment and during the last 15 generations of the experiment. Although the average values of this modulus decreased from 1.00 to 0.75, that change was not significant. Likewise among the 10 LH populations, four showed increases in the modulus of their leading eigenvalue, indicating a decrease in stability. Increased stability did not generally evolve in the laboratory populations.

One interpretation of these results is that the intensity of density-dependent selection was simply too small to expect to see improvements in stability. Over the first 20 generations, we measured the larval feeding rates in each of the 20 populations. This character has been documented to increase under crowded larval conditions primarily due to its effect on larval competitive ability (Joshi and Mueller, 1988; Mueller *et al.*, 1993). We found rapid differentiation between the HL and LH populations independent of the ancestral source of the flies (e.g., CU versus UU). In all cases the feeding rates were greater in the LH populations. Thus, the environments we imposed led to strong responses in density-dependent natural selection. But such a selection did not have measurable effects on the stability of the population dynamics. It may be that stability changes very slowly as a result of density-dependent selection. But any such change must be so slow that even after 68 generations demonstrable effects of density-dependent selection cannot be detected.

In these experiments we purposely kept the total population size at about 1,000 adults or larger to prevent inbreeding depression. In *Drosophila,* inbreeding is likely to cause reductions in female fecundity and that may have the effect of stabilizing the population dynamics. In fact, we have suggested elsewhere that the apparent evolution of more stable dynamics in Nicholson's blowfly experiments (observed by Stokes *et al.* [1988]) may have been a consequence of inbreeding during the severe population blowfly bottlenecks (Mueller *et al.*, 2000).

Prasad *et al.* (2003) studied *Drosophila* populations selected for rapid development time and their controls. The rapid developing lines show reductions in female fecundity of about 35% compared to their controls. When both selected and control populations are maintained in the HL and LH environments, the rapid developing lines show reduced population fluctuations consistent with more stable dynamics.

Population cycles may be due to a variety of causes other than density-dependent population regulation. Indeed many cycles in nature are thought to be a consequence of multispecies interactions like those between predators and prey (see Turchin 2003, for a recent review). Future experimental work on population stability will benefit from an examination of evolution in multi-species communities. An illustration of this type of work is Yoshida *et al.* (2003). They studied the consequences of evolution on the predator-prey cycles of an experimental rotifer-algal community. They demonstrated that the characteristics of the predator-prey cycle changed as the algae evolved defenses against predation.

III. POPULATIONS WITH AGE STRUCTURE

A. Age-Specific Mortality Rates

1. Theory

The theory of natural selection in age-structured populations is well developed (Norton, 1928; Charlesworth, 1994). The basic demographic parameters needed for this theory are the probabilities of an individual surviving to age$-x$, $l(x)$, and the number of newborns produced by individuals age-x that survive to the first age class, $m(x)$. Populations with these demographic parameters will grow exponentially at a rate r that can be determined from the equation,

$$\sum e^{-rx} l(x)m(x) = 1, \tag{3}$$

where the summation is over all age classes. The theory of natural selection supposes that these demographic parameters vary between genotypes. Thus, in a simple single-locus setting the demographic parameters for genotype A_iA_j are $l_{ij}(x)$ and $m_{ij}(x)$. The fitness of this genotype at genetic equilibrium is,

$$w_{ij} = \sum e^{-\hat{r}x} l_{ij}(x)m_{ij}(x), \tag{4}$$

where $\hat{r}$ is the rate of exponential growth for the equilibrium population. For positive values of $\hat{r}$, fitness will be most strongly affected by the survival and fertility values at early ages. As pointed out by Hamilton (1966), mutations increasing survival at earlier ages would be most strongly selected for.

These ideas have been used to develop evolutionary explanations for the general observation that age-specific mortality rates generally increase at an exponential rate with increasing age, which can also be equated to senescence. One specific theory of senescence, called antagonistic pleiotropy (Medawar, 1952; Williams, 1957), suggests that natural selection will often view as favorable an allele that increases survival or fertility early in life even if it also has a deleterious (pleiotropic) effect late in life. A second theory, called mutation accumulation, suggests that populations will harbor a collection of deleterious alleles that affect age-specific survival and fertility (Medawar, 1952; Edney and Gill, 1968). The frequency of these deleterious alleles will be at an equilibrium dictated by the force of selection and the rate of mutation. Since the force of selection against deleterious alleles will be weak when the alleles are expressed late in life, the frequency of late-acting deleterious alleles is expected to be higher than the frequency of mutations affecting early fitness. Both the genetic theories of aging are supported by the formal theories of selection in age-structured populations. They are not even logically incompatible. Assessing the importance of these two theories is an empirical question.

2. *Laboratory Experiments*

There are a variety of possible tests of the evolutionary theories of aging. One interesting test, which can only be carried out in a laboratory setting, would be to reverse the normal way selection acts; that is, make reproduction late in life more valuable than reproduction early in life. With natural selection acting in this fashion, there should be improvements in late life fertility and declines in later mortality rates. Rose (1984) performed this experiment with *D. melanogaster*. Control populations, called Bs, reproduced at the end of a two-week egg-to-adult life cycle. The experimental populations, called Os, were cultured from eggs laid by older females. The age of these older females was progressively increased until its present level of 10 weeks from egg. This process has lead to genetic changes in the O populations that have doubled their longevity relative to the B populations and decreased mortality rates (Fig. 2). The decreased mortality rates are not due to a simple age-independent reduction in overall mortality. The actual rate of increase in mortality with age in the O populations is one-third the rate seen in the B populations. Thus, the actual rate of aging has been genetically reduced in the long-lived O populations. Rose (1984) also documented a decline in the early fecundity of O females relative to B females but an increase in late life fecundity. This type of trade-off is consistent with the

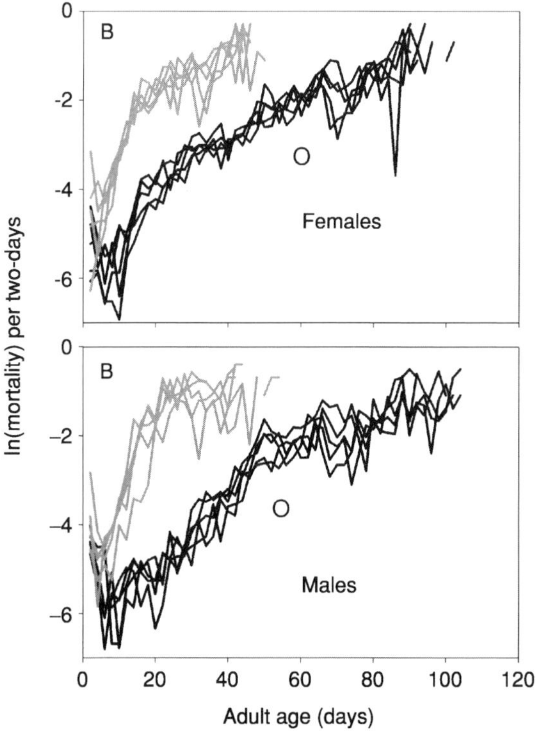

Figure 2 The natural log of mortality rates in females (top) from the B (gray) and O (black) populations and males (bottom) from the B and O populations.

action of alleles that have antagonistic pleiotropic effects on longevity and fertility.

If there are naturally occurring deleterious alleles that contribute to senescence, then their frequencies could be increased by further reducing the strength of selection against these alleles. We can accomplish this in the laboratory by permitting reproduction only at very young ages. This effectively makes late-acting deleterious alleles neutral (assuming they have no pleiotropic effects on early survival or fertility). The probability that these neutral alleles will be fixed is independent of the population size but the speed with which they move to fixation will depend on the population size. Thus, if we also make the populations small we should see a relatively rapid increase in these deleterious alleles. Mueller (1987) maintained populations under these conditions and in fact was able to document accelerated senescence due to naturally occurring deleterious alleles.

The general paradigm employed by Rose and his colleagues has been replicated many times in independent laboratories. Some of these studies have corroborated the basic results that selection for late-life fitness increases longevity but at a cost of decreased early female fecundity (Luckinbill *et al.*, 1984; Partridge *et al.*, 1999). Occasionally, selection for late-life fitness has not resulted in a measurable decline in early fecundity (Partridge and Fowler, 1992; Roper *et al.*, 1993). Some of these differences can be understood by recognizing that the laboratory is not a single environment and even subtle differences in selection protocols can significantly affect the course of evolution (Leroi *et al.*, 1994a,b).

B. Mortality-Rate Plateaus

1. Theory

Benjamin Gompertz (1825) made a seminal contribution to demography by suggesting that age-specific mortality, $u(x)$, might be modeled by the simple exponential relationship:

$$u(x) = Ae^{\alpha x}. \quad (5)$$

The two parameters of the Gompertz equation, A and α, measure age-independent and dependent sources of mortality respectively. The Gompertz equation can provide very good statistical fits to mortality or survival curves for a wide variety of species. For that reason this model has become a standard equation for summarizing the kinetics of biological mortality.

In 1992, two papers appeared that questioned the Gompertzian view of demography (Carey *et al.*, 1992; Curtsinger *et al.*, 1992). In these studies it appeared that at the most advanced ages, observed mortality rates failed to increase exponentially or, for that matter, increase at all. After the existence of these mortality plateaus was confirmed in a number of very different organisms, like fruit flies, yeast, wasps, humans, and nematodes, an explanation for their existence was sought (Brooks *et al.*, 1994; Vaupel *et al.*, 1998).

One hypothesis posits that at the time aging commences each individual's chance of dying is described by the Gompertz equation, but that there is heterogeneity between individuals for the values of A, α, or both. This variability may be due to genetic or environmental differences. When there is variation in A of magnitude σ^2, the average mortality of individuals age x is given by (Vaupel *et al.*, 1979):

$$\bar{u}(x) = \frac{A\exp(\alpha x)}{1 + \sigma^2 A\alpha^{-1}[\exp(\alpha x) - 1].} \tag{6}$$

This model predicts mortality plateaus if σ^2 is sufficiently large. Variation can also be incorporated in α, although analytic results are more complicated for that model.

A second explanation for mortality plateaus is that they arise as a natural consequence of selection in age-structured populations and genetic drift (Mueller and Rose, 1996; Charlesworth, 2001). Under this theory, survival shows an exponential increase at early ages due to the exponential weighting of fitness as in equation 3. At very advanced ages the strength of selection is so weak that random genetic drift dominates changes in allele frequencies (Fig. 3). Thus, mortality remains high but there is no tendency for these rates to change with age. In Fig. 3 we demonstrate how in a large population, since the effects of drift are weaker, the onset of the mortality plateau is displaced to later ages. A corollary of this theory is that the age at which the plateau starts should be a function of the age-specific strength of natural selection. If selection remains strong, even at advanced ages, the plateau in mortality rates should be accordingly later in life than the case when selection is only strong in early life. Charlesworth (2001) has developed analytical models of mutation accumulation that also predict the existence of mortality plateaus, thus supporting the general conclusions of Mueller and Rose (1996).

Tests of the heterogeneity theory are more difficult to design since the sources of heterogeneity are left vague by most theories. We have attempted to test some aspects of the heterogeneity theory but will not present the details of those results here (Mueller *et al.*, 2003).

2. *Laboratory Experiments*

The evolutionary theory of late-life mortality plateaus has been experimentally tested by Rose *et al.* (2002). The general approach was to compare populations that had been subjected to different, long-term, age-specific

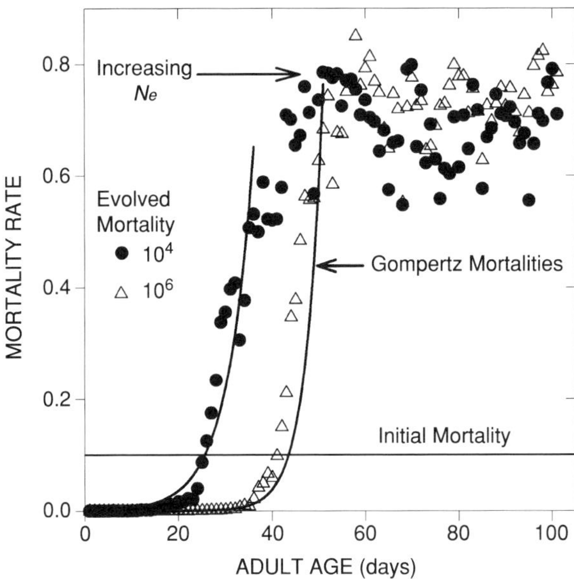

Figure 3 The evolution of mortality plateaus. In a population with no pattern of age-specific mortality (initial mortality) evolution results in the increase of mortality with age at approximately Gompertian rates until later in life. At that point drift takes over and mortality simply remains high but without an age-specific pattern (see Mueller and Rose, 1996, for additional details of these simulations).

selection regimes: reproduction early in life versus reproduction later in life. These populations comprised five sets of replicated stocks: B_{1-5}, O_{1-5}, CO_{1-5}, ACO_{1-5}, and NRO_{1-5}. The ACO and B populations have an early age of last reproduction (9 and 14 days, respectively), the CO populations have an intermediate last age of reproduction (28 days), and the O populations have a late last age of reproduction (70 days). The five NRO populations were derived from their respective O populations and have a last age of reproduction of 14 days. The NRO culture procedure was like that of the O populations, except that the flies were placed in cages at about 10 days from the egg stage with egg collection at 14 days after feeding on yeast. Except for the NRO group, these populations were each maintained for more than 100 generations at effective population sizes of $\geq$1000.

Using maximum likelihood techniques (Mueller *et al.*, 1995), we fit a two-stage Gompertz model to the cohort survival data collected from these populations. This model assumes that, up until an age called the breakday, mortality is described by the Gompertz equation. After the breakday, mortality is constant. The value of the breakday is determined from maximum likelihood techniques. The specific prediction from the evolutionary theories

Table 1 The mean longevity and breakday for populations subjected to different regimes of age-specific survival (Rose *et al.*, 2002). The pairs of selection treatments to be compared are arranged side-by-side, e.g., compare B to O, ACO to CO, and so on.

	Early reproduction	Later reproduction
	B	O
Male longevity**	20.6	52.3
Male breakday**	23.6	58.0
Female longevity**	20.8	48.2
Female breakday**	24.0	68.4
	ACO	CO
Male longevity**	26.2	44.2
Male breakday**	42.6	58.6
Female longevity**	23.5	37.2
Female breakday**	40.6	57.0
	NRO	O
Male longevity**	41.8	53.3
Male breakday**	48.2	68.6
Female longevity**	39.2	50.4
Female breakday*	54.6	67.8

*$p < 0.1$; **$p < 0.01$.

of late-life mortality plateaus is that the onset of the plateau, or the breakday, should be later in those populations experiencing strong selection for reproduction later in life. The mean longevities and breakdays for the three different pairs of selection treatments are shown in Table 1. In each case, selection for late-life reproduction increased both the average longevity and the age at which the mortality plateau begins (breakday). Except for one case, these differences are all statistically significant at $p < 0.01$.

Several studies have examined the impact of different types of heterogeneity on mortality plateaus (Carey *et al.*, 1995; Khazaeli *et al.*, 1995). Using highly inbred lines of *Drosophila*, Fukui *et al.* (1993) demonstrated that the existence of late-life mortality plateaus does not require genetic variability. Thus, for the heterogeneity theory of late-life mortality plateaus to remain viable there must be sufficient environmentally generated variability to cause such plateaus. Khazaeli *et al.* (1995) exposed *Drosophila* to variable levels of desiccation and initially interpreted their results as supporting the existence of heterogeneity. However, Curtsinger and Khazaeli (1997) later retracted that interpretation noting that the normal heat shock response of *Drosophila* made it impossible to unambiguously interpret their experimental results. The experimental results of Carey *et al.* (1995) with medflies led him to conclude that environmentally induced variation is not the likely cause of mortality-leveling in later life (Carey, 2003). While the sources of environmental variation are effectively infinite, the inability to readily demonstrate causation of plateaus from identifiable sources of variation places the heterogeneity theory of mortality plateaus in a precarious position.

C. Fecundity Plateaus

1. Theory

The theory of fitness in age-structured populations (equation 3) controls selection on fecundity as well as survival. If the arguments presented to explain the existence of late-life mortality are essentially correct, then these same forces ought to be shaping the pattern of female fecundity in late life. That is, at sufficiently advanced ages selection for fecundity will be so weak that it cannot compete with drift. At that point, all ages become interchangeable and there should be no pattern in late-life fecundity other than it being very low (Rauser *et al.*, 2003). A well-developed formal theory about the evolution of late-life fecundity does not yet exist. This is clearly an area for additional future research. However, as with mortality plateaus, we expect that the onset of a fecundity plateau will respond to age-specific selection in much the same way mortality plateaus respond.

2. Laboratory Experiments

Female fecundity in large cohorts of flies from three ACO populations (ACO_{1-3}) was followed at daily intervals (Rauser *et al.*, 2003). The results showed a rapid decline in fecundity as the flies aged, followed by an extended period during which small but relatively unchanging numbers of eggs per female were produced (Fig. 4).

We tested the evolutionary theory that predicts level fecundity at late ages by fitting the late-life fecundity data from the three ACO populations to a two-stage linear model, which most simply summarizes our *a priori* predictions. Since our focus was on fecundity at late ages we did not try to model the age-specific fecundity patterns at early or mid-life. Our model predicts linearly declining egg numbers at advanced ages until an age at which egg numbers stop declining and plateau at low values. Specifically, this model shows that female fecundity at age x is

$$f(x) = \begin{cases} a_0 + a_1 x \text{ when } & x \leq a_2 \\ a_0 + a_1 a_2 \text{ when } & x > a_2, \end{cases} \qquad (7)$$

where a_2 is the breakday or the age at which the fecundity plateau begins. We also estimated the plateau height, which was significantly different than zero

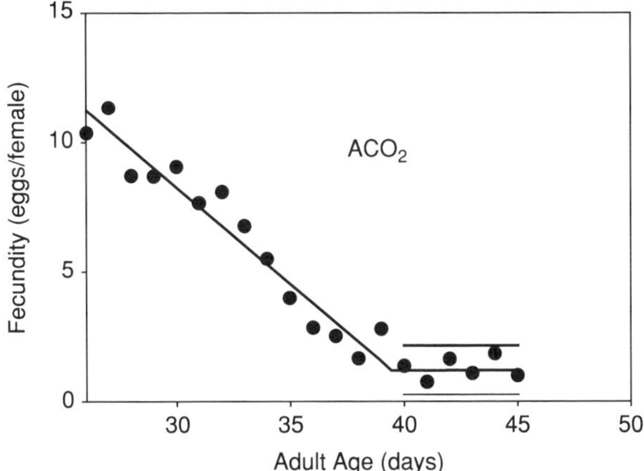

Figure 4 Late-life fecundity in the ACO_2 population. The line shows a two-stage linear model fit by non-linear least squares to the observed fecundity values (circles). In addition to the parameters for the two lines, the age of plateau onset is fit by the regression technique. A 95% confidence interval on the plateau regression is also shown to emphasize that female fecundity in the plateau is demonstrably >0.

for all three populations (Fig. 4). This result suggests that fecundity is level and non-zero at late ages as the evolutionary theory predicts.

It is important to note that the two-stage linear model does not necessarily need to support the existence of late-life fecundity plateaus. For instance, the breakday could be estimated to be at the last day females were alive, which would not be consistent with the existence of a plateau. In addition, the height of the plateau could simply be zero. This latter result is not inconsistent with an evolutionary model, but does not constitute strong support for the evolutionary model of late-life. Nevertheless, when we apply these objective criteria to our large datasets they support the overall visual impression that female fecundity reaches a non-zero plateau later in life.

IV. DISCUSSION

The use of experimental systems of *Drosophila* has contributed to our understanding of density-dependent and age-specific natural selection and population dynamics. Laboratory experiments have repeatedly allowed us to test the general theories of ecology. The use of experimental systems is not always straightforward and requires attention to details like the source of the experimental populations, effective population sizes, effects of mutant alleles, and adaptation to the laboratory environment (Rose *et al.*, 1996). However, the great strength of laboratory experimental systems is their power when testing a theory. Oftentimes, these experimental populations can be maintained in a way that eliminates confounding factors that are present in field systems. Ultimately this makes the interpretation of laboratory experiments simpler.

Recently a number of papers have discussed the limitations of using laboratory populations for the study of evolutionary problems (Gibbs, 1999; Hoffmann and Harshman, 1999; Sgro and Partridge, 2000; Harshman and Hoffmann, 2000; Hoffmann *et al.*, 2001; Linnen *et al.*, 2001). For instance, Sgro and Partridge (2000) point out that "Conditions in the laboratory are different from those in the field and could alter both the way that genes affecting life-history traits are expressed ... and the resulting genetic correlations between them." Implicit in this statement is the idea that there are two environments: the laboratory and the field. In reality there is no single definable entity called the field. For fruit flies, the field may be almost anywhere in the world. Even if one were to focus on a specific locality, as Sgro and Partridge (2000) did, we do not expect the conditions of temperature, humidity, competitors, and disease in these localities to be precisely the same from one year to the next.

Suppose we could study life-history evolution in the field environment of Montpellier, France, in 1996. We would still not be sure that the genetic correlations and gene expression observed in France would be the same as in Irvine, California, in 2004. At least in the laboratory we can declare in a fairly precise way the quantitative state of the environment; that is, we know what the temperature, humidity, light regime, food resources, and biological competitors were during the entire experiment. In this sense, we have vastly more information about our laboratory environments than we do about any particular field environment and thus the potential to control whichever variables that matter for the evolutionary and ecological phenomena we are interested in.

The fact that the laboratory is different from any field environment should be expected. Hoffmann *et al.* (2001) document a decline in starvation and desiccation resistance of laboratory-adapted flies compared to recently collected flies from Montpellier, France. In retrospect, the laboratory must either be more or less stressful with respect to water balance compared to a particular field environment. It is unlikely to be precisely the same. Hoffmann *et al.* (2001) determined that wild caught flies are 46% more desiccation-resistant than laboratory-adapted flies. Hoffmann *et al.* (2001) suggest these declines in stress resistance reduce the usefulness of laboratory-selected populations since "... then much of the response to selection for increased stress resistance may be achieved by moving the laboratory populations back to resistance levels that approach those of recently founded populations" (Hoffmann *et al.*, 2001). However, when long established laboratory populations of *Drosophila* are selected for desiccation resistance they improve by a factor of more than 300% not a mere 46% (Phelan *et al.*, 2003). This result also shows that the concern that laboratory populations must necessarily and quickly lose genetic variation for these stress resistant traits in benign laboratory environments is unfounded. It is true that if laboratory populations are kept at chronically small sizes, loss of variation is virtually inevitable. But the use of laboratory culture procedures that maintain large effective population sizes can minimize this effect.

One must always be careful about extrapolating conclusions obtained under one set of environmental conditions to a broader set of environments. However, it seems to us that we can determine which aspects of the environment matter only by performing experiments in which these environmental variables are carefully manipulated.

The complications of natural populations, like heterogeneous environments, migration, meteorological disaster, and so on, should not be viewed as barriers to laboratory studies but as challenges. Many interesting problems that can be addressed by laboratory populations still exist. For instance, there is still very little experimental work on the population dynamics of age-structured populations. It is unclear how age-structure affects population stability. Ultimately, we are optimistic that with the recognition of the many important contributions that can be made by experimental ecology, those employing its techniques will increase in number.

ACKNOWLEDGMENTS

We thank two anonymous referees for their useful comments on the manuscript. Dr. Rauser was supported by a GAANN fellowship and a Grant-in-Aid of Research from Sigma Xi.

REFERENCES

Barclay, H.J. and Gregory, P.T. (1981) An experimental test of models predicting life-history characteristics. *Am. Nat.* **117**, 944–961.

Barclay, H.J. and Gregory, P.T. (1982) An experimental test of life-history evolution using *Drosophila melanoagaster* and *Hyla regilla*. *Am. Nat.* **120**, 26–40.

Borash, D.J., Gibbs, A.G., Joshi, A. and Mueller, L.D. (1998) A genetic polymorphism maintained by natural selection in a temporally varying environment. *Am. Nat.* **151**, 148–156.

Boyce, M.S. (1984) Restitution of *r*- and *K*-selection as a model of density-dependent natural selection. *Annu. Rev. Ecol. Syst.* **15**, 427–447.

Brooks, A.G., Lithgow, G.J. and Johnson, T.E. (1994) Mortality rates in a genetically heterogeneous population of *Caenorhabditis elegans*. *Science* **263**, 668–671.

Gadgil, M. and Solbrig, O.T. (1972) The concept of r and K selection: Evidence from wild flowers and some theoretical considerations. *Am. Nat.* **106**, 14–31.

Carey, J.R. (2003) *Longevity*. Princeton University Press, Princeton, NJ.

Carey, J.R., Liedo, P., Orozco, D. and Vaupel, J.W. (1992) Slowing of mortality rates at older ages in large medfly cohorts. *Science* **258**, 457–461.

Carey, J.R., Liedo, P., Orozco, D. and Vaupel, J.W. (1995) Mortality dynamics of density in the Mediterranean fruit fly. *Exp. Gerontol.* **30**, 605–629.

Charlesworth, B. (1994) *Evolution in Age-Structured Populations*, 2nd ed. Cambridge University Press, London.

Charlesworth, B. (2001) Patterns of age-specific means and genetic variances of mortality rates predicted by the mutation-accumulation theory of ageing. *J. Theor. Biol.* **210**, 47–65.

Curtsinger, J.W., Fukui, H.H., Townsend, D.R. and Vaupel, J.W. (1992) Demography of genotypes: Failure of the limited life-span paradigm in *Drosophila melanogaster*. *Science* **258**, 461–463.

Curtsinger, J.W. and Khazaeli, A. (1997) A reconsideration of stress experiments and population heterogeneity. *Exp. Gerontol.* **32**, 727–729.

Edney, E.B. and Gill, R.W. (1968) Evolution of senescence and specific longevity. *Nature* **220**, 281–282.

Fukui, H.H., Xiu, L. and Curtsinger, J. (1993) Slowing of age-specific mortality rates in *Drosophila melanogaster*. *Exp. Gerontol.* **28**, 585–599.

Gatto, M. (1993) The evolutionary optimality of oscillatory and chaotic dynamics in simple population models. *Theor. Popul. Biol.* **43**, 310–336.

Gibbs, A.G. (1999) Laboratory selection for the comparative physiologist. *J. Exp. Biol.* **202**, 2709–2718.

Gompertz, B. (1825) On the nature of the function expressive of the law of human mortality, and on a new mode of determining the value of contingencies. *Phil. Trans. Roy. Soc. (London)* **115**, 513–585.

Hamilton, W.D. (1966) The moulding of senescence by natural selection. *J. Theor. Biol.* **12**, 12–45.

Harshman, L.G. and Hoffmann, A.A. (2000) Laboratory selection experiments using *Drosophila*: What do they really tell us? *Trends Ecol. Evol. Biol.* **15**, 32–36.

Hoffmann, A.A., Hallas, R., Sinclair, C. and Partridge, L. (2001) Rapid loss of stress resistance in *Drosophila melanogaster* under adaptation to laboratory culture. *Evolution* **55**, 436–438.

Hoffmann, A.A. and Harshman, L.G. (1999) Desiccation and starvation resistance in *Drosophila*: Patterns of variation at the species, population and intrapopulation levels. *Heredity* **83**, 637–643.

Joshi, A. and Mueller, L.D. (1988) Evolution of higher feeding rate in *Drosophila* due to density-dependent natural selection. *Evolution* **42**, 1090–1093.

Khazaeli, A.A., Xiu, L. and Curtsinger, J.W. (1995) Stress experiments as a means of investigating age-specific mortality in *Drosophila melanogaster*. *Exp. Gerontol.* **30**, 177–184.

Leroi, A.M., Chen, W. and Rose, M.R. (1994a) Long term laboratory evolution of a genetic life-history trade-off in *Drosophila melanogaster*. 2. Stability of genetic correlations. *Evolution* **48**, 1258–1268.

Leroi, A.M., Chipppindale, A.K. and Rose, M.R. (1994b) Long term laboratory evolution of a genetic life-history trade-off in *Drosophila melanogaster*. 1. The role of genotype-by-environment interaction. *Evolution* **48**, 1244–1257.

Linnen, C., Tatar, M. and Promislow, D. (2001) Cultural artifacts: a comparison of senescence in natural, laboratory-adapted and artificially selected lines of *Drosophila melanogaster*. *Evol. Ecol. Res.* **3**, 877–888.

Luckinbill, L.S., Arking, R., Clare, M.G., Cirocco, W.C. and Buck, S.A. (1984) Selection for delayed senescence in *Drosophila melanogaster*. *Evolution* **38**, 996–1003.

MacArthur, R.H. and Wilson, E.O. (1967) *The Theory of Island Biogeography*. Princeton University Press, Princeton, NJ.

McNaughton, S.J. (1975) r- and K-selection in Typha. *Am. Nat.* **109**, 251–261.

Medawar, P.B. (1952) *An Unsolved Problem of Biology*. H. K. Lewis, London.

Mueller, L.D. (1985) The evolutionary ecology of *Drosophila*. *Evol. Biol.* **19**, 37–98.

Mueller, L.D. (1987) Evolution of accelerated senescence in laboratory populations of *Drosophila*. *Proc. Natl. Acad. Sci. USA* **84**, 1974–1977.

Mueller, L.D. (1997a) Theoretical and empirical examination of density-dependent selection. *Annu. Rev. Ecol. Syst.* **28**, 269–288.

Mueller, L.D. and Ayala, F.J. (1981b) Trade-off between *r*-selection and *K*-selection in *Drosophila* populations. *Proc. Natl. Acad. Sci. USA* **78**, 1303–1305.

Mueller, L.D. and Ayala, F.J. (1981) Dynamics of single-species population growth: Stability or chaos? *Ecology* **62**, 1148–1154.

Mueller, L.D., Drapeau, M.D., Adams, C.S., Hammerle, C.W., Doyal, K.M., Jazayeri, A.J., Ly, T., Beguwala, S.A., Mamidi, A.R. and Rose, M.R. (2003) Statistical tests of demographic heterogeneity theories. *Exp. Gerontolo.* **38**, 373–386.

Mueller, L.D., Graves, J.L., Jr., and Rose, M.R. (1993) Interactions between density-dependent and age-specific selection in *Drosophila melanogaster*. *Funct. Ecol.* **7**, 469–479.

Mueller, L.D., Guo, P.Z. and Ayala, F.J. (1991) Density-dependent natural selection and trade-offs in life history traits. *Science* **253**, 433–435.

Mueller, L.D. and Huynh, P.T. (1994) Ecological determinants of stability in model populations. *Ecology* **75**, 430–437.

Mueller, L.D. and Joshi, A. (2000) *Stability in Model Populations*. Princeton University Press, Princeton, NJ.

Mueller, L.D., Joshi, A. and Borash, D.J. (2000) Does population stability evolve? *Ecology* **81**, 1273–1285.

Mueller, L.D., Nusbaum, T.J. and Rose, M.R. (1995) The Gompertz equation as a predictive tool in demography. *Exp. Gerontol.* **30**, 553–569.

Mueller, L.D. and Rose, M.R. (1996) Evolutionary theory predicts late-life mortality plateaus. *Proc. Natl. Acad. Sci. USA* **93**, 15249–15253.

Norton, H.T.J. (1928) Natural selection and Mendelian variation. *Proc. Lond. Math. Soc.* **28**, 1–45.

Partridge, L. and Fowler, K. (1992) Direct and correlated responses to selection on age at reproduction in *Drosophila melanogaster*. *Evolution* **46**, 76–91.

Partridge, L., Prowse, N. and Pignatelli, P. (1999) Another set of responses and correlated responses to selection on age at reproduction in *Drosophila melanogaster*. *Proc. R. Soc. Lond. B* **266**, 255–261.

Phelan, J.P., Archer, M.A., Beckman, K.A., Chippindale, A.K., Nusbaum, T.J. and Rose, M.R. (2003) Breakdown in correlations during laboratory evolution I. Comparative analyses of *Drosophila* populations. *Evolution* **57**, 527–535.

Prasad, N.G., Dey, S., Shakarad, M. and Joshi, A. (2003) The evolution of population stability as a by-product of life-history evolution. *Proc. R. Soc. Lond. B* **270** (Suppl. 1), S84–S86.

Rauser, C.L., Mueller, L.D. and Rose, M.R. (2003) Aging, fertility, and immortality. *Exp. Gerontol.* **38**, 27–33.

Reznick, D.N. (1989) Life history evolution in guppies. 2. Repeatability of field observations and the effects of season on life histories. *Evolution* **43**, 1285–1297.
Reznick, D.N. and Bryga, H. (1987) Life history evolution in guppies. 1. Phenotypic and genetic changes in an introduction experiment. *Evolution* **41**, 1370–1385.
Reznick, D.N. and Bryga, H. (1996) Life history evolution in guppies (*Poecillia reticulata*: Poeciliidae). V. Genetic basis of parallelism in life histories. *Am. Nat.* **147**, 339–359.
Reznick, D.N., Bryga, H. and Endler, J.A. (1990) Experimentally induced life-history evolution in a natural population. *Nature* **346**, 357–359.
Reznick, D., Bryant, M.J. and Bashey, F. (2002) *r*- and *K*-selection revisited: The role of population regulation in life-history evolution. *Ecology* **83**, 1509–1520.
Reznick, D.N. and Endler, J.A. (1982) The impact of predation on life history evolution in Trinidadian guppies (*Poecilia reticulata*). *Evolution* **36**, 160–177.
Reznick, D.N., Rodd, F.H. and Cardenas, M. (1996) Life history evolution in guppies (*Poecillia reticulata*: Poeciliidae). IV. Parallelism in life history phenotypes. *Am. Nat.* **147**, 319–338.
Reznick, D.N., Shaw, F.H., Rodd, F.H. and Shaw, R.G. (1997) Evaluation of the rate of evolution in natural populations of guppies (*Poecillia reticulata*). *Science* **275**, 1934–1937.
Roper, C., Pignatelli, P. and Partridge, L. (1993) Evolutionary effects of selection on age at reproduction in larval and adult *Drosophila melanogaster*. *Evolution* **47**, 445–455.
Rose, M.R. (1984) Laboratory evolution of postponed senescence in *Drosophila melanogaster*. *Evolution* **38**, 1004–1010.
Rose, M.R., Drapeau, M.D., Yazdi, P.G., Shah, K.H., Moise, D.B., Thakar, R.R., Rauser, C.S. and Mueller, L.D. (2002) Evolution of late-life mortality in *Drosophila melanogaster*. *Evolution* **56**, 1982–1991.
Rose, M.R., Nusbaum, T.J. and Chippindale, A.K. (1996) Laboratory evolution: The experimental wonderland and the Cheshire cat syndrome. In: *Adaptation* (Ed. by M.R. Rose and G.V. Lauder), pp. 221–241. Academic Press, San Diego, CA.
Roughgarden, J. (1971) Density-dependent natural selection. *Ecology* **52**, 453–468.
Sgro, C.M. and Partridge, L. (2000) Evolutionary responses of the life history of wild-caught *Drosophila melanogaster* to two standard methods of laboratory culture. *Am. Nat.* **156**, 341–353.
Sheeba, V. and Joshi, A. (1998) A test of simple models of population growth using data from very small populations of *Drosophila melanogaster*. *Curr. Sci.* **75**, 1406–1410.
Sokolowski, M.B., Pereira, H.S. and Hughes, K. (1997) Evolution of foraging behavior in Drosophila by density-dependent selection. *Proc. Natl. Acad. Sci. USA* **94**, 7373–7377.
Stearns, S.C. (1976) Life-history tactics: a review of the ideas. *Q. Rev. Biol.* **51**, 3–47.
Stearns, S.C. (1977) The evolution of life history traits: A critique of the theory and a review of the data. *Annu. Rev. Ecol. Syst.* **8**, 145–171.
Stokes, T.K., Gurney, W.S.C., Nisbet, R.M. and Blythe, S.P. (1988) Parameter evolution in a laboratory insect population. *Theor. Pop. Biol.* **34**, 248–265.
Taylor, C.E. and Condra, C. (1980) r- and K-selection in *Drosophila pseudoobscura*. *Evolution* **34**, 1183–1193.
Thompson, W.R. (1948) Can economic entomology be an exact science? *Canadian Entomologist* **80**, 49–55.
Turchin, P. (2003) *Complex Population Dynamics: A Theoretical/Empirical Synthesis*. Princeton University Press, Princeton, NJ.

Turelli, M. and Petry, D. (1980) Density-dependent selection in a random environment: An evolutionary process that can maintain stable population dynamics. *Proc. Natl. Acad. Sci. USA* **77**, 7501–7505.

Vaupel, J.W., Carey, J.R., Christensen, K., Johnson, T.E., Yashin, A.I., Holm, N.V., Iachine, I.A., Kannisto, V., Khazaeli, A.A., Liedo, P., Longo, V.D., Zeng, Y., Manton, K.G. and Curtsinger, J.W. (1998) Biodemographic trajectories of longevity. *Science* **280**, 855–860.

Vaupel, J.W., Manton, K.G. and Stallard, E. (1979) The impact of heterogeneity in individual frailty on the dynamics of mortality. *Demography* **16**, 439–454.

Williams, G.C. (1957) Pleiotropy, natural selection, and the evolution of senescence. *Evolution* **11**, 398–411.

Yoshida, T., Jones, L.E., Ellner, S.P., Fussmann, G.F. and Hairston, N.G., Jr., (2003) Rapid evolution drives ecological dynamics in a predator-prey system. *Nature* **424**, 303–306.

Nonlinear Stochastic Population Dynamics: The Flour Beetle *Tribolium* as an Effective Tool of Discovery

ROBERT F. COSTANTINO, ROBERT A. DESHARNAIS, JIM M. CUSHING, BRIAN DENNIS, SHANDELLE M. HENSON AND AARON A. KING

> When observation and theory collide, scientists turn to carefully designed experiments for resolution. Their motivation is especially high in the case of biological systems, which are typically far too complex to be grasped by observation and theory alone. The best procedure, as in the rest of science, is first to simplify the system, then to hold it more or less constant while varying the important parameters one or two at a time to see what happens.
>
> —Edward O. Wilson (2002)

I. INTRODUCTION

Prior to the seminal work of R.M. May in the 1970s, the prevailing paradigm viewed the unpredictable fluctuations in population time-series data as random effects due to environmental variability and/or measurement errors.

ADVANCES IN ECOLOGICAL RESEARCH VOL. 37 0065-2504/05 $35.00

DOI: 10.1016/S0065-2504(04)37004-2

In the absence of environmental variability, according to this view, population numbers would either equilibrate or settle into regular periodic oscillations. May's (1974) suggestion that simple deterministic rules might explain the complex fluctuations observed in animal abundances led to an intense search for chaos in extant population data. The results of the search were suggestive, but equivocal, and May's hypothesis remained the subject of lively debate (Zimmer, 1999; Perry *et al.*, 2000).

We took a different approach. Our interdisciplinary research team composed of statisticians, mathematicians, and biologists gathered together in the early 1990s to document *experimentally* the occurrence of nonlinear dynamic phenomena in biological populations. We began with the idea that nonlinear theory yields testable hypotheses concerning changes in the dynamical behaviors of populations. For example, in the case of the quadratic map (sometimes called the "logistic" map), changes in the intrinsic growth rate lead to a sequence of dynamical behaviors from equilibria to periodic cycles, to aperiodic chaotic behavior. Our thought was that a sequence of changes in dynamical behavior, which is a common feature of nonlinear models, could be tested, in principle, under controlled laboratory conditions. This would provide a connection between theory and data that was missing from ecology.

From the very beginning of our collaboration, fundamental questions greeted us at every turn as we looked at historical time-series data and at data collected in our laboratories. Over the years we struggled to combine deterministic concepts such as equilibria, cycles, saddle nodes, bifurcations, basins of attraction, multiple attractors, resonance, and chaos with observations. What would a stable equilibrium, let alone chaos, look like in a population? Could a saddle node be invoked as an explanation for different transient behaviors of time series among replicate populations? Is chaos even possible if we consider discrete-state population models? Is it useful to consider populations as discrete-state stochastic systems?

In ecological theory, a central (and abiding) problem is to situate deterministic theory in the context of biological systems where important demographic events are probabilistic. Chance variation, in such fundamental biological processes as the number of offspring per adult and the chance of an individual surviving to adulthood, is a part of population dynamics. Probabilistic variation enters the overall research effort in the statistical methods associated with model identification, parameter estimation, and model validation. Chance events are also a component of the interpretation of population behavior; probabilistic variation is essential to the explanation of ecological time-series data. We expand on these points.

First, a mathematical population model, built and tested as a serious scientific hypothesis, must be somehow connected to data. A probabilistic version of the model must be constructed to account for inevitable deviations of data from the predictions of the deterministic model. Demographic/

environmental variability must be modeled in order to construct an appropriate estimating function for the model parameters (based on the likelihood or conditional sums of squares, for instance). Statistical diagnostic procedures should be used to evaluate the uncertainty component of the model.

Second, chance events interact with deterministic forces to produce emergent dynamic behaviors. The deterministic skeleton fixes the geometry of state space, providing a stage for the transient dance of stochasticity. Chance events allow the system to visit (and re-visit) the various deterministic entities on the stage, including unstable invariant sets, which under strict deterministic theory would have little or no impact on population time series. Ecological time series can display a stochastic mix of many of the dynamic features of the skeleton, including multiple attractors, transients, unstable invariant sets (such as saddles and unstable manifolds) and lattice effects. Stochasticity enlarges the repertoire of time-series orbits; each population, even in a set of laboratory replicates, may display a unique sequence of population abundances.

In this chapter, we expand upon the message that in order to understand population fluctuations, deterministic *and* stochastic forces must be viewed as an integral part of the ecological system. We begin by explaining our models, both animal and mathematical. We then discuss how we estimated model parameters and validated the model. Next, with the parameterized model in hand, we present an overview of some of the nonlinear phenomena and related topics that we have documented in our experimental system: chaotic dynamics, population outbreaks, saddle nodes, phase switching, lattice effects, the anatomy of chaos and, finally, mechanistic models of stochasticity.

II. ANIMAL MODEL

Based on our experience with the flour beetle *Tribolium castaneum* (Herbst), we made it our choice as the animal model (Costantino and Desharnais, 1991). Nevertheless, in the spring of 1992, it was far from clear if the beetle system would meet the demands for an in-depth investigation of nonlinear population dynamics. Several features made the beetle attractive. Much of the biology is understood and the life cycle is sufficiently complicated leading to rich dynamic possibilities; moreover, there is a consensus that cannibalism plays a major role in beetle dynamics (Mertz, 1972). Laboratory protocols to culture and manipulate the insects in a controlled setting are well established. Bi-weekly census counts can be accurate and may be taken over many reproductive cycles in a relatively short period of time. The animals are routinely cultured in half-pint milk bottles containing 20 g of a standard medium; these incubators are maintained at a constant temperature with relative humidity and amount of light. In general, environmental variation is minimized. However, environmental variability can be imposed on the system, for example, by altering the size of the habitat following a census.

III. DETERMINISTIC SKELETON

Small. That was the word of caution for building the model. Keep the number of parameters small. Keep the number of state variables small. A model with an over-abundance of parameters and state variables (relative to the amount of data available) will be difficult to analyze mathematically and impossible to test statistically.

To construct the mathematical model, we proposed deterministic equations for the prediction of measurable state variables from one census time to the next. The state variables were chosen to be the numbers of larvae, L, pupae, P, and adult beetles, A. The resulting larval-pupal-adult (LPA) model is a system of three difference equations. The deterministic LPA model predicts the numbers of animals in each stage at time $t+1$ given the actual number of animals in each stage at time t:

$$\begin{aligned} L_{t+1} &= bA_t \exp\left(-\frac{c_{ea}}{V}A_t - \frac{c_{el}}{V}L_t\right) \\ P_{t+1} &= (1-\mu_l)L_t \\ A_{t+1} &= P_t \exp\left(-\frac{c_{pa}}{V}A_t\right) + (1-\mu_a)A_t. \end{aligned} \tag{1}$$

Here L_t is the number of feeding larvae at time t, P_t is the number of non-feeding larvae, pupae and callow adults at time t, and A_t is the number of sexually mature adults at time t. The unit of time is two weeks, which is the approximate amount of time spent in each of the L and P classes under experimental conditions. The average number of larvae recruited per adult per unit time in the absence of cannibalism is $b > 0$, and the fractions μ_a and μ_l are the adult and larval probabilities of dying from causes other than cannibalism in one time unit. In *T. castaneum*, larvae and adults eat eggs and adults eat pupae. The exponential expressions represent the fractions of individuals surviving cannibalism in one unit of time, with cannibalism coefficients c_{ea}/V, c_{el}/V, $c_{pa}/V > 0$. Habitat size V has units equal to the volume occupied by 20 g of flour, the amount of medium routinely used in our laboratory.

IV. STOCHASTIC MODELS

The LPA model (1) is not complete. The model must include a probabilistic portion that specifies how the variability in the data was determined. There are many ways of formulating a stochastic model. We present two possibilities.

Ecologists have drawn a distinction between environmental stochasticity (chance variation from extrinsic sources affecting many individuals) and demographic stochasticity (intrinsic chance variation in individual births and deaths) in populations (May, 1974; Shaffer, 1981; Simberloff, 1988). Environmental noise can be modeled as being additive on the log scale, while demographic noise is additive on the square root scale (Dennis *et al.*, 1995, 2001).

In our early explorations of the LPA model we used an environmental stochastic version (2) to describe the dynamics of larvae, pupae, and adults (Dennis *et al.*, 1995):

$$\begin{aligned} L_{t+1} &= bA_t \exp\left(-\frac{c_{ea}}{V}A_t - \frac{c_{el}}{V}L_t + E_{1t}\right) \\ P_{t+1} &= (1-\mu_l)L_t \exp(E_{2t}) \\ A_{t+1} &= \left(P_t \exp\left(-\frac{c_{pa}}{V}A_t\right) + (1-\mu_a)A_t\right)\exp(E_{3t}). \end{aligned} \tag{2}$$

Later experiments and analyses supported a demographic stochastic LPA model (3) (Dennis *et al.*, 2001):

$$\begin{aligned} L_{t+1} &= \left[\sqrt{bA_t \exp\left(-\frac{c_{ea}}{V}A_t - \frac{c_{el}}{V}L_t\right)} + E_{1t}\right]^2 \\ P_{t+1} &= \left[\sqrt{(1-\mu_l)L_t} + E_{2t}\right]^2 \\ A_{t+1} &= \left[\sqrt{P_t \exp\left(-\frac{c_{pa}}{V}A_t\right) + (1-\mu_a)A_t} + E_{3t}\right]^2. \end{aligned} \tag{3}$$

In models (2) and (3), $\mathbf{E}_t = [E_{1t}, E_{2t}, E_{3t}]'$ is a random vector and is assumed to have a trivariate normal distribution with a mean vector of $\mathbf{0}$ and a variance-covariance matrix of $\mathbf{\Sigma}$. Covariances among E_{1t}, E_{2t}, and E_{3t} at any given time t are assumed (and represented by off-diagonal elements of $\mathbf{\Sigma}$), but we expect the covariances between times to be small by comparison. Thus, we assume that $\mathbf{E}_0$, $\mathbf{E}_1$, and so on are uncorrelated. In rare instances where a large negative E_{it} causes the term in square brackets to be negative, the value of that life stage is set to zero.

V. PARAMETER ESTIMATION AND MODEL VALIDATION

A mathematical model is a scientific hypothesis expressed in the peculiarly precise, quantitative language of mathematics. How is the hypothesis to be evaluated? How are model parameters to be estimated? How is a statistical test of the model constructed? To what extent can the precision of the mathematical model be translated into strong statistical tests? More generally, how are theory and data to be joined? Finding working solutions to these problems has been an important feature of our research effort.

The stochastic construction represented by models (2) and (3) have a number of statistical advantages. First, written on the logarithmic scale for (2) and on the square root scale for (3), the stochastic models are seen to be a type of multivariate, nonlinear, autoregressive (NLAR) model:

$$\boldsymbol{W}_{t+1} = \boldsymbol{h}(\boldsymbol{W}_t) + \boldsymbol{E}_t.$$

Here $\boldsymbol{W}_t$ is the column vector of the logarithmic-transformed state variables, $W_t = [\ln L_t, \ln P_t, \ln A_t]'$, or square root-transformed, $\boldsymbol{W}_t = [\sqrt{L_t}, \sqrt{P_t}, \sqrt{A_t}]'$, state variables; $\boldsymbol{h}(\boldsymbol{W}_t) = [g\{bA_t \exp(-(c_{ea}/V)A_t - (c_{el}/V)L_t\}, g\{1 - \mu_l)L_t\}, g\{P_t \exp(-(c_{pa}/V)A_t + (1 - \mu_a)A_t\}]'$ is a column vector of functions, where $g(\cdot)$ is either a logarithmic or square root transformation and $\boldsymbol{E}_t$ has a multivariate normal $(\boldsymbol{0}, \boldsymbol{\Sigma})$ distribution. Development of statistical methods (estimation, testing, and evaluation) for NLAR models has received much attention (Tong, 1990). Second, this stochastic construction preserves the dynamical properties of the deterministic model (1) through one-step predictions (conditional expected values). Third, changes in the underlying deterministic forms can easily be accommodated in the stochastic construction. Fourth, the model is easy to simulate and, finally, parameter estimates are straightforward to compute.

The stochastic construction has biological advantages as well. The models allow for covariance of fluctuations in larvae, pupae, and adults in a given time period, as described by the covariance of elements in $\boldsymbol{E}_t$. Autocovariances of the noise elements through time, however, are not expected to be important compared to the covariances between the elements within a time, provided the underlying dynamics (deterministic skeleton) are specified correctly. The different scales of variability for larvae, pupae, and adults are accounted for through the parameters on the main diagonal of the variance-covariance matrix $\boldsymbol{\Sigma}$.

The stochastic LPA models (2) and (3) provide an explicit likelihood function. A likelihood function gives the probability that under the proposed stochastic model the random outcome would be the observed time series. A likelihood function is a fundamental tool in statistical inference (Stuart and Ord, 1991) and represents the crucial connection between data and model. Data for a particular *Tribolium* population are a realization of the joint stochastic variables L_t, P_t, and A_t. The data take the form of a trivariate time series: (l_0, p_0, a_0), (l_1, p_1, a_1), ..., (l_q, p_q, a_q). Let $\boldsymbol{w}_t$ denote the column vector of transformed observations at time $t : \boldsymbol{w}_t = [\ln l_t, \ln p_t, \ln a_t]'$ or $\boldsymbol{w}_t = [\sqrt{l_t}, \sqrt{p_t}, \sqrt{a_t}]'$. Suppose $\boldsymbol{\theta}$ denotes the unknown parameters in the functions in $\boldsymbol{h}(\cdot)$, i.e., the parameters in the deterministic model equations. Additional unknown parameters are in the variance-covariance matrix $\boldsymbol{\Sigma}$. The likelihood function $L(\boldsymbol{\theta}, \boldsymbol{\Sigma})$ is given by

$$L(\boldsymbol{\theta}, \boldsymbol{\Sigma}) = \prod_{t=1}^{q} p(\mathbf{w}_t|\mathbf{w}_{t-1}),$$

where $p(\boldsymbol{w}_t|\boldsymbol{w}_{t-1})$ is the joint transition probability density function, that is, the joint probability density function for $\boldsymbol{W}_t$ conditional on $\boldsymbol{W}_{t-1} = \boldsymbol{w}_{t-1}$ and evaluated at $\boldsymbol{w}_t$. The maximum likelihood (ML) estimates of parameters in $\boldsymbol{\theta}$ and $\boldsymbol{\Sigma}$ are those values that jointly maximize $L(\boldsymbol{\theta}, \boldsymbol{\Sigma})$ or, equivalently, ln $L(\boldsymbol{\theta}, \boldsymbol{\Sigma})$. The ML estimates of parameters in $\boldsymbol{\theta}$ must be obtained numerically for any particular dataset. We have found that maximizing the log-likelihood using the Nelder-Mead simplex algorithm is convenient, reliable and easy to program (Olsson and Nelson, 1975; Press *et al.*, 1986). The ML estimates for the stochastic LPA models have desirable statistical properties. The ML estimates are asymptotically unbiased, asymptotically efficient and asymptotically normally distributed. However, the properties of ML estimates do not hold if the model is a poor description of the underlying stochastic mechanisms producing the data. In particular, if the noise vector $\boldsymbol{E}_t$ does not have a multivariate normal distribution or is correlated through time, then the ML estimates could be biased. Since we aim to identify dynamic behavior by estimating where the parameters in $\boldsymbol{\theta}$ are in parameter space, an alternative estimation method that yields more robust parameter estimates is useful.

The method of conditional least squares (CLS) was also used for estimation of the parameters. The CLS methods relax many distributional assumptions about the noise variables in the vector $\boldsymbol{E}_t$ (Klimko and Nelson, 1978; Tong, 1990). The CLS estimates are consistent (converge to the true parameters as sample size increases), even if $\boldsymbol{E}_t$ is non-normal and autocorrelated, provided the stochastic model has a stationary distribution. In the LPA model, CLS estimates reduce to three univariate cases because any given parameter does not appear in more than one model equation. The CLS estimates are based on the sum of squared differences between the value of the variable observed at time t and its expected (or one-step forecast) value, given the observed state of the system at time $t - 1$. For the LPA model, there are three such conditional sums of squares:

$$Q_1(\boldsymbol{\theta}_1) = \sum_{t=1}^{q} \left\{ g(l_t) - g\left[b a_{t-1} \exp\left(-\frac{c_{ea}}{V} a_{t-1} - \frac{c_{el}}{V} l_{t-1} \right) \right] \right\}^2$$

$$Q_2(\boldsymbol{\theta}_2) = \sum_{t=1}^{q} \{ g(p_t) - g[(1 - \mu_l) l_{t-1}] \}^2$$

$$Q_3(\boldsymbol{\theta}_3) = \sum_{t=1}^{q} \left\{ g(a_t) - g\left[p_{t-1} \exp(-\frac{c_{pa}}{V} a_{t-1}) + (1 - \mu_a) a_{t-1} \right] \right\}^2.$$

Here $\boldsymbol{\theta}_1 = [b, c_{el}, c_{ea}]', \boldsymbol{\theta}_2 = \mu_l$, and $\boldsymbol{\theta}_3 = [c_{pa}, \mu_a]'$ are the parameter vectors from the model equations. The conditional sums of squares are constructed on the logarithmic or square root scales because that is the scale on which we assume noise is additive. Three separate numerical minimizations are required, one for each of the above sums of squares. We find the Nelder-Mead simplex algorithm convenient. The estimates of the parameters in the variance-covariance matrix of $\boldsymbol{E}_t$ are then found from the sums of squares and cross-products matrix constructed using the conditional residuals.

Model evaluation procedures center on the residuals defined as the differences of the logarithmic (or square root) state variables and their one-step (estimated) expected values:

$$\boldsymbol{e}_t = \boldsymbol{w}_t - \hat{\boldsymbol{h}}(\boldsymbol{w}_{t-1}).$$

Here $\boldsymbol{e}_t$ is a vector of residuals for $g(l_t)$, $g(p_t)$, $g(a_t)$ in a population at time t, and $\hat{\boldsymbol{h}}$ denotes the functions in $\boldsymbol{h}$ evaluated at the ML parameter estimates. If the model fits, then $\boldsymbol{e}_1, \boldsymbol{e}_2, \ldots, \boldsymbol{e}_q$ should behave approximately like uncorrelated observations from a trivariate normal distribution. Departures of the residuals from normality can be investigated using graphical procedures such as quantile-quantile plots and tested using the Lin-Mudholkar statistic (Tong, 1990). Autocorrelations of residuals indicate a relationship between successive prediction errors and thus might suggest a systematic lack of fit between model and data. First and second (or higher) order autocorrelations of residuals can be computed and tested for significance.

In Dennis *et al.* (1995), we applied a maximum likelihood procedure to the data of Desharnais and Costantino (1980) and conducted rigorous diagnostics for the evaluation of the stochastic model (2). A single set of parameter values from the control cultures was able to describe the dynamics of nine demographically manipulated cultures, even though none of the data from these manipulated populations were used to obtain the parameter estimates. The observed time series for a representative replicate together with the one-step predictions are graphed in Fig. 1. The solid lines connect the observed census data (closed circles). Dashed lines connect the observed numbers at time t with the forecast (open circles) at time $t + 1$. The accuracy of a particular forecast can be judged by comparing the predictions at time $t + 1$ with the number of animals actually observed at time $t + 1$. In general, the graph reveals a close association between the one-step forecast and the census data. Parameter estimation placed the population in the region of parameter space that corresponds to stable 2-cycles of the deterministic skeleton of the LPA model.

Similar statistical methods for parameter estimation and model validation were applied to the data from two other experiments described in this chapter (Dennis *et al.*, 1997, 2001). In both cases, the diagnostic analyses of the residuals supported the LPA model.

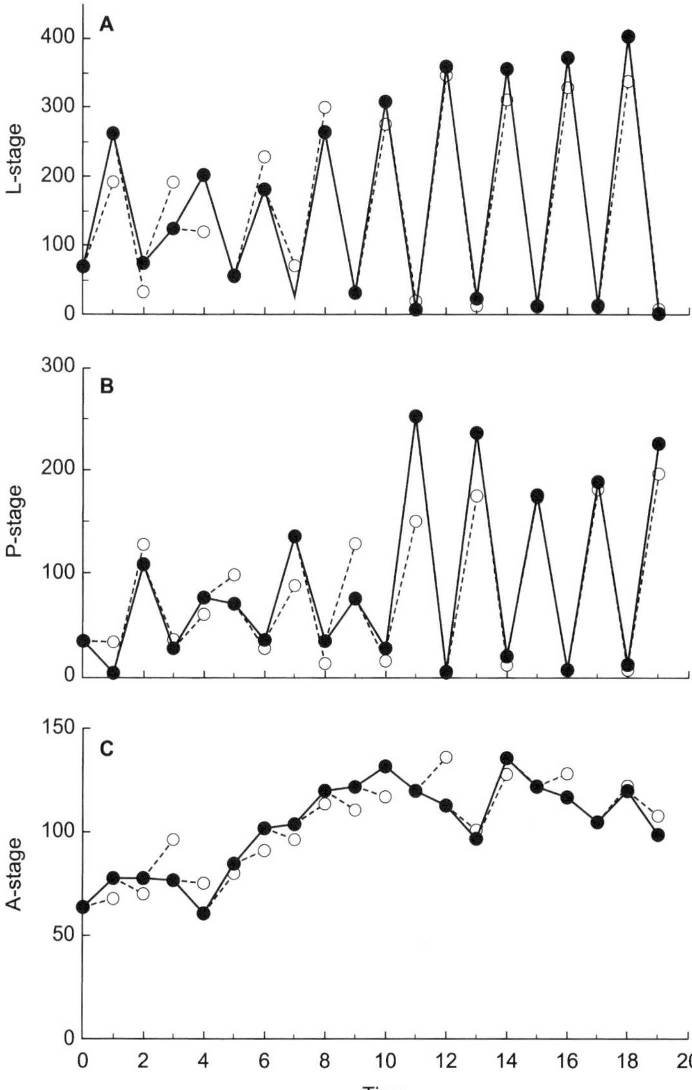

Figure 1 Time-series data (closed circles) and one-step forecasts (open circles) for the control replicate A from the experiment of Desharnais and Costantino (1980). Solid lines connect the observed census data. Dashed lines connect the observed numbers at time t with the one-step forecast at time $t + 1$. The maximum likelihood parameter estimates are: $b = 11.67$, $c_{el} = 0.0093$, $c_{ea} = 0.0110$, $c_{pa} = 0.0178$, $\mu_l = 0.5129$ and $\mu_a = 0.1108$.

VI. EXPERIMENTAL CONFIRMATION OF NONLINEAR DYNAMIC PHENOMENA

The laboratory system—beetles in a bottle together with the deterministic and stochastic models—was devised to test basic ecological hypotheses in isolation from confounding factors. With a parameterized and validated model for laboratory cultures of *T. castaneum* in hand, we were ready to open a new phase of our research in which experiments are focused directly on phenomena such as bifurcation sequences, equilibria, cycles, stable and unstable manifolds, and chaos.

Recognizing that we are part of a continuing tradition of using flour beetles in ecological research, it seems appropriate to recall the philosophy of laboratory research stated 50 years ago by pioneering ecologist Thomas Park (1955):

"Research in laboratory population ecology should take its orientation from some phenomenon known or suspected to occur in nature and known or suspected to have significant ecological consequences. Its objective is not to erect an indoor ecology but, rather, to illustrate conceptually the general problem to which it is addressed. The research is thus the handmaiden of field investigation; not the substitute. Findings derived from such studies are *models* of selected events in natural environments. The models, though not simple, are simplified; they are under a regimen of planned control, and their intrinsic interactions are likely to be intensified. To this extent they are unrealistic. But they remain, nonetheless, quantitative *biological* models and their unrealistic aspects may be a virtue instead of a vice. This is to say, they can contribute to the maturation of ecology, at least until that time when they are no longer needed."

We begin our presentation of selected nonlinear phenomena documented in our laboratories by describing an experiment based on a model-predicted sequence of transitions in dynamic behavior occurring in response to changes in the adult death rate.

VII. BIFURCATIONS IN THE DYNAMIC BEHAVIOR OF POPULATIONS

Nonlinear mathematical models can undergo sudden transitions in dynamic behavior in response to changes in parameter values. Specifically, the long-term attractors of model trajectories—stable points, stable cycles, loops, strange attractors—can exhibit abrupt changes in identity and stability when model parameter values are altered. The anatomy and taxonomy of these changes is the focus of the bifurcation theory in nonlinear dynamics.

An important biological point is that a given nonlinear model may forecast a unique parade of dynamic behaviors in response to parameter change. If a population rate quantity—birth rate, death rate, migration rate and so on—could be manipulated experimentally, the resulting observed population responses would serve as a test of the model bifurcation sequence.

The LPA model predicts such a sequence of transitions in dynamic behavior in response to changing values of the adult death rate parameter μ_a. A bifurcation (or final-state) diagram is a plot of the asymptotic dynamical behaviors as a function of adult mortality (Fig. 2). For very small values of mortality there is a stable fixed point. As μ_a increases, a period doubling bifurcation to stable 2-cycles occurs. With further increases in μ_a, a surprising reversal to an interval of stable equilibria occurs. An increase of μ_a to values near one results in another destabilization of equilibria and, in this case, a bifurcation to invariant loop attractors in phase space and quasi-periodic time series (called a Neimark-Sacker bifurcation).

To test the bifurcation prediction we manipulated adult mortality rates in beetle cultures (Costantino *et al.*, 1995; Dennis *et al.*, 1997). Rigorous

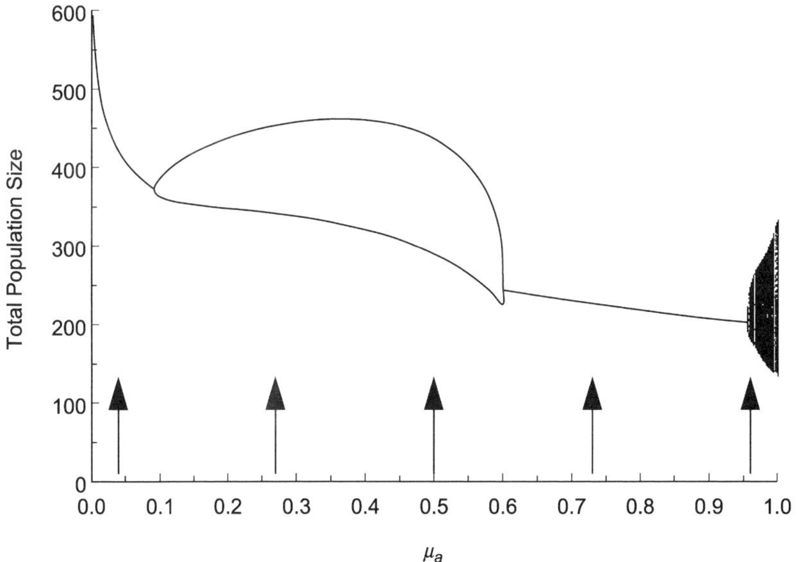

Figure 2 Bifurcation diagram for the LPA model (1) with adult mortality μ_a as the bifurcation parameter. The maximum likelihood parameter estimates are $b = 7.48$, $c_{ea} = 0.0091$, $c_{pa} = 0.0041$, $c_{el} = 0.0120$ and $\mu_l = 0.2670$. The arrows indicate those values of μ_a at which experiments were conducted.

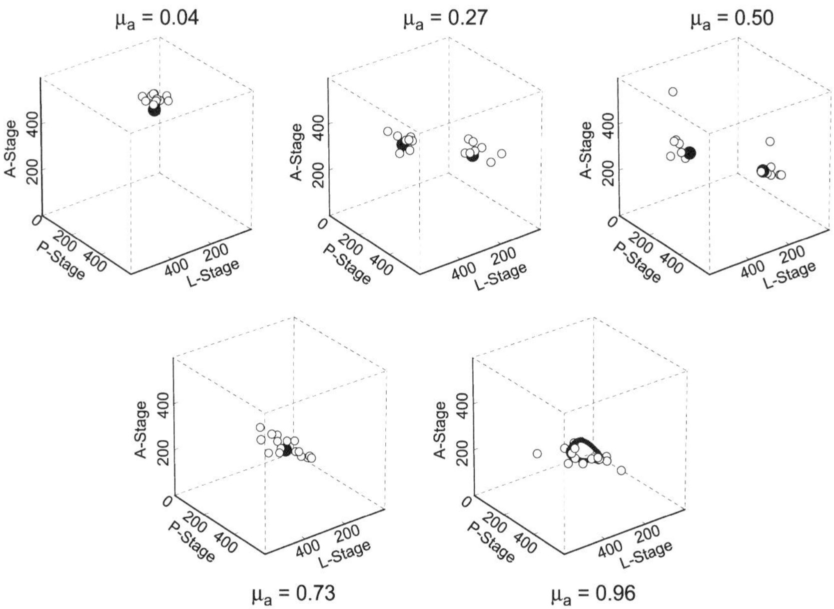

Figure 3 Phase space plots of the data (SS genetic strain) obtained from the bifurcation experiment with $\mu_a = 0.04$ (equilibrium), $\mu_a = 0.27$, 0.50 (2-cycles), $\mu_a = 0.73$ (equilibrium), and $\mu_a = 0.96$ (invariant loop).

statistical verification of the predicted shifts in dynamic behavior provided convincing evidence that the observed transitional changes did indeed correspond to those forecast by the mathematical model (Fig. 3).

VIII. A SECOND BIFURCATION EXPERIMENT: THE HUNT FOR CHAOS

Ecologists searched for chaos in historical data. This involved various statistical methods for the analysis of existing time series (e.g., Ellner and Turchin, 1995; Perry *et al.*, 2000; Turchin and Ellner, 2000). Our approach was experimental. We focused not on a particular time series, but rather on a collection of time series taken from treatments designed to lie along a route to chaos. Thus, any claim that chaotic dynamics influenced the populations would not rest on a single dataset, but would be supported by the dynamics observed across an entire sequence of predicted bifurcations.

Thus, our hunt for chaos took the form of a second-transition experiment whose protocol was based upon an LPA model-predicted sequence of bifurcations (Costantino *et al.*, 1997). This sequence occurs in the model when the adult death rate is $\mu_a = 0.96$ and the adult cannibalism of pupae c_{pa} is

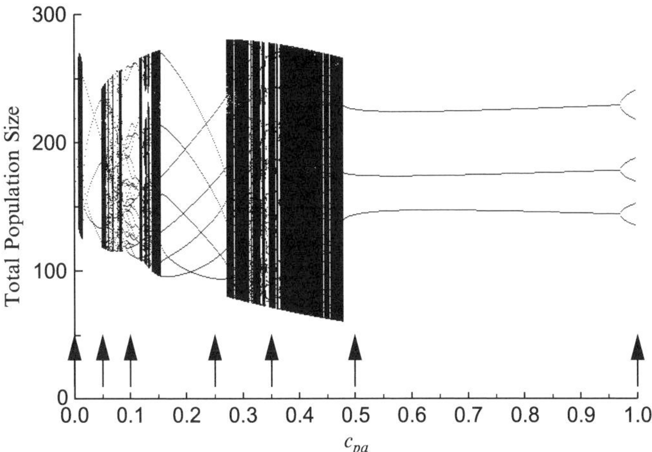

Figure 4 Bifurcation diagram for the LPA model (1) using c_{pa} as the bifurcation parameter. The adult death rate was set experimentally at $\mu_a = 0.96$. The other parameter values are $b = 10.45$, $\mu_l = 0.2000$, $c_{el} = 0.01731$ and $c_{ea} = 0.01310$. The arrows indicate those c_{pa} values at which experiments were performed.

increased (by manipulating adult recruitment). The sequence (Fig. 4) begins at $c_{pa} = 0$ with quasi-periodic oscillations around an invariant loop. With further increases in c_{pa} the predicted dynamics pass through a complicated array of aperiodic attractors and period-locking windows (where the motion around the loop is exactly periodic) until finally chaotic and strange attractors dominate. For sufficiently large values of c_{pa}, a distinctive cycle of period three which bifurcates to a six-cycle near $c_{pa} = 1$ is predicted.

Based on these predictions, the "hunt for chaos experiment" was designed with adult mortality rate at $\mu_a = 0.96$. The adult recruitment rate was manipulated so that it would equal $P_t \exp(-c_{pa} A_t)$, with values of c_{pa} set at 0.00, 0.05, 0.10, 0.25, 0.35, 0.50, and 1.00. There was also an unmanipulated control treatment.

The data from these experiments, together with the predicted deterministic attractors and stochastic realizations of model (3) using the estimated parameter values, are plotted in phase space in Fig. 5. The pattern of changes in dynamics and variability from treatment to treatment are well-captured by the model, from the stable point equilibrium of the control, to the irregular behavior of $c_{pa} = 0.35$, to the strong periodic signals in the $c_{pa} = 0.50$ and 1.00 treatments.

These results illustrate two important messages. Biological systems can undergo transitions between different types of deterministic behaviors in response to changing conditions. Moreover, these transitions might be predictable by means of suitable stochastic versions of the models (Dennis *et al.*, 2001).

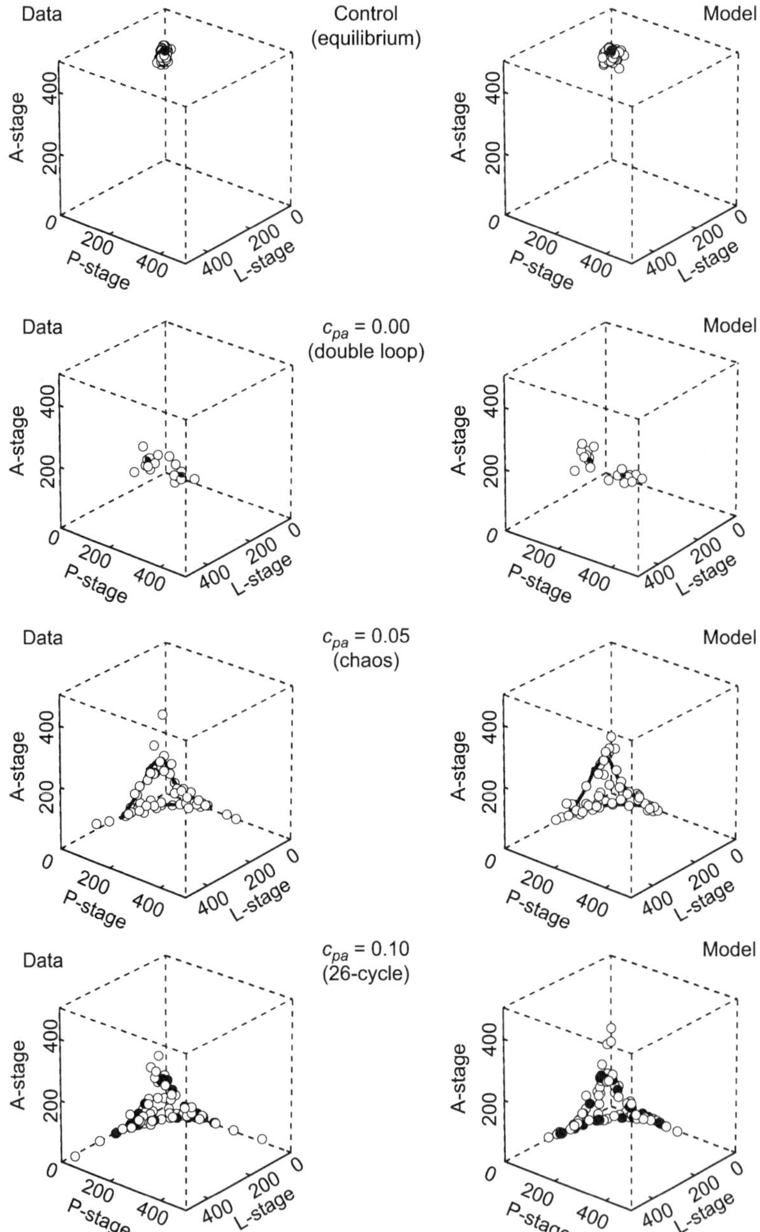

Figure 5 (*continued*)

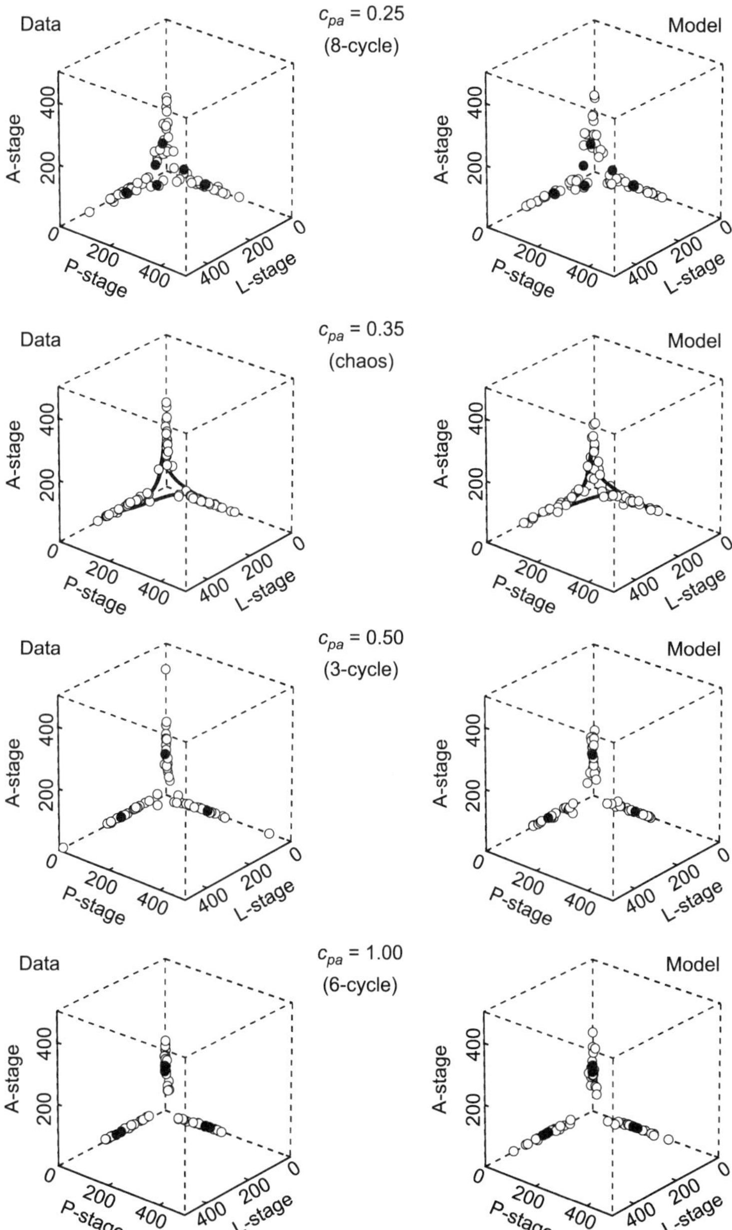

Figure 5 Phase-space plots for the data and stochastic model (3) of the experiment associated with the bifurcation diagram in Fig. 4. The model simulations used $\mu_a = 0.007629$ for the controls, $\mu_a = 0.96$ for the treatments where adult mortality was manipulated and the c_{pa} value given in the figure together with the conditional least squares parameter estimates $b = 10.45$, $c_{el} = 0.01731$, $c_{ea} = 0.01310$, $\mu_l = 0.2000$

IX. CHAOS AND POPULATION OUTBREAKS

Sensitivity to initial conditions is a key characteristic of chaos. This property led to suggestions on how small perturbations might be used to influence the dynamics of chaotic systems. One idea is that by "nudging" the parameters or state variables at points in the trajectory where the system is particularly sensitive to perturbations, one might produce a desired effect, which may be large relative to the perturbation applied. Several authors discuss this method for population control in ecology (Doebeli, 1993; Hawkins and Cornell, 1999; Shulenburger *et al.*, 1999; Sole *et al.*, 1999), but provide no experimental test of the procedure. In this section, we show how small demographic perturbations in adult numbers can be used to dampen large chaotic fluctuations in the densities of larvae (Desharnais *et al.*, 2001).

The chaotic strange attractor for $\mu_a = 0.96$ and $c_{pa} = 0$ has regions of differing sensitivities to initial conditions (Fig. 6A). For each of the 2,000 points on the attractor we computed the three eigenvalues of the Jacobian matrix of the deterministic LPA model. Each point was shaded according to λ_t, the logarithm of largest modulus of the three eigenvalues: light gray for negative values, dark gray for moderate positive values, and black for large positive values. These numbers, which ranged from -1.03 to 3.92, are the local Liapunov exponents for one step in the orbit (Bailey *et al.*, 1997). They measure the effect of small perturbations on the population trajectory. Values of $\lambda_t > 0$ indicate regions of phase space where nearby trajectories diverge in the next time step; values of $\lambda_t < 0$ occur in regions where nearby trajectories converge. The black coloration in Fig. 6A indicates a "hot" region of the attractor where larval and adult numbers are small and numbers of pupae are large. Small perturbations in this region can have a large effect on the population.

We closely studied orbits of simulated populations and noticed that a difference of a few adults in the "hot region" resulted in widely divergent trajectories. This led to the identification of two rules which we subsequently used in the experimental protocol. The first or "in-box" rule forecasts a reduction in larval numbers with small perturbations in the number of adult beetles: if the life stage vector $[L_t, P_t, A_t]$ is such that $L_t \leq 150$ and $A_t \leq 3$ then three adults are added to the culture; otherwise no perturbation is made. We developed a second or "out-box" rule as a control to demonstrate that it is the dynamics associated with the "hot spots" on the chaotic attractor that are responsible for the reduction in larval numbers and not simply the fact that adults were added to the culture. Under this rule, if the life stage vector

---

and the following variance-covariance estimates: $\hat{\sigma}_{11} = 1.621$, $\hat{\sigma}_{12} = -0.1336$, $\hat{\sigma}_{13} = -0.01339, \hat{\sigma}_{22} = 0.7375, \hat{\sigma}_{23} = -0.0009612$ and $\hat{\sigma}_{33} = 0.01212$ for the controls and $\hat{\sigma}_{11} = 2.332, \hat{\sigma}_{12} = 0.007097, \hat{\sigma}_{22} = 0.2374$ and $\hat{\sigma}_{13} = \hat{\sigma}_{23} = \hat{\sigma}_{33} = 0$ for the treatments where adult mortality and recruitment were manipulated.

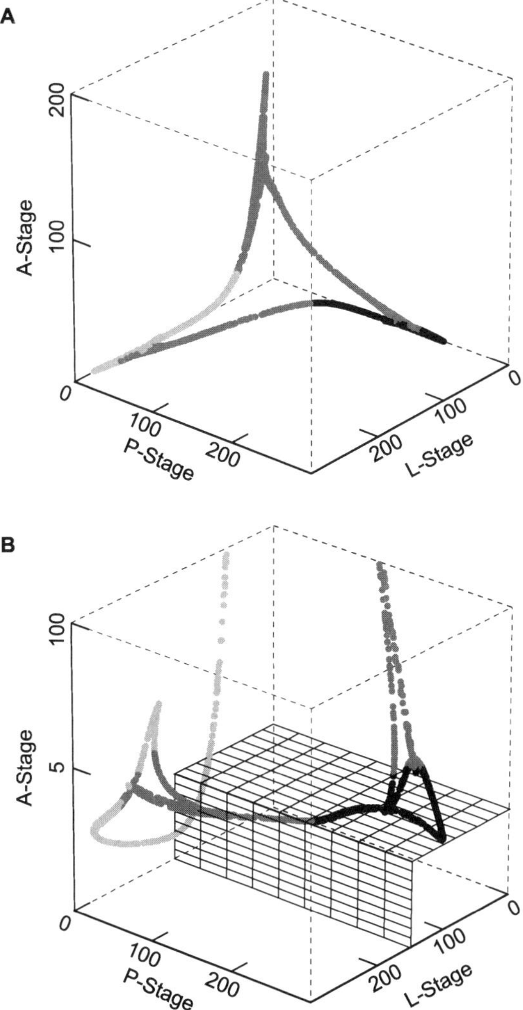

Figure 6 Chaotic attractor of the deterministic LPA model in phase space. The pictures were generated from 2000 iterations of the LPA model after the initial transients disappeared. The attractor is shaded according to λ_t, the logarithm of largest moduli of the three eigenvalues of the Jacobian matrix of the LPA model (1) evaluated at the point (L_t, P_t, A_t) using $\mu_a = 0.96$, $c_{pa} = 0.35$, and the conditional least squares parameter estimates listed in the caption of Fig. 5. The colors range from light gray for negative values ($\lambda_t < 0$), to dark gray for moderate values ($0 \leq \lambda_t \leq 3$), to black for large positive values ($\lambda_t > 3$). (A) The full attractor has "hot regions" (black) where the trajectories show strong divergence and "cold regions" (light gray) where trajectories converge. (B) The axis scale for adult numbers is changed to magnify the base of the attractor. The grids show the boundary of the "in-box" and "out-box" regions used in the experimental design described in the text. The "hot spots" of the attractor fall mostly within this box.

$[L_t, P_t, A_t]$ is such that $L_t > 150$ or $A_t > 3$ then three adults are added to the culture; otherwise no manipulation is made. The regions where the in-box and out-box perturbations are applied are represented in Fig. 6B.

We conducted an experimental evaluation of the predicted perturbation responses by establishing nine laboratory populations of the RR strain of the flour beetle *T. castaneum*. As in the study described in the previous section, we experimentally set the adult mortality rate at 0.96 and manipulated the adult recruitment rate so that it would equal $P_t \exp(-c_{pa} A_t)$ with $c_{pa} = 0.35$. Three of the populations formed an experimental control treatment where no perturbations were applied for the duration of the experiment. For the six remaining cultures, the aforementioned procedure was continued for 132 weeks; however, at week 134, and thereafter for a total of 78 weeks, in addition to manipulating μ_a and c_{pa}, we applied the in-box perturbation rule to three populations and the out-box rule to three populations until week 210 after which we stopped the in-box and out-box perturbations and maintained the cultures for another 54 weeks.

Predicted and observed time series for larval numbers are shown in Fig. 7. The panels on the left side of the figure show realizations from the stochastic version of the LPA model (3) with parameters estimated from a previous study (Dennis *et al.*, 2001). The panels on the right side are for one representative replicate population from each of the three experimental treatments. Both the simulated and observed populations in the unperturbed control treatment (Fig. 7A and 7B) show large chaotic fluctuations in larval numbers similar to those observed in previous studies (Costantino *et al.*, 1997; Dennis *et al.*, 2001). The in-box perturbations, which were designed to decrease the amplitude of the fluctuations in insect numbers, had the desired effect. The model and experimental populations in the in-box treatment exhibit large amplitude fluctuations prior to the in-box perturbations (solid symbols in Fig. 7C and 7D), but these oscillations dampened dramatically after the in-box perturbations were applied (open symbols in Fig. 7C and 7D). On the other hand, as predicted by the model, the out-box populations continued to exhibit large amplitude fluctuations in larval numbers during the out-box perturbations (Fig. 7E and 7F). This was despite the fact that, in accordance with the experimental protocol, the out-box perturbations were applied more often than the in-box perturbations. This demonstrates that the dampening effect of the in-box treatment was due to the timing of the perturbations to coincide with the occurrence of life-stage numbers in a sensitive region of phase space (box in Fig. 6B). During the final 54 weeks of the experiment, the LPA model prediction was for the oscillatory amplitudes of the cultures to return to the levels attained prior to the application of the perturbations. That behavior was observed (Fig. 7).

Can small perturbations be used to influence the dynamics of natural ecosystems? The question has yet to be addressed by field ecologists.

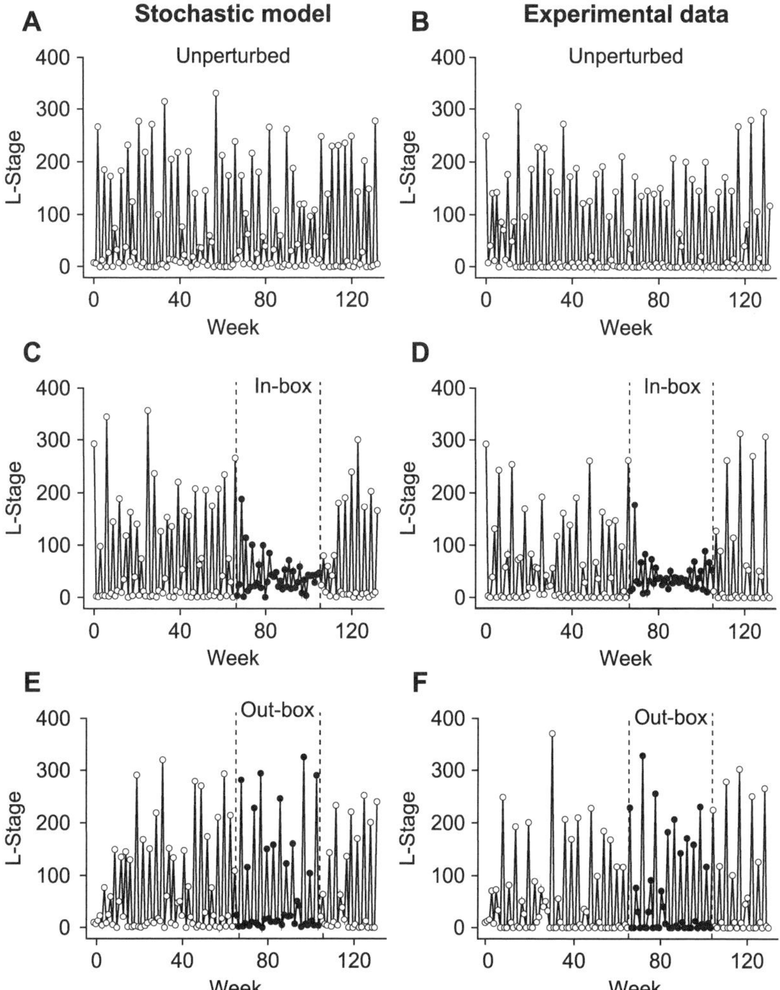

Figure 7 Time series of larval numbers for the stochastic model and experimental data under three conditions: unperturbed treatment, in-box treatment, and out-box treatment. The simulations are from the demographic stochastic model (3) with $\mu_a = 0.96$, $c_{pa} = 0.35$, and the conditional least squares parameter and variance-covariance estimates listed in the caption of Fig. 5.

Certainly the introduction of a handful of individuals of a non-native species into a region can have wide ranging effects. However, the possibility of changing the dynamics of an abundant species in the field with small perturbations has not been explored to our knowledge. Here we have made the step from theoretical possibility to laboratory demonstration. For a similar approach to be effective in field populations, a model of the dynamics of the system is required which can be used to make accurate predictions. Such models will come from careful studies of the mechanisms that determine ecological change (Kendall *et al.*, 1999; Perry *et al.*, 2000; Turchin and Ellner, 2000; Turchin, 2003).

X. BACK IN THE SADDLE (NODE) AGAIN

Random events can frequently and repeatedly produce visits near *unstable* equilibria, cycles or other invariant sets. Such visitations can result in the influence of unstable invariant sets on the dynamics of a population (Cushing *et al.*, 1998). This is particularly true for unstable invariant sets that are not repelling. In the higher dimensional systems typically found in ecology, unstable invariant sets often are associated with attracting regions in phase space; that is, there are points in phase space whose orbits approach the unstable set. These points constitute the "stable manifold" of the invariant set. Moreover, points near, but not on this stable manifold are temporarily drawn toward the unstable set before being repelled away towards an attractor. Unstable invariant sets of this type are called "saddles." Nonlinear systems, particularly those with complex dynamics, are generally replete with saddles.

Since random events can cause deviations away from an attractor and place an orbit near an unstable set or its stable manifold, a population's dynamics become a mixture of influences from both stable and unstable sets. Indeed, the predicted stationary distribution of most stochastic models covers all (or most) of the feasible phase space and, therefore, such a mixture is theoretically certain to happen. The resulting temporal patterns are then a matter of the relative strengths of the influences due to the unstable and stable invariant sets.

Here we present evidence for the influence of a saddle observed in a replicate of the $c_{pa} = 0.05$ treatment of the hunt for chaos experiment. In Fig. 8, plots of the larval, pupal, and adult time series of the deterministic model (1), stochastic realizations of model (3), and the data are presented. The deterministic time series displays no influence of the presence of the unstable equilibrium (Fig. 8A). The stochastic model forecasts that the time series of some stochastic realizations and some experimental orbits will visit the unstable equilibrium and remain relatively nonoscillatory at low levels before returning to the asymptotic stable attractor. One such stochastic realization is presented in Fig. 8B. A saddle fly-by occurs in the interval marked by the double ended arrow. The experimental data show a saddle fly-by in the interval from time step $t = 8$ to $t = 17$.

In the four panels of Fig. 9, the deterministic model-predicted invariant loop (black closed curve) and the data orbit (open symbols) are presented in

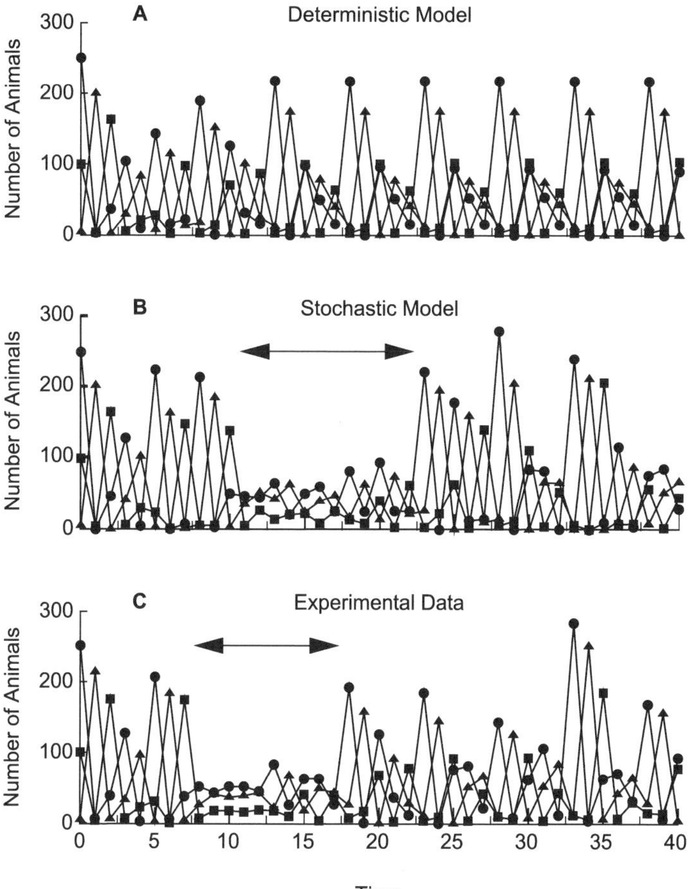

Figure 8 The time series of larval (circles), pupal (triangles), and adult (squares) numbers. (A) Deterministic model which reveals no influence of the presence of the unstable equilibrium. (B) Stochastic model (3) with a saddle fly-by displayed in the interval marked by the double ended arrow. (C) Experimental data with a saddle fly-by seen in the interval from $t = 8$ to $t = 17$. The deterministic and stochastic model simulations used $\mu_a = 0.96$, $c_{pa} = 0.05$, and the conditional least squares parameter estimates listed in the caption of Fig. 5. The stochastic model also used the variance-covariance estimates listed in the caption of Fig. 5.

phase space. In the first panel, the data for time steps 0 to 8 (weeks 0 to 16) show a temporal motion around the loop. However, at time step 8 there is a perturbation away from the loop which places the data point near the model-predicted *unstable* equilibrium (solid circle). From time steps 8 to 13 (weeks 16 to 26) the data orbit stays clustered very near this predicted

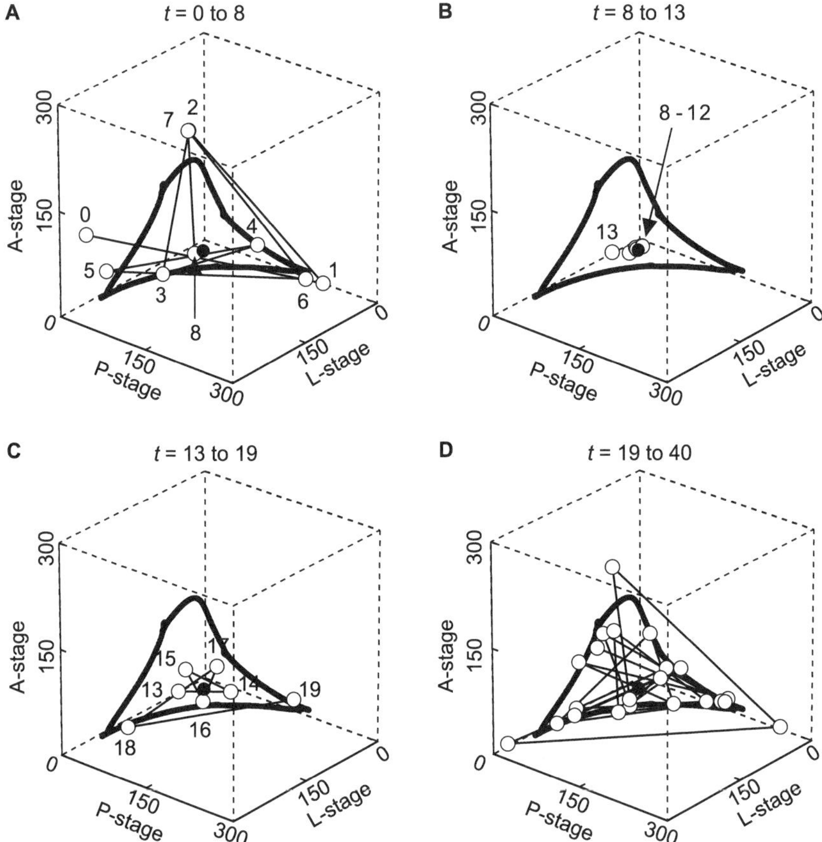

Figure 9 The model-predicted invariant loop (closed curve) and unstable equilibrium (closed circle) for the hunt experiment treatment $c_{pa} = 0.05$ are shown together (in phase-space) with the data orbit of one replicate (open circles). Lines connect the data values through time.

equilibrium. Indeed, if the interval of observation was restricted to the first 13 time steps of the experiment, an inaccurate conclusion of equilibrium dynamics might be made. The equilibrium, however, is unstable and the data orbit, from time steps 13 to 19, displays a "star-like" rotational motion in state space as the data orbit leaves the vicinity of the saddle equilibrium and returns to the stable invariant loop. This geometrically distinctive path is predicted by the deterministic LPA model (1). The linearization at the equilibrium has a conjugate pair of complex eigenvalues $re^{\pm i\theta}$ of magnitude $r \approx 1.265 > 1$ and polar angle $\theta \approx 2.576$ (and a third real positive eigenvalue $\lambda \approx 0.3945 < 1$). This complex eigenvalue implies a rotational motion away from the saddle of approximately $2\pi/\theta \approx 2.439$ radians (139.8°) per step, the motion occurring approximately in a plane parallel to that spanned by the eigenvectors $(L, P, A) \approx (1, -1.116, 0.4860)$ and $(1, -0.3526, -0.2817)$. In the fourth panel, the data return to the model-predicted quasi-periodic motion around the invariant loop.

For the deterministic LPA model, only those time series whose orbits pass near the stable manifold will be strongly influenced by the saddle node. Moreover, once the population has reached the stable attractor it will stay there forever. This is true neither for stochastic LPA model time series nor for experimental observations. Chance events can cause a population to land near the stable manifold and come under the influence of the saddle node. This might even reoccur on several occasions in a time series, and several 'fly-bys' of the saddle would then be present in the data (Cushing *et al.*, 2003). The stochastic component of the dynamics, therefore, can account for different transient behaviors of time series in identically replicated experimental populations.

XI. PHASE SWITCHING IN POPULATION CYCLES

Populations often exhibit temporal oscillations, and sometimes these oscillations cause a phase shift. A common phenomenon observed in oscillating *Tribolium* cultures is a change of phase in which, for example, a high-low periodic pattern "chicken-steps" (skips) to a low-high pattern. Fig. 10A displays larval numbers for two of the control replicates from the experiments of Desharnais and Costantino (1980). The cultures, shown to be oscillating with period 2 (Dennis *et al.*, 1995), display phase shifts in both replicates which eventually lead to asynchronous oscillations. Two other examples appear in Fig. 10, one from the experiment of Costantino *et al.* (1995) involving a two-cycle (Fig. 10B) and another from the experiment of Costantino *et al.* (1997) involving a three-cycle (Fig. 10C).

We hypothesize that phase shifts correspond to stochastic jumps between basins of attraction in an appropriate phase space which associates the different phases of a periodic cycle with distinct attractors (Henson *et al.*, 1998).

At the maximum likelihood LPA parameters estimated from the control replicates, reported in the Desharnais and Costantino (1980), the model admits an unstable fixed-point with coordinates (rounded to the nearest beetle):

$$[L, P, A] = [124, 60, 97].$$

This fixed point is stable in some directions and unstable in others, it is a saddle node as described in the previous section. The model also predicts two

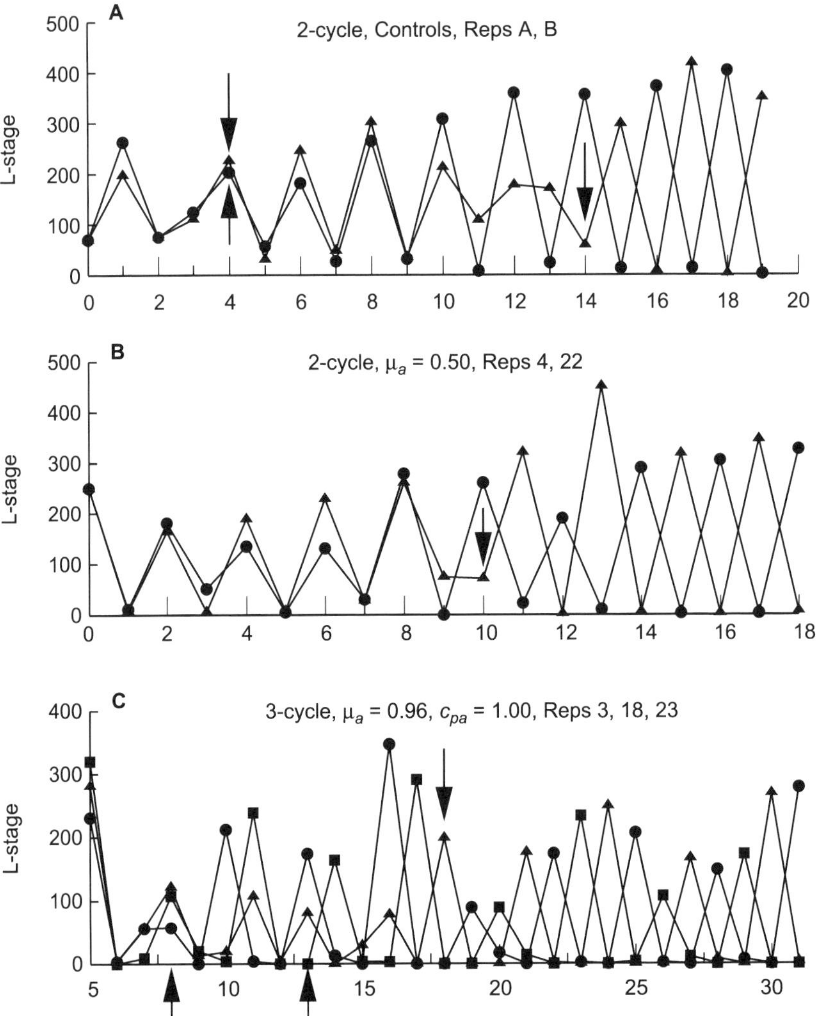

Figure 10 Time series data showing phase switching in population cycles. (A) Replicate A (circles) changes phase at time step 4 while replicate B (triangles) changes phase at time step 4 and 14. After time step 14 both replicates from the experiment of Desharnais and Costantino (1980) are asynchronous. (B) Replicate 4 (circles) does not change phase. Replicate 22 (triangles) changes phase at time step 10. After time step 10 both replicates from the experiment of Costantino *et al.* (1995) are asynchronous. (C) Replicates 3, 18, and 23 from the experiment of Costantino *et al.* (1997) show three-cycle dynamics (transient time steps 0 to 4 are omitted for clarity). Replicate 23 (squares) does not shift phase. Replicate 3 (circles) changes phase at time step 8. Replicate 18 (triangles) changes phase at time step 13 and 18. After time step 18 all three replicates are out of phase.

locally stable two-cycle solutions: one which alternates between the stage vectors

$$[L, P, A] = [18, 158, 106]$$
$$[L, P, A] = [325, 9, 118]$$

and the other, which traverses the same vectors in the opposite phase

$$[L, P, A] = [325, 9, 118]$$
$$[L, P, A] = [18, 158, 106].$$

Because they "live" on the same attractor {[18, 158,106], [325, 9, 118]}, the two different 2-cycles previously listed are indistinguishable when plotted in phase space. However, these 2-cycles do determine different phases for each component. For example, the first cycle determines a low-high oscillation in the larval component, while the second determines a high-low oscillation in the larval component. In order to differentiate between these out-of-phase 2-cycles as separate attractors with distinct basins of attraction, we turn to the composite of the LPA model.

The "composite LPA model" (the composite map, whose solutions correspond to even time steps of solutions of the LPA model) identifies the above 2-cycles as two different fixed point attractors given by the stage vectors

$$[L, P, A]_\Delta = [18, 158, 106] \text{ and } [L, P, A]_\mathrm{O} = [325, 9, 118]$$

(Note the subscripts Δ and O are used to label the two attractors.) The saddle point of the LPA map (labeled with the subscript +) is also a saddle point

$$[L, P, A]_+ = [124, 60, 97]$$

of the composite map.

The basins of attraction of the two stable fixed points of the composite LPA model are set in three-dimensional phase space and are computed numerically. In this particular example, the basins are fairly simple sets. Throughout a large portion of phase space, they are separated by a two-dimensional surface (containing the saddle) that forms part of the "basin boundary." Initial conditions on one side of the boundary lead to composite map solutions that approach$[L, P, A]_\Delta$, while initial conditions on the other side generate composite map solutions approaching $[L, P, A]_\mathrm{O}$. Solutions starting on the basin boundary near the saddle point tend to the saddle $[L,P,A]_+$ (locally, the boundary is the "stable manifold" of the unstable saddle). Indeed, near the saddle, the stable manifold of this unstable entity forms the watershed geometrical feature of phase space. Near the origin, however, the basin boundary becomes much more complicated; this does not concern us.

Deterministic and stochastic time series generated by the LPA model for the L-stage are displayed in Fig. 11 using the initial condition [70, 35, 64] of

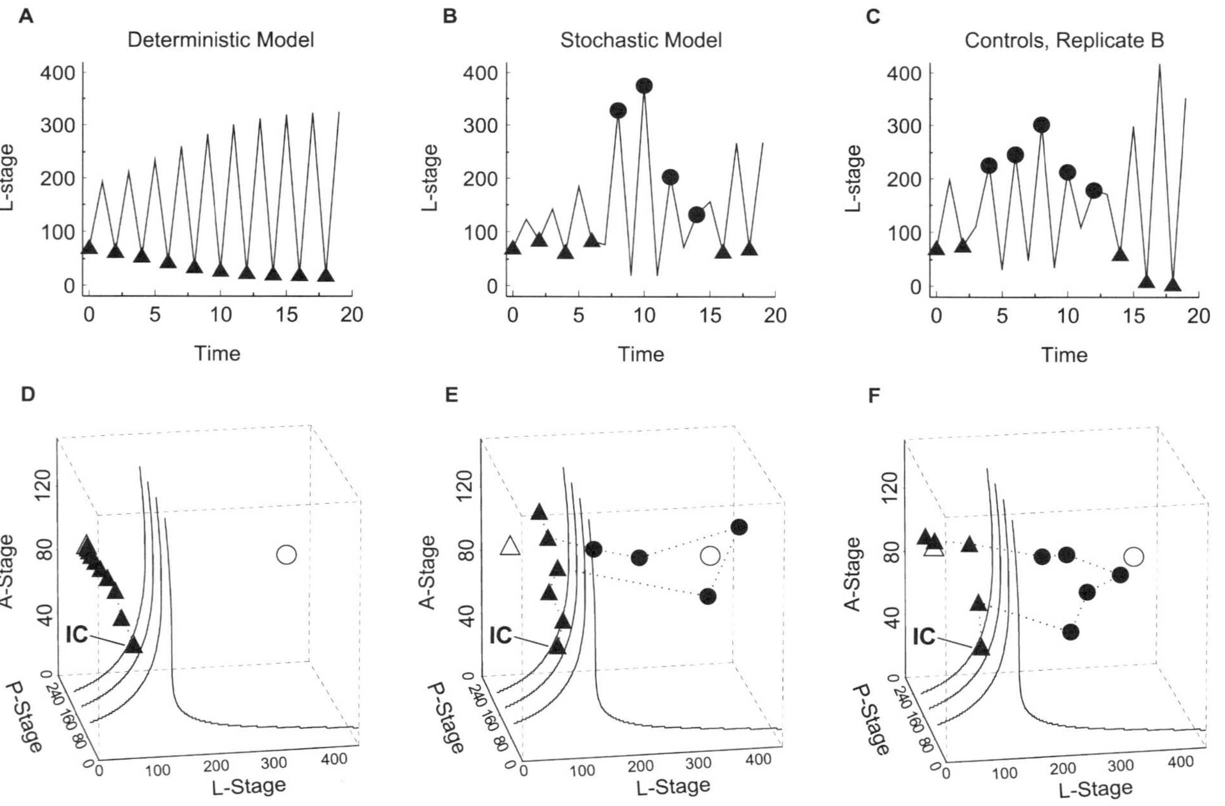
A
Deterministic Model
L-stage
Time
B
Stochastic Model
L-stage
Time
C
Controls, Replicate B
L-stage
Time
D
A-Stage
P-Stage
L-Stage
IC
E
A-Stage
P-Stage
L-Stage
IC
F
A-Stage
P-Stage
L-Stage
IC

the experiment previously described. The deterministic time series approaches the 2-cycle shown in Fig. 11A. In composite phase space, the corresponding solution of the composite LPA map approaches the fixed point $[L,P,A]_\Delta = [18, 158, 106]$ (Fig. 11D). The stochastic model L-stage time series, on the other hand, shifts phase at time $t = 7$ and again at $t = 15$ (Fig. 11B). In composite phase space, these phase changes occur exactly when the basin boundary is crossed (Fig. 11E). The data for replicate B from Desharnais and Costantino (1980) are shown in Fig. 11C and 11F. As the LPA model predicts, phase shifting occurs in the data time series precisely when the data cross the model-predicted basin boundary in composite phase space.

Deterministic attractors alone do not account for the phase-shifting mechanism proposed. Indeed, in some situations attractors may be of little interest as final states. Other invariant sets such as basin boundaries and saddle points, along with stochasticity, play a key role and may lead to data dominated by transient rather than asymptotic dynamics.

XII. LATTICE EFFECTS

The discovery that simple deterministic population models can display complex aperiodic fluctuations such as chaos (May, 1974) inspired decades of empirical and theoretical work in ecology (Hastings *et al.*, 1993; Dennis *et al.*, 2001). The resulting mathematical models of population dynamics almost invariably employ a continuous state space. That is, variables representing population densities in these models are real-valued. But animals, and, for many practical purposes, plants are individuals. More realistic models, therefore, would cast population densities as discrete variables,

---

Figure 11 Model predictions and the data. Time series for the deterministic model (1) were generated for the L-stage using maximum likelihood parameter estimates $b = 11.67$, $\mu_l = 0.5129$, $c_{pa} = 0.0178$, $c_{ea} = 0.0110$, $c_{el} = 0.0093$, $\mu_a = 0.1108$ and the initial condition [70, 36, 64]. Time series for the stochastic model (2) were generated for the L-stage using the same parameter estimates and initial condition as for the deterministic model and the variance-covariance estimates $\hat{\sigma}_{11} = 0.2771$, $\hat{\sigma}_{12} = 0.02792, \hat{\sigma}13 = 0.009796, \hat{\sigma}_{22} = 0.4284, \hat{\sigma}_{23} = -0.008150$ and $\hat{\sigma}_{33} = 0.01112$. (A) The deterministic time series approaches a 2-cycle. (D) In composite phase space, the corresponding solution of the composite LPA map approaches the fixed point $[L, P, A]_\Delta = [18, 158, 106]$. (B) The stochastic model L-stage time series, on the other hand, shifts phase at time $t = 7$ and again at $t = 15$. (E) In composite phase space, these phase changes occur exactly when the basin boundary is crossed. The data for replicate B from the experiment of Desharnais and Costantino (1980) are shown in panels (C) and (F). As the LPA model predicts, phase shifting occurs in the data time series precisely when the data cross the model-predicted basin boundary in composite phase space.

with state space a discrete lattice of numbers. As long as the population size is bounded, deterministic models of the latter type have finitely many possible states and hence display only periodic cycles. In particular, discrete-state deterministic models with bounded dynamics cannot display chaos.

Approximating population size with continuous-state models is commonly justified by the assumption that population numbers remain so large that the discrete state space lattice is sufficiently fine (May, 1974). However, the deterministic dynamics of associated discrete-state and continuous-state models can be quite different even for very large population sizes, so that the effects caused by the discreteness of animal densities (*lattice effects*) cannot always be ignored (Jackson, 1989).

As we have repeatedly emphasized, ecological systems are invariably stochastic. Discrete-state models, when perturbed by stochasticity, can recover the deterministic dynamics of the underlying continuous state space. The dynamics of such a model are a blend of the dynamics predicted by the deterministic continuous-state model and the cyclic dynamics predicted by the deterministic discrete-state model (Henson *et al.*, 2001; King *et al.*, 2002, Henson *et al.*, 2003; King *et al.*, 2004).

As it turns out, lattice effects are not theoretical oddities arising from simple population models. We were able to verify the existence of lattice effects in the chaotic treatments of the "hunt for chaos" experiment described earlier in this chapter (Henson *et al.*, 2001). We present one example in detail.

Figure 12A shows a chaotic attractor of the LPA model. The data from the experimental treatment corresponding to this attractor clearly exhibit the temporal and phase space patterns of the predicted chaotic dynamics (King *et al.*, 2004). However, the data also reveal a near 6-cycle pattern not predicted by the LPA model. We show that this unexpected 6-pattern is in fact a lattice effect.

Suppose that, in order to simulate dynamics on a whole integer lattice, we integerize the LPA model as follows. Since we manipulated the experimental parameters μ_a and c_{pa} by adding or subtracting integer numbers of adults, we can more accurately describe the experimental protocol by replacing the A-equation in the LPA model by an A-equation in which recruitment and survival are integer quantities. In addition, the survival/recruitment processes for the other state variables are fundamentally integer processes. One possible deterministic lattice model for the experiment in question, and the one used in Henson *et al.* (2001) is

$$\begin{aligned} L_{t+1} &= \operatorname{int}\left[bA_t \exp\left(-\frac{c_{ea}}{V}A_t - \frac{c_{el}}{V}L_t\right)\right] \\ P_{t+1} &= \operatorname{int}[(1-\mu_l)L_t] \\ A_{t+1} &= \operatorname{int}\left[P_t \exp\left(-\frac{c_{pa}}{V}A_t\right)\right] + \operatorname{int}[(1-\mu_a)A_t]. \end{aligned} \tag{4}$$

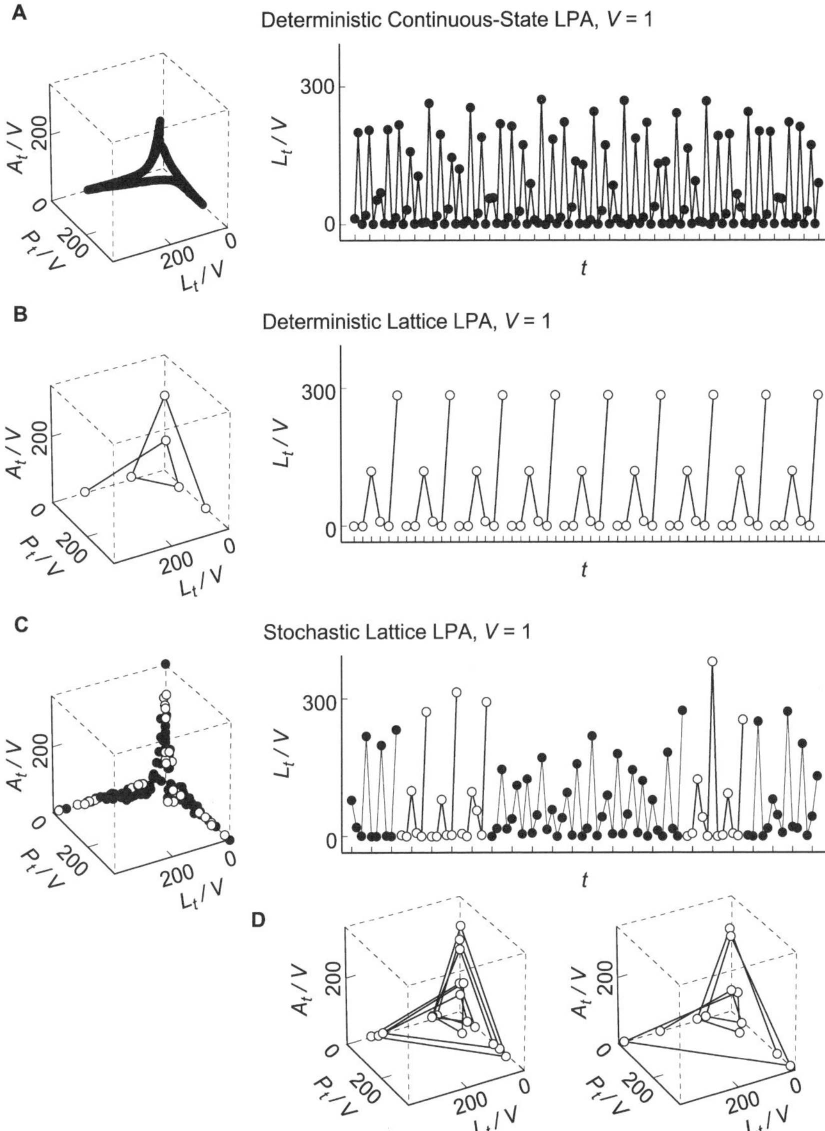

Figure 12 Density dynamics of the LPA models with $b = 10.67$, $\mu_l = 0.1955$, $\mu_a = 0.96$, $c_{el} = 0.01647$, $c_{ea} = 0.01313$, and $c_{pa} = 0.35$. For the demographic stochastic model (3), the variance and covariance entries of the matrix Σ were taken to be $\sigma_{11} = 2.332$, $\sigma_{22} = 0.2374$, and $\sigma_{12} = \sigma_{21} = 0$. (A) The chaotic attractor of LPA model (1). (B) A six-cycle attractor of the lattice LPA model with $V = 1$ (on the order of 107 lattice points). (C) A stochastic realization with $V = 1$ exhibits a mixture of patterns, with intermittent patterns that resemble the lattice 6-cycle in panel (B) interspersed among episodes that resemble the chaotic attractor in panel (A). (D) The 6-cycle patterns in (C) are shown in phase space.

This is a discrete state (or "lattice") LPA model. When $V = 1$, the lattice model predicts a six-cycle attractor (Fig. 12B).

One stochastic version of the integerized LPA model results from adding demographic variability on the square root scale to the two unmanipulated life-stage equations, namely to the equations for the larval and pupal stages:

$$\begin{aligned} L_{t+1} &= \text{int}\left[\left(\sqrt{bA_t \exp\left(-\frac{c_{ea}}{V}A_t - \frac{c_{el}}{V}L_t\right)} + E_{1t}\right)^2\right] \\ P_{t+1} &= \text{int}\left[\left(\sqrt{(1-\mu_l)L_t} + E_{2t}\right)^2\right] \\ A_{t+1} &= \text{int}\left[P_t \exp\left(-\frac{c_{pa}}{V}A_t\right)\right] + \text{int}[(1-\mu_a)A_t], \end{aligned} \tag{5}$$

where E_{1t} and E_{2t} are random normal variables with mean zero and variance-covariance matrix $\mathbf{\Sigma}$. In the rare cases in which a large negative E causes a negative value inside a square, we set the right-hand side of that equation equal to zero. Equation (5) is a stochastic discrete state LPA model.

When $V = 1$, time series generated by the stochastic lattice model resembles the chaotic attractor; however, the lattice effect 6-pattern episodically recurs (Fig. 12C).

The 6-pattern forecast by the stochastic lattice model is clearly evident in the three experimental replicates. Figure 13A shows the larval time series data from one replicate. The intermittently occurring 6-pattern is also seen in the phase space representation of the data (Fig. 13B).

Lattice effects can dramatically alter the predictions of ecological models, especially in systems for which the continuous-state deterministic dynamics are complex. In deterministic models, discretizing state space can replace a complicated continuous-state attractor with a simpler lattice attractor. Yet the continuous-state dynamics remain important inasmuch as they continue to shape the transient behavior on the lattice. In the presence of demographic variability, the system is influenced by both transients and attractors, and thus displays episodes which alternately resemble the dynamics of the continuous-state and lattice models. We emphasize that such lattice effects are not only found in relatively coarse lattices or in small populations: indeed, in our experimental study of chaotic population dynamics, lattice effects were important even with 10^7 lattice points.

XIII. ANATOMY OF CHAOS

Chaos is a mathematical concept. In reality, populations are stochastic, discrete-state systems. In the previous section we saw that, although discrete-state deterministic systems cannot exhibit chaos, discrete-state stochastic systems can exhibit a dynamic blend of lattice effects and what appears to

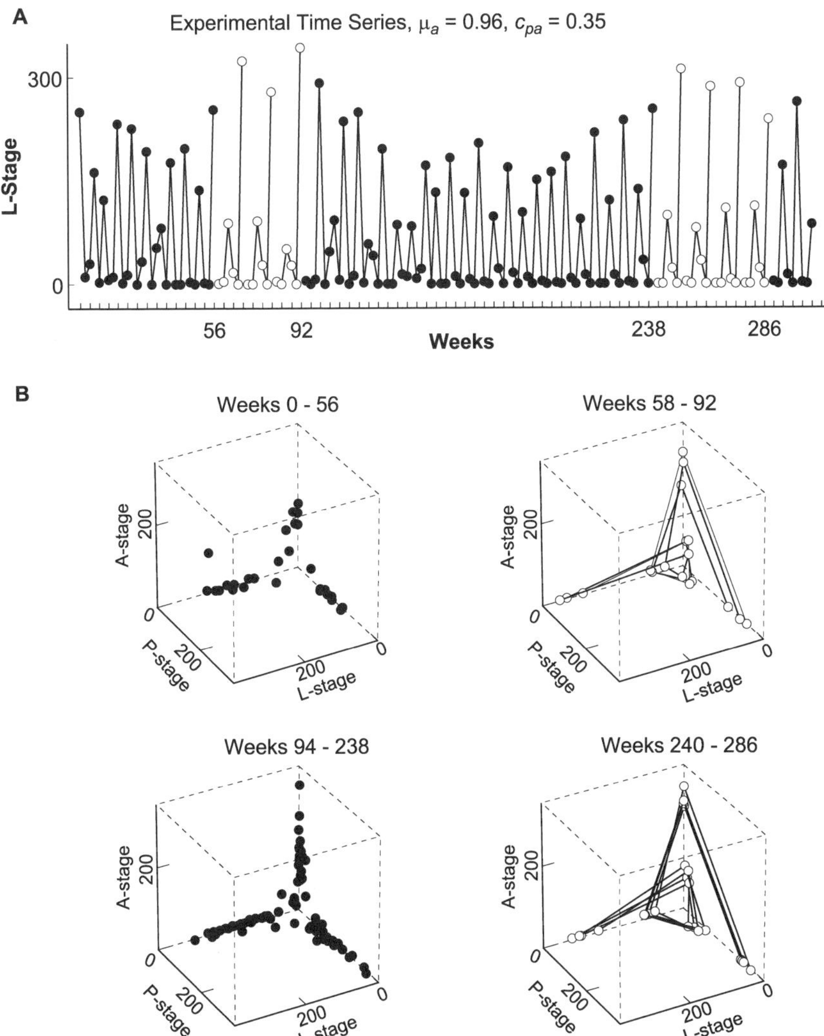

Figure 13 A 304 week data time series obtained from one replicate of the *Tribolium* experiment (Dennis *et al.*, 2001) where adult mortality and recruitment were manipulated to be $\mu_a = 0.96$ and $c_{pa} = 0.35$. (A) Selected temporal episodes that resemble the lattice model six-cycle shown in Fig. 12B are displayed as open circles. The remaining data points (closed circles) resemble the chaotic time series. (B) The selected temporal episodes in (A) are shown in phase space (on the order of 10^7 lattice points). Compare the 6-pattern episodes (open circles) to the 6-cycle lattice attractor in Fig. 12B.

be chaos. But can the chaotic signal be quantified? What, in fact, do we mean when we say a discrete-state stochastic system is "chaotic"?

At the heart of chaos is the concept of sensitivity to initial conditions: a small perturbation can have a big effect. In populations, stochastic perturbations are not necessarily small, and they occur often. It is tempting to conclude that all fine structures associated with a chaotic mathematical model would be washed out by noise in experimental data. This is not necessarily the case (Cushing *et al.*, 2001).

Mathematically speaking, chaotic attractors are composed of infinitely many periodic orbits of saddle stability-type. Thus, chaotic dynamics exhibit continual fly-bys—not of saddle nodes as discussed in a previous section—but of saddle cycles. In time series, these fly-bys appear as recurrent episodes of near-periodic dynamics. Sensitivity to initial conditions rearranges the recurrent episodes but does not destroy them. Scientists studying oscillatory chemical reactions, electroencephalographic recordings, and epidemiological case reports have all noted the appearance of recurrent near-periodic episodes in putatively chaotic dynamical systems (Kendall *et al.*, 2003; Lathrop and Kostelich, 1989; Schaffer *et al.*, 1993; So *et al.*, 1996, 1997; Xu, 2002).

Identification of cyclic episodes in time series requires much data. For the oscillatory chemical reactions and electroencephalographic recordings previously mentioned, it was possible to resolve these fine structures clearly because of the wealth of data. In the epidemiological study, however, the identification of these patterns was confined to cycles of periods one and two, even though the dataset afforded by measles case reports in major cities are extensive by ecological standards (Schaffer *et al.*, 1993). There is, in fact, a dearth of long ecological time series. The eight-year long *Tribolium* dataset ($\approx$ 70 generations) from the "hunt for chaos" experiment represents a unique opportunity to examine the signal of chaos as it is manifest in biological populations.

We have developed the following hypothesis. Populations, being discrete-state stochastic systems, should display episodes of lattice cycles interspersed with episodes of chaotic signal. The chaotic signal itself should exhibit recurrent episodes of cyclic behavior. Chaotic population time series, therefore, should be a mixture of cycles predicted by both the discrete-state and continuous-state models, woven together by stochasticity. In this section we present a tool to test this hypothesis.

For the chaotic ($c_{pa} = 0.35$) treatment in the "hunt for chaos" experiment there are a large number of model-predicted periodic orbits. They are of two types: (1) saddle-cycles embedded in the continuous-state LPA model attractor, and (2) lattice cycles of the discrete-state LPA model. Although there are infinitely many of the first type, the level of demographic variability and the length of the data time series puts a limit on our ability to distinguish among these cycles. Thus, we focus our attention on a dominant period-11 saddle-cycle. At the same parameter estimates, the discrete-state LPA model has precisely nine periodic orbits. These orbits fall into three groups based on their periodicity: 3-, 6- and 8-cycles. Cycles within each of the groups are very similar, with none differing from any other by more than 30 animals. Figure 14 shows the period-11 saddle-cycle and the lattice 3-, 6- and 8-cycles.

We can quantify fly-bys of periodic orbits using a measure of the "distance" between data points and periodic orbits. We generalize the notion of distance using a quantity we call the *lag-metric comparison*, or LMC. Essentially, the LMC measures the average distance in state space between the data and each phase of the model cycle. To be precise, given a length-N sequence of data vectors, $d = \{d_t\}_{t=0}^{N-1}$, and a model of period-T cycle, $\{m_t\}_{t=0}^{\infty}(m_{t+T} = m_t)$, we define the LMC of d and m at lag s and time t by the formula

$$LMC(s,t) = \frac{1}{T}\sum_{q=0}^{T-1}||d_t - m_{t+s-q}||$$
$$s = 0, ..., T-1, \quad t = T-1, ..., N-1,$$

where $||x|| = |x_1| + |x_2| + |x_3|$ is a norm on the three-dimensional state space. In Fig. 15, we plot $LMC(s,t)$ against t directly.

Figure 15 shows plots of the LMC of the data from replicate 13 of the "hunt for chaos" experiment against time. At any time t a low value of the LMC indicates that the data lie close to the model-predicted T-cycle and have done so over the course of the preceding T time units; a high LMC value indicates poor correspondence between model cycle and data. Plotted against time, the LMC appears as a "braid" with one strand for each phase. Time intervals over which the data trajectory closely follows the model cycle ("cycle episodes" for short) appear as unplaited portions of the braid. Tightly plaited portions indicate lack of correspondence between data and the particular model cycle in question.

Viewing the complete 424-week replicate 13 data series using the LMC, we see that, initially, the population trajectory follows the saddle 11-cycle. During this same interval, the lattice 8-cycle is also evident. This is not surprising, since these two cycles lie close together. After about week 50, a 6-cycle episode is identified. Over weeks 110 to 134, another 8-cycle/11-cycle episode appears, which is followed by a 3-cycle episode over weeks 150 to 172. At around week 200, a 130-week 6-cycle episode begins. Weeks 328 to 356 display another 11-cycle/8-cycle episode. Finally, it should be noted that a 1-cycle episode (i.e., an equilibrium fly-by) is evident in these data around week 364.

Overall, the pattern we observe in the data is one of transient but recurrent near-periodic episodes, each traceable to a model-predicted periodic orbit. From this perspective, the principal role of stochasticity is to move the system from one cycle to another. The set of patterns observed, as well as

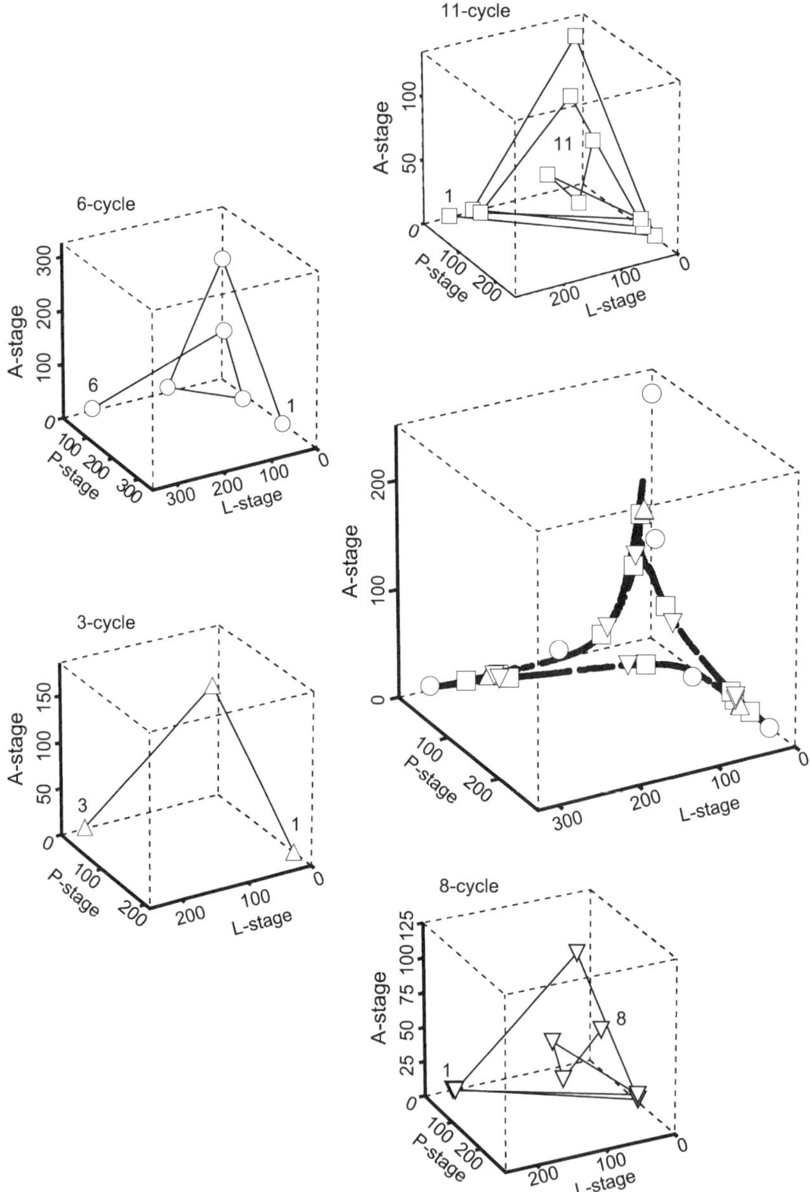

Figure 14 Model-predicted continuum and lattice cycles. The four small phase-space graphs depict the 11-cycle from the continuous-state model (1) and the 8-, 6-, 3-cycles from the discrete-state deterministic model (4). In the central graph, all of the cycles are superimposed on the chaotic attractor of the deterministic model. The graphs were generated by using the conditional least-squares parameter estimates $b = 10.45$, $c_{ea} = 0.01310$, $c_{el} = 0.01731$, and $\mu_a = 0.2000$ with $\mu_a = 0.96$, $c_{pa} = 0.35$ both set experimentally.

their relative prominence in the mixture, is a prediction of the mathematical model. It is worth pointing out that this method of quantifying the influence of chaos in population data is based on the fine structure of the dynamics. By contrast, commonly-used measures, such as the "stochastic Liapunov exponent" are long-time averages which can measure only the gross properties of an entire system (Desharnais *et al.*, 1997; Dennis *et al.*, 2003).

XIV. MECHANISTIC MODELS OF THE STOCHASTICITY

We have presented three stochastic LPA models in this chapter: the lognormal model (2), the square root model (3), and the integerized square root model (5). Although these models have been useful, stochasticity was introduced to the deterministic skeleton through the addition of biologically unspecified random variables. A next step is to associate more carefully the uncertainty to biological features such as reproduction and survival.

A simple stochastic model of this type for the population dynamics of the beetle uses the binomial and Poisson distributions to characterize the aggregation of demographic events within the life stages (Dennis *et al.*, 2001; Henson *et al.*, 2003; Desharnais *et al.*, 2005). The Poisson-binomial (PB) model is

$$\begin{aligned}
L_{t+1} &\sim Poisson\left[ba_t \exp\left(-\frac{c_{ea}}{V}a_t - \frac{c_{el}}{V}l_t\right)\right] \\
P_{t+1} &\sim Binomial[l_t, (1-\mu_l)] \\
R_{t+1} &\sim Binomial\left[p_t, \exp\left(-\frac{c_{pa}}{V}a_t\right)\right] \\
S_{t+1} &\sim Binomial[a_t, (1-\mu_a)] \\
A_{t+1} &= R_{i+1} + S_{t+1},
\end{aligned} \tag{6}$$

where L_{t+1} is the number of feeding larvae, P_{t+1} is the number of non-feeding larvae, pupae, and callow adults, R_{t+1} is the number of sexually mature adult recruits, S_{t+1} is the number of surviving mature adults, and l_t, p_t, r_t, and s_t are the respective abundances observed at time t. The total number of mature adults A_{t+1} is given by $R_{t+1} + S_{t+1}$, and $a_t = r_t + s_t$ is the total number of mature adults observed at time t. Here "$\sim$" means "is distributed as."

The PB model (6) has purely demographic variability. The L-stage is a compound process: a random number of potential recruits are produced with conditional mean ba_t, and each potential recruit subsequently undergoes a survival process in which the conditional survival probability exp $(-c_{el}l_t - c_{ea}a_t)$ depends on the system state variables l_t and a_t. We assume that the number of potential recruits has a Poisson ba_t distribution, and that

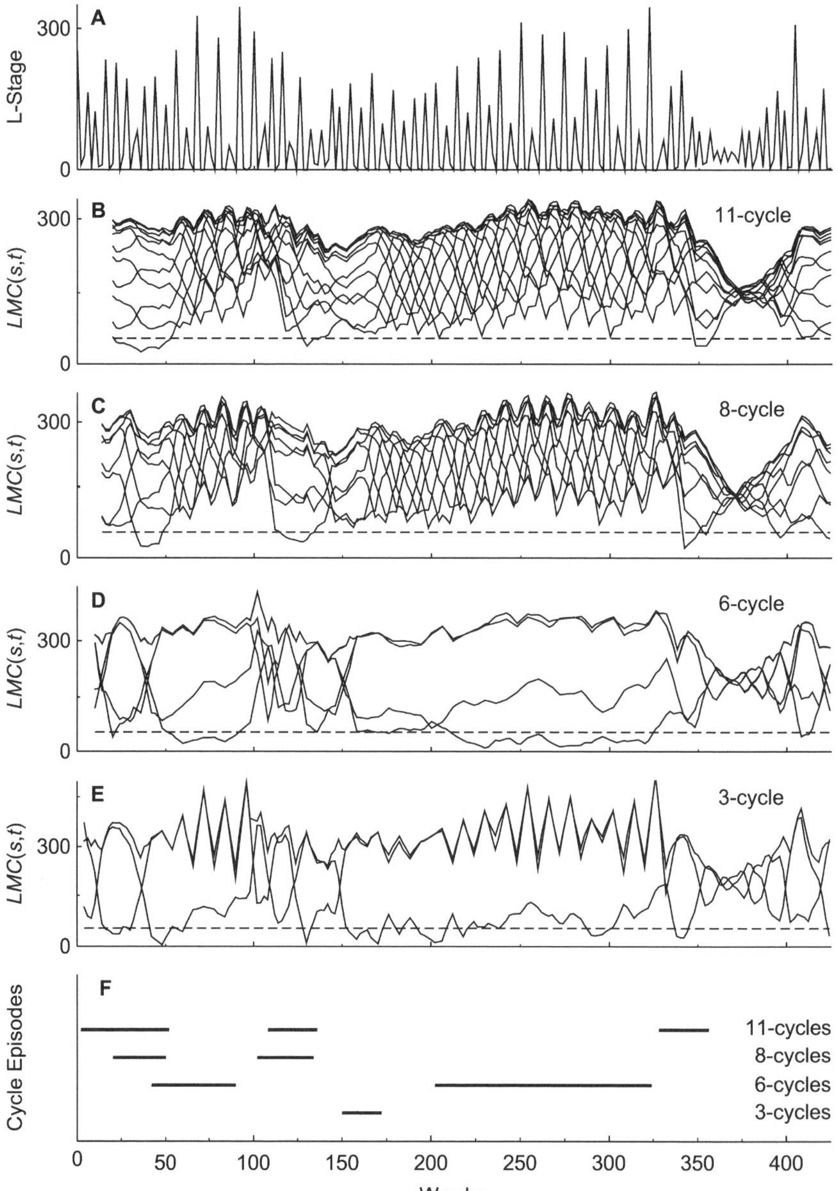

Figure 15 Lag-metrics in the data. (A) Raw time series data. For clarity, only the L-stage numbers are shown. LMC with respect to the model-predicted cycles: (B) continuous-state model saddle 11-cycle; (C) discrete-state model 8-cycle; (D) discrete-state model 6-cycle; and (E) discrete-state model 3-cycle. During intervals for which the "braid" appears tightly plaited, the data bear little or no resemblance to

the number of subsequent survivors has a binomial distribution. The conditional distribution of L_{t+1} given $L_t = l_t$ and $A_t = a_t$ becomes a Poisson distribution with mean $ba_t \exp(-c_{el}l_t - c_{ea}a_t)$.

The distribution of P_{t+1} given $L_t = l_t$ has a binomial $(l_t, (1 - \mu_l))$ distribution. The A-stage equation is the sum of two survival processes: recruits from the P-stage, denoted R_{t+1}, and surviving adults, denoted S_{t+1}. We assume that R_{t+1} given $P_t = p_t$ and $A_t = a_t$ has a binomial $(p_t, \exp(-c_{pa}\, a_t))$ distribution. The P-stage survival probability is the nonlinear function $\exp(-c_{pa}a_t)$. S_{t+1} is assumed to have a binomial $(a_t, (1 - \mu_a))$ distribution.

In the PB model there are no noise variances and covariances to be estimated. The stochastic model has the same number of parameters as the deterministic LPA model (1). State space is discrete. The PB model is a stochastic lattice (integer valued) model. The assumption of demographic variability seems appropriate for laboratory cultures of beetles grown under standard conditions. We have used this type of model to study competition between two species of flour beetle where competitive exclusion is a common outcome (Desharnais *et al.*, 2005).

XV. BEYOND BEETLES

The rarity of designed manipulations and replications in natural systems makes rigorous testing of models difficult. Models of natural systems must necessarily be evaluated on the basis of biological plausibility, how well they describe the data and, when possible, how well they predict new data. Models of natural systems retain the status of hypotheses and are used only tentatively as building blocks in theories about population abundance patterns. Our laboratory documentation of nonlinear phenomena such as saddles nodes, phase switching, bifurcations, lattice effects, and chaos suggests that these phenomena may be worthy hypotheses to incorporate into investigations by field ecologists.

What are the prospects for finding this sort of fine-structure in the dynamics of natural systems? Clearly, we have exploited some special features of the *Tribolium* system. Foremost among these are: a very detailed

---

the corresponding model-predicted cycle. Unplaited portions of the braid correspond to intervals for which the data closely resemble the model cycle. As shown in panel (F), we identified T-cycle episodes by setting the threshold number of animals $\theta = 55$ (dashed line) and threshold duration $K = 12$ for all model-predicted cycles. Thus, to be identified as a T-cycle episode in panel (F), non-equilibrium patterns were required to be in evidence for 24 consecutive weeks (more than seven generations), a very stringent requirement. The effects of varying θ and/or K on the episodes identified can be readily seen from inspection of the LMC plots in panels (B) and (E).

understanding of the life history and population biology of the organism, the ability to observe all essential state variables, the absence of measurement error, the ability to essentially eliminate environmental variability, and the isolation of each population from interaction with other populations. In systems with different properties, the attainable resolution may be coarser. We hope our research using the *Tribolium* model system, however, will raise expectations for how quantitatively precise the model/data fit can be in population biology.

Ecologists can be encouraged that simple nonlinear models can help unlock substantial gains in understanding population systems. Keys to transforming nonlinear models from scientific caricatures to testable scientific hypotheses are: incorporating demographic/environmental variability as well as the deterministic signal in biologically based models, explicitly connecting models and data, including statistics in the mathematical analysis, rigorously evaluating model performance, and effectively combining biology, mathematics, and statistics in an interdisciplinary approach.

Our use of a laboratory population system served the purpose that laboratory experiments have always served to isolate factors and to rigorously attribute cause. We were interested in whether the concepts from nonlinear dynamics—cycles, multiple attractors, chaos—could ever advance beyond the status of hypotheses and be convincing explanations of population fluctuations. We were interested in whether a mathematical population model could ever be considered reliable scientific knowledge. The laboratory allowed us to manipulate conditions, perform a census of each population and replicate, so that key predictions of the model could be tested.

Our results strengthen the relevance of mathematical modeling in population ecology. Not only was a mathematical model *useful* in describing population patterns, it was *essential* for understanding the experimental results. Nonlinear dynamic concepts, combined with stochasticity, *are* the explanations of the phenomena that we documented. In addition, advanced statistical modeling techniques were *required* for connecting model and data. Throughout much of ecology, mathematical models have been no more than simplified teaching concepts, not to be taken seriously, and statistics has been a set of recipes for data analysis. Herein we have displayed a population system in which mathematical modeling and mathematical statistics form an integral part of the theories themselves.

ACKNOWLEDGMENTS

This research was supported, in part, by National Science Foundation grants DMS-9625576, DMS-9616205, DMS-9981374, DMS-9973126, DMS-9981458, DMS-9981423, DMS-0210474, DMS-0414212.

REFERENCES

Bailey, B.A., Ellner, S. and Nychka, D.W. (1997) Chaos with confidence: Asymptotics and applications of local Lyapunov exponents. In: *Nonlinear Dynamics and Time Series: Building a Bridge Between the Natural and Statistical Sciences* (Ed. by C. Cutler and D.T. Kaplan), pp. 115–133. American Mathematical Society, Providence, RI.

Costantino, R.F., Cushing, J.M., Dennis, B. and Desharnais, R.A. (1995) Experimentally induced transitions in the dynamic behavior of insect populations. *Nature* **375**, 227–230.

Costantino, R.F., Desharnais, R.A., Cushing, J.M. and Dennis, B. (1997) Chaotic dynamics in an insect population. *Science* **275**, 389–391.

Costantino, R.F. and Desharnais, R.A. (1991) *Population Dynamics and the Tribolium Model: Genetics and Demography*. Springer-Verlag, New York, NY.

Cushing, J.M., Dennis, B., Desharnais, R.A. and Costantino, R.F. (1998) Moving toward an unstable equilibrium: Saddle nodes in population systems. *J. Anim. Ecol.* **67**, 298–306.

Cushing, J.M., Henson, S.M., Desharnais, R.A., Dennis, B., Costantino, R.F. and King, A.A. (2001) A chaotic attractor in ecology: Theory and experimental data. *Chaos Solitons Fractals* **12**, 219–234.

Cushing, J.M., Costantino, R.F., Dennis, B., Desharnais, R.A. and Henson, S.M. (2003) *Chaos in Ecology: A Study of Nonlinear Systems*. Academic Press, New York, NY.

Dennis, B., Desharnais, R.A., Cushing, J.M. and Costantino, R.F. (1995) Nonlinear demographic dynamics: Mathematical models, statistical methods, and biological experiments. *Ecol. Monogr.* **65**, 261–281.

Dennis, B., Desharnais, R.A., Cushing, J.M. and Costantino, R.F. (1997) Transitions in population dynamics: Equilibria to periodic cycles to aperiodic cycles. *J. Anim. Ecol.* **66**, 704–729.

Dennis, B., Desharnais, R.A., Cushing, J.M., Henson, S.M. and Costantino, R.F. (2001) Estimating chaos and complex dynamics in an insect population. *Ecol. Monogr.* **71**, 277–303.

Dennis, B., Desharnais, R.A., Cushing, J.M., Henson, S.M. and Costantino, R.F. (2003) Can noise induce chaos? *Oikos* **102**, 329–340.

Desharnais, R.A. and Costantino, R.F. (1980) Genetic analysis of a population of *Tribolium*. VII. Stability: Response to genetic and demographic perturbations. *Can. J. Genet. Cytol.* **22**, 577–589.

Desharnais, R.A., Costantino, R.F., Cushing, J.M. and Dennis, B. (1997) Estimating chaos in an insect population. *Science* **276**, 1881–1882.

Desharnais, R.A., Costantino, R.F., Cushing, J.M., Henson, S.M. and Dennis, B. (2001) Chaos and population of insect outbreaks. *Ecol. Lett.* **4**, 229–235.

Desharnais, R.A., Edmunds, J., Costantino, R.F. and Henson, S.H. (2005) Species competition: Uncertainty on a double invariant loop. *J. Diff. Equations and Applications* (in press).

Doebeli, M. (1993) The evolutionary advantage of controlled chaos. *Proc. Royal Soc. London B* **254**, 281–285.

Ellner, S. and Turchin, P. (1995) Chaos in a noisy world: New methods and evidence from time-series analyses. *Am. Nat.* **145**, 343–375.

Hastings, A., Holm, C.L., Ellner, S., Turchin, P. and Godfray, H.C.J. (1993) Chaos in ecology: Is mother nature a strange attractor? *Annu. Rev. Ecol. Syst.* **24**, 1–33.

Hawkins, B.A. and Cornell, H.V. (1999) *Theoretical Approaches to Biological Control*. Cambridge University Press, New York, NY.

Henson, S.M., Cushing, J.M., Costantino, R.F., Dennis, B. and Desharnais, R.A. (1998) Phase switching in population cycles. *Proc. R. Soc. Lond. B* **265**, 2229–2234.

Henson, S.M., Costantino, R.F., Cushing, J.M., Desharnais, R.A. and King, A.A. (2001) Lattice effects observed in chaotic dynamics of experimental populations. *Science* **294**, 602–605.

Henson, S.M., King, A.A., Costantino, R.F., Cushing, J.M., Dennis, B. and Desharnais, R.A. (2003) Explaining and predicting patterns in stochastic population systems. *Proc. R. Soc. Lond. B* **270**, 1549–1553.

Jackson, E.A. (1989) *Perspectives of Nonlinear Dynamics*, Vol. 1, Cambridge University Press Cambridge, England 216–219

Kendall, B.E., Briggs, C.J., Murdock, W.W., Turchin, P., Ellner, S.P., McCauley, E., Nisbet, R.M. and Wood, S.N. (1999) Why do populations cycle? A synthesis of statistical and mechanistic modeling approaches. *Ecology* **80**, 1789–1805.

Kendall, B.E., Schaffer, W.M., Olsen, L.F., Tidd, C.W. and Jorgensen, B.L. (1993) In: *Predictability and Nonlinear Modelling in Natural Sciences and Economics* (Ed. by J. Grassman and G. van Straten), pp. 184–203. Kluwer, Dordrecht, The Netherlands.

King, A.A., Desharnais, R.A., Henson, S.M., Costantino, R.F. and Cushing, J.M. (2002) Random perturbations and lattice effects in chaotic population dynamics. *Science* **297**, 2163.

King, A.A., Costantino, R.F., Cushing, J.M., Henson, S.M., Desharnais, R.A. and Dennis, B. (2004) Anatomy of a chaotic attractor: Subtle model-predicted patterns revealed in population data. *Proc. Nat. Acad. Sci. USA* **101**, 408–413.

Klimko, L.A. and Nelson, P.I. (1978) On conditional least squares estimation for stochastic processes. *Annals Statistics* **6**, 629–642.

Lathrop, D.P. and Kostelich, E.J. (1989) Characterization of an experimental strange attractor by periodic orbits. *Phys. Rev. A* **40**, 4028–4031.

May, R.M. (1974) Biological populations with nonoverlapping generations: Stable points, stable cycles, and chaos. *Science* **186**, 645–647.

Mertz, D.B. (1972) The *Tribolium* model and the mathematics of population growth. *Ann. Rev. Ecol. Syste.* **3**, 51–78.

Olsson, D.M. and Nelson, L.S. (1975) The Nelder-Mead simplex procedure for function minimization. *Technometrics* **17**, 45–51.

Park, T. (1955) Experimental competition in beetles, with some general implications. In: *The Numbers of Man and Animals* (Ed. by J.B. Craig and N.W. Pirie), pp. 69–82. Oliver & Boyd, London.

Perry, J.N., Smith, R.H., Woiwod, I.P. and Morse, D.R. (Eds.) (2000) *Chaos in Real Data: The Analysis of Nonlinear Dynamics from Short Ecological Time Series*. Kluwer, Dordrecht, the Netherlands.

Press, W.H., Flannery, B.P., Teukolsky, S.A. and Vetterling, W.T. (1986) *Numerical recipes: The art of scientific computing*. Cambridge University Press, Cambridge, England.

Schaffer, W.M., Kendall, B.E., Tidd, C.W. and Olson, L.F. (1993) Transient periodicity and episodic predictability in biological dynamics. *IMA J. Math. Appl. Med. Biol.* **10**, 227–247.

Shaffer, M.L. (1981) Minimum population sizes for species conservation. *BioSciences* **31**, 131–134.

Shulenburger, L., Ying-Cheng, L., Yalcinkaya, T. and Holt, R.D. (1999) Controlling transient chaos to prevent species extinction. *Phys. Lett.* **66**, 1123–1125.

Simberloff, D. (1988) The contribution pf population and community ecology to conservation science. *Annu. Rev. Ecol. Syst.* **19**, 473–511.

So, P., Ott, E., Schiff, S.J., Kaplan, D.T., Sauer, T. and Grebogi, C. (1996) Detecting unstable periodic orbits in chaotic experimental data. *Phys. Rev. Lett.* **76**, 4705–4708.

So, P., Ott, E. Sauer, Gluckman, B.J., Grebogi, C. and Schiff, S.J. (1997) Extracting unstable periodic orbits from chaotic time series data. *Phys. Rev. E* **55**, 5398–5417.

Sole, R.V., Gamarra, J.G.P., Ginovart, M. and Lopez, D. (1999) Controlling chaos in ecology: From deterministic to individual based models. *Bull. Math. Biol.* **61**, 1187–1207.

Stuart, A. and Ord, J.K. (1991) Kendall's Advanced Theory of Statistics, Vol. 2: Classical inference and relationships, 5th ed. Griffin, London, England.

Tong, H. (1990) *Nonlinear Time series: A Dynamical System Approach.* Oxford University Press, Oxford, England.

Turchin, P. (2003) *Complex Population Dynamics: A Theoretical/Empirical Synthesis.* Princeton University Press, Princeton, NJ.

Turchin, P. and Ellner, S.P. (2000) Living on the edge of chaos: Population dynamics of Fennoscandian voles. *Ecology* **81**, 3099–3116.

Wilson, E.O. (2002) *The Future of Life.* Alfred Knopf, New York, NY.

Xu, D., Li, Z., Bishop, S.R. and Galvanetto, U. (2002) Estimation of periodic-like motions of chaotic evolutions using detected unstable periodic patterns. *Pattern Recognition Letters* **23**, 245–252.

Zimmer, C. (1999) Life after chaos. *Science* **284**, 83–86.

Population Dynamics in a Noisy World: Lessons From a Mite Experimental System

TIM G. BENTON AND ANDREW P. BECKERMAN

I. SUMMARY

Both density dependence and variability in the environment are ubiquitous for biological systems. Theory has repeatedly indicated that environmental noise and density dependence interact, sometimes in unpredictable ways. Phenomenological models may capture the dynamical processes, but often give little insight into the way organisms may "filter" an environmental signal into dynamical response. Laboratory models provide ideal systems in which to explore the effects of the current environment on future population size, as well as being proving grounds for time series analysis, because they can experimentally be used to concurrently explore the biology and the population dynamics. Here we give an overview of ongoing studies of the soil mite *Sancassania berlesei* living in a stochastic environment generated by random supplies of food. We show: (1) how environmental noise affects the mean, variance, and temporal pattern of the population dynamics, (2) how much biological knowledge can be extracted by time series analysis, in the presence or absence of knowledge of the stochastic forcing, and (3) how the stochastic forcing interacts with the population density and age structure

ADVANCES IN ECOLOGICAL RESEARCH VOL. 37 0065-2504/05 $35.00

DOI: 10.1016/S0065-2504(04)37005-4

to affect per capita food, and how this, in turn, changes growth rates and influences life history decisions, such as age and size at maturity, reproductive allocation, and so on. The biological mechanisms indicate that a given sequence of environmental states may lead to quite different population dynamics depending on subtleties of the initial age structures and density. Ultimately, understanding (or predicting) population dynamics, becomes a question of understanding the inter-relationship of the environment and the life history.

II. INTRODUCTION

A major challenge of ecology is to understand the mechanisms that determine population size, in order to be able to predict future population dynamics and apply this knowledge to managing populations. In recent years, theory has repeatedly indicated that environmental noise can interact with deterministic density dependence to give rise to population dynamics which are not solely "deterministic dynamics with noise" but which may be qualitatively different (Bjørnstad and Grenfell, 2001). This qualitative difference may surface through several routes. (1) For example, the addition of noise can excite a system which is close to a bifurcation, such that it exhibits the dynamic behavior that it would show if the parameters of the model placed the system on the other side of the bifurcation (Nisbet and Gurney, 1976; Greenman and Benton, 2003). (2) Noise can cause the dynamics to move between different attractors. There are a number of situations of this type. For example, noise can move a system between coexisting attractors—such as point equilibria and cycles—so that the dynamics intermittently exhibit both types of behavior (Nisbet and Onyiah, 1994; Greenman and Benton, 2004); or noise can move a system towards a point that may be an attractor in some dimensions, but a repellor in others. In this case, such as a saddle point, the attractor is unstable and noise will cause the system to move away again (Cushing *et al.*, 1998). (3) Noise can also cause a resonance whereby density-dependent interactions between cohorts introduce delays in the density-dependent regulation, such that long-term, multigenerational trends result (Bjørnstad *et al.*, 1999a). (4) Noise can cause systems which are at superficially similar states to diverge through time. One class of such behavior occurs with systems that are chaotic, where there is well-known sensitivity to initial conditions. This also occurs in systems which are close to a bifurcation and undergoing stochastic excitation (Greenman and Benton, 2003). The other class of behavior is discussed in the following text and explains how the differences in initial conditions (e.g., age structure) cause differing responses to the same environmental state (Coulson *et al.*, 2001), such that, in a spatially replicated system, the correlation between populations is always less than the correlation in the environmental noise (Grenfell *et al.*, 1998; Greenman and Benton, 2001; Benton *et al.*, 2001a).

Another source of noise, in addition to environmental noise (i.e., random fluctuations in the environment), is the random way that individuals progress through their life history. This is called demographic stochasticity, and arises through the "discreteness of individuality": an organism can be male or female, dead or alive, but the "vital rate" (gender ratio, survival, fecundity) may be non-integer. In large populations, demographic stochasticity can be ignored because it is averaged across a large number of individuals, but not so in small populations: if there is one individual with a survival rate, s, of 0.75, then it will either survive (s = 1.0) or not (s = 0); in statistical terms, the survival rate will be distributed as a binomial distribution, with a mean of 0.75. This integer arithmetic can have important dynamical consequences, both in terms of increasing extinction rates (Caswell, 2001; Fox and Kendall, 2002), and also adding sufficient noise to cause movement between nearby attractors (Henson *et al.*, 2003).

Even if environmental noise does not create a profound change in the dynamics (such as switching between attractors), it will inevitably cause changes in the population dynamics by causing changes in the life history. As these changes work their way through the life cycle, lags are introduced between cause (environmental state) and effect (population dynamics). Management of populations in the natural world will often require population predictions to be made, which will become more precise as our understanding of the biological effects of environmental noise increases. From a theoretical perspective, "environmental noise" is variability in some environmental factor which causes random variability in the population dynamics. Thus, in a population model, noise may be a simple additive "error" term in an equation (e.g., births = a*$\log_e$(population density) + ε, where ε is typically drawn from a normal distribution with mean of 0). However, from a biological perspective, the ε term is effectively shorthand for "other factors we do not know about." Different environmental states may cause predictable changes in the life histories of the organisms, which, in turn, may cause predictable changes in the population dynamics. Thus, the organism may act as a biological filter, modifying an environmental signal into a population dynamics signal. Therefore, it may be possible, if the biology of the organism is well understood, to remove the ε term from the model and have a deterministic relationship between the biology and the environment (in our example this may be births = a*population density + b*food + c*temperature). Understanding the mapping of environment onto dynamics does not alter the fact that the environment may still be a random variable and so dynamics still will be unpredictable, but the mechanistic understanding should mean that the stochasticity in the model is appropriate and will interact with the biology in a reliable way. Hence, the predictions of such a model would be more likely to be qualitatively accurate.

Teasing apart the relationship between environmental variability and population dynamics is inherently complex. Field studies can generate time series, but the time series already incorporate environmental variability and there is rarely any true replication. Models are sensitive to assumptions and there may be little biological understanding of whether or not the assumptions are good or bad. Laboratory systems, therefore, may provide an important step in linking modelling to field systems. This is because such systems can allow replicated, population-level experiments to investigate how the environmental state maps onto the population dynamics. In turn, this allows the characterization of the "deterministic dynamics"—from populations kept under near-constant conditions—as well as the dynamics under stochastic forcing. Time series analysis can then be used to infer the biological mechanisms creating the population dynamics (the "top-down approach"). At the other scale, behavioral and life-history experiments can dissect the way in which individuals respond to a given environment, and so the properties of the "biological filter" can be constructed from first principles (the "bottom-up approach").

In this chapter, we report the results of ongoing studies of a soil mite laboratory system. Following a brief review of the biology of the mites and the experimental methods this chapter is divided into three sections. The first section is phenomenological, describing the impact of noise on population dynamics. We show that there is no simple mapping of the environment onto dynamics (alternatively, populations subject to the same environmental sequence do not respond in the same way). The second section uses time series analysis to try to reconstruct important aspects of the biology driving the dynamics, and ask the question "if we know the environmental stochasticity, how much does it improve our ability to detect the underlying biology?" In the third section, we report experiments at the life-history level which provide insight into the way that the organisms respond to the environment. In particular, we show that there is considerable plasticity in the life history of the organisms (and the way they respond to density), which means that the observed life history is the result of a complex interplay between current conditions, past conditions within the organisms' lifetimes, and the conditions experienced by previous generations. Understanding the plasticity in the life history allows considerable insight into the way that population dynamics may be shaped by environmental heterogeneity. In contrast, time-series analysis and simple analytical models may not capture the necessary biology and, therefore, may be unlikely to be able to predict population dynamics without a considerable degree of uncertainty.

A. The Mite Model System

The soil mite *Sancassania berlesei* (Michael), is common in soil, poultry litter, and stored food products. Populations of *S. berlesei* were collected from an agricultural manure heap (composted poultry waste) in 1996 and 1998 and have been kept in stock ever since (stock cultures number c1-2.5 $\times$ 10^5 individuals).

The life cycle consists of five stages, beginning with eggs (length: 0.16 $\pm$ SD 0.01 mm), continuing through a six-legged larvae (length: 0.22 $\pm$ 0.01 mm), a protonymph, tritonymph, and then to adulthood (female length at maturity: 0.79 $\pm$ 0.17 mm, range 0.47 to 1.17, n = 64; males: 0.72 $\pm$ 0.11 mm, range 0.55 to 1.02, n = 39). As indicated by the standard deviations of the adult lengths, there is considerable plasticity in the life history (discussed in detail in a following section) and much of it is governed by intake rates of food. An individual's intake rate is a function of a number of factors: population density, stage structure, and the amount of food supplied and its spatial configuration; together these factors create the individual's competitive environment. Clumped food allows some individuals to monopolize food and gain large amounts of resources. Which animals gain access to clumped food is likely to be a combination of body size (adults can push their way onto a food clump and exclude juveniles) and lottery (the first there may be harder to displace). Uniformly spread food allows a more equal distribution of food among the population.

Eggs hatch 2 to 5 days after being laid; large eggs hatch earlier than small eggs. Juveniles can mature from as little as 4 days after hatching to 50+ days after hatching (Beckerman *et al.*, 2003), depending on food and density. The longevity of the adults can also vary from c10 to c50 days. Thus, total longevity varies from 3 weeks (high food, low density) to 7+ weeks (low food, high density). Fecundity is related to resources, and so body size, and inversely related to survival.

Cultures are kept at constant 24 °C in three unlit incubators. Food is supplied in the form of powdered or granulated yeast, the granules of which are sieved to reduce size variation. One granule averaged 1.5 mg $\pm$ SD 0.35. Culture and experiment vessels are glass tubes 20 mm in diameter and 50 mm in height. These are half-filled with plaster of Paris which when kept moist maintains humidity in the tubes. The tops of the tubes are sealed with a circle of filter paper held in place by the tubes' cap with ventilation holes cut into it. The mites are censused using a Leica MZ8 binocular microscope and a hand counter. In each tube, a sampling grid is etched into the plaster surface to facilitate more accurate counting and observation. All adults are counted in the tube, but juveniles and eggs are counted in a randomly chosen quarter. Measurements of individuals and eggs were made from digital images captured from the microscope and measured using ImageJ 1.28u software (http://rsb.info.nih.gov/ij).

In the population experiments, we add stochasticity in the timing and amount of food supplied, while trying to maintain other factors as close to constant as possible. Our rationale for this is that many natural environmental factors will either vary the absolute food supply (e.g., the weather), the requirement for food (e.g., temperature), or the availability of food (e.g., patchiness, territoriality, inter-specific competition). Each treatment supplied food at the same mean daily rate (equivalent to one ball per day), but the variance changed. The algorithm we developed was to supply yeast randomly within each window of time, such that over the window length, w (days), the cultures received a total w balls of yeast, supplied in random amounts on randomly selected days within the window. Thus, a low variance in food supply had a window of 7 days, so the weekly food regime could vary from 1,1,1,1,1,1,1 to 7,0,0,0,0,0,0, whereas the high variance food supply had a window of 21 days, giving a potential period without food of 20 days in each window. The distributions of food supply are described in more detail elsewhere (Benton *et al.*, 2002). Populations were also maintained on constant food regimes to act as contrasts to those in the variable environments.

The underlying population dynamics were varied by: (1) varying the spatial distribution of food—granules or powder—and thereby changing the competition and distribution of food among individuals. Powdered food gives equilibrium dynamics, granulated food gives decaying oscillations (Fig. 1). (2) Imposing delayed density-dependent mortality (culling 20% of the number of adults present one week previously) produces population cycles with a period of 21 days (Fig. 1).

B. The Effect of Noise on Population Dynamics

In this section we describe the way that the population dynamics are altered by temporal variation in resources and discuss the mechanisms that may underlie the population patterns.

1. Means and Variances

When mites receive food supplied at random over time, it leads to variation in the population dynamics (compare Figs. 1 and 2). The distribution of population sizes is best described by a gamma or negative binomial distribution (Dennis and Costantino, 1988; Benton *et al.*, 2002). These two distributions are similar with the gamma being a continuous time version of the discrete time negative binomial, and are the sum of independent random

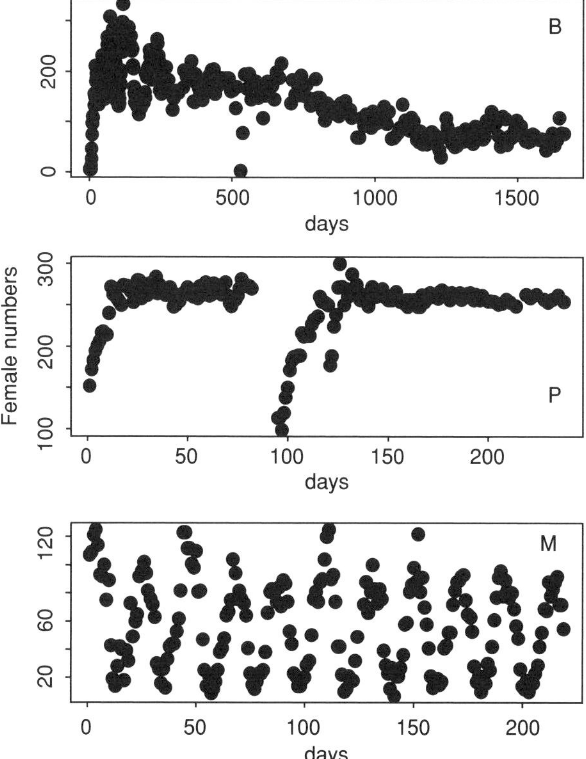

Figure 1 Time series from mites kept under "near constant" food conditions. B = long-term culture fed on 1 ball of yeast per day. The reduction in numbers soon after day 500 reflected a perturbation due to culturing by an unfamiliar technician. The long-term decline between days 700 and 1,000 reflects a change in the average size of balls of yeast, due to a change in the size of sieve we used to standardize grain size. P = culture fed on powdered yeast. Again, the perturbation at about day 95 reflects the care of the cultures by an unfamiliar technician. Following the perturbation, the population returns to equilibrium. M = a culture with imposed 20% adult mortality based on counts one week beforehand. This delayed density dependence creates 21-day cycles in the dynamics.

variables, each with a geometric (negative binomial) or exponential (gamma) distribution. The distribution typically changes shape as the variation increases (Fig. 3) (Benton *et al.*, 2002), becoming more skewed and with a smaller mean (Table 1), although the details depend on the stage class considered (Table 1).

The relationship between the population mean and the population variance is one that has given rise to considerable interest over the years. Typically, there is a power-law relationship, with the slope of the l-variance

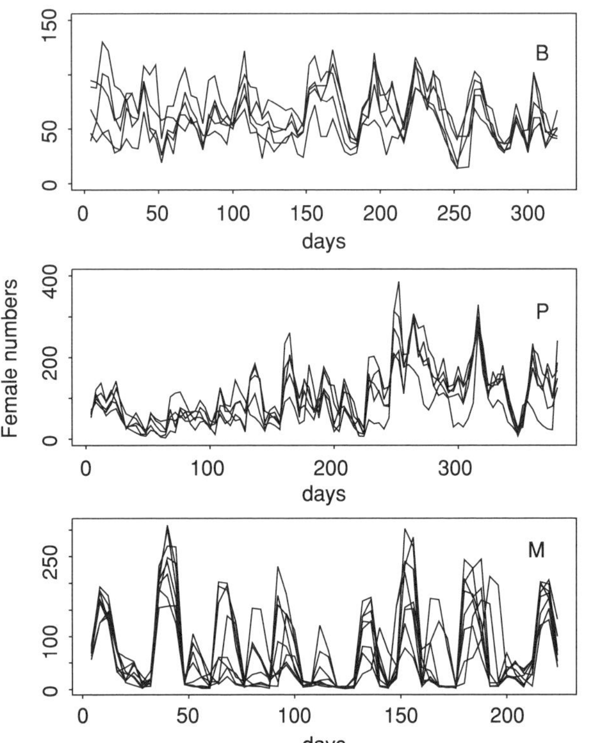

Figure 2 Dynamics of populations supplied with food randomly within windows of 14 days. B = 5 cultures fed on balls, P = 5 cultures fed on powder, M = 8 cultures with imposed 20% adult mortality delayed by 1 week. The cultures within each treatment group were all fed the same random regime of food (correlation between food supplies = 1), and counted every 4 days.

versus l-mean relationship being between 1 and 2 (Taylor, 1961; Taylor *et al.*, 1978; Taylor and Woiwod, 1982). For populations that are fluctuating around the equilibrium, according to a gamma or log-normal distribution, the slope of the relationship should equal 2. The rationale for this follows. Consider a species whose populations fluctuate with mean μ and variance σ^2, due to random variations in the environment. If populations occur in different habitats, then the carrying capacities will change but the ratio of σ to μ (the coefficient of variation [CV]) may stay the same. If this assumption holds, as the carrying capacity, K, varies between populations, the means should scale as $K\mu$ and the standard deviations as $K\sigma$, hence the variances as $K^2\sigma^2$. Plotted on a log-log graph, the slope of the variance-mean relationship should then be 2. That the value from empirical relationships is typically <2 has been ascribed to a number of factors including

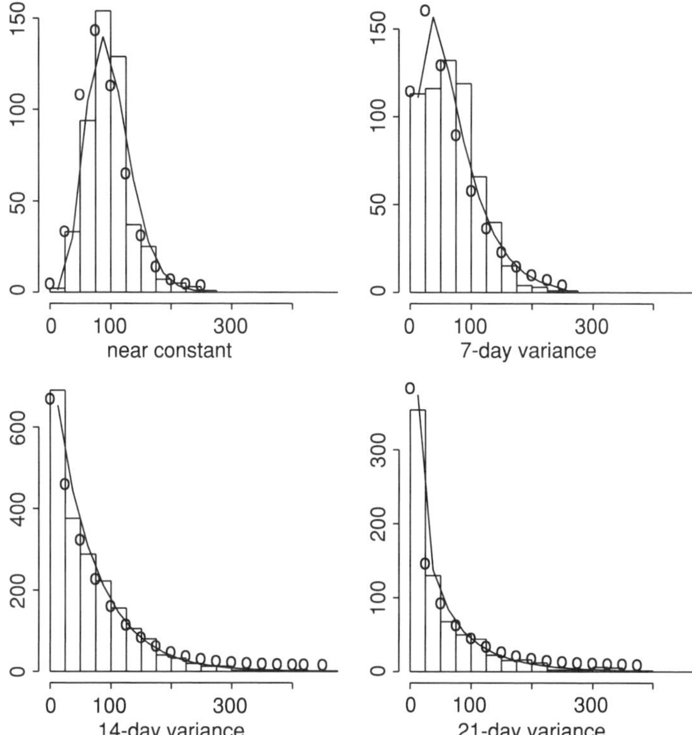

Figure 3 The distribution of egg counts under different amounts of environmental variance. The egg counts from a randomly selected quarter of each culture tube were pooled across replicate time series of mite populations within different environmental variances (with the daily mean supply the same = 1 ball of yeast per day) are plotted. As the variance increases the mean decreases and skew increases (Table 1). The lines represent the expected numbers in each class for a negative binomial distribution fitted to the data. The shape parameter, k, = 8.024, 1.880, 1.029, 0.654 for the C, 7-, 14- and 21-day treatments, with means of 98.1, 68.0, 66.5 and 52.5, respectively. The sample sizes reflect different numbers of replicates (C = 5, 7-day = 10, 14-day = 25 and 21-day = 10) and different lengths of the time series.

demographic stochasticity (Anderson *et al.*, 1982), trophic interactions (Kilpatrick and Ives, 2003), spatial heterogeneity (Keeling, 2000), non-linear dynamics (Perry, 1994), and sampling error (Titmus, 1983). Another simple explanation is the incorrect assumption that the CV remains the same across different populations. If the population is resource limited and the variance in resources changes across space, then the slope of the mean-variance relationship can actually assume any value, including negative ones. The exact value depends on the relative way that the means and variances vary (R.J.H. Payne and T.G. Benton, unpublished manuscript). Within groups of

Table 1 Descriptive statistics for counts of adults, juveniles, and eggs kept under different regimes of stochasticity. The mean and 95% CI were estimated from bootstrapping in a stratified way, with tubes as strata. Other statistics estimated from pooled samples. Constant, 5 tubes × 77 counts per tube, n = 385. 7-day, 10 tubes × 62 counts, n = 610. 14-day, 10 tubes × 60 counts, 5 tubes × 94 counts and 10 tubes × 98 counts, n = 2050. 21-day, 5 tubes × 61 counts, 5 × 85, n = 730. Egg and juvenile counts multiplied by 4 to give estimates of numbers in each tube. Constant = 1ball of food per day; "n-day variability" is n balls at random over each period of n days, with the variance in the food supply increasing with n.

		Constant	7-Day Variability	14-Day Variability	21-Day Variability
Adults	Mean	150.1	127.1	121.2	95.3
	95% CI for mean	147.4–152.9	124.3–129.9	119.1–123.0	91.8–98.8
	SD	44.9	52.5	60.7	65.0
	Skewness	0.39	0.69	1.02	1.60
Juveniles	Mean	662.9	458.8	504.4	531.0
	95% CI for mean	641.9–682.0	445.1–475.0	492.7–519.9	506.4–557.0
	SD	221.0	183.0	303.3	357.0
	Skewness	0.83	1.26	1.61	1.39
Eggs	Mean	392.3	272.0	266.5	212.0
	95% CI for mean	380.9–405.5	259.5–285.9	254.7–278.1	193.5–232.8
	SD	121.7	171.3	262.2	275.4
	Skewness	0.71	0.66	1.93	2.55

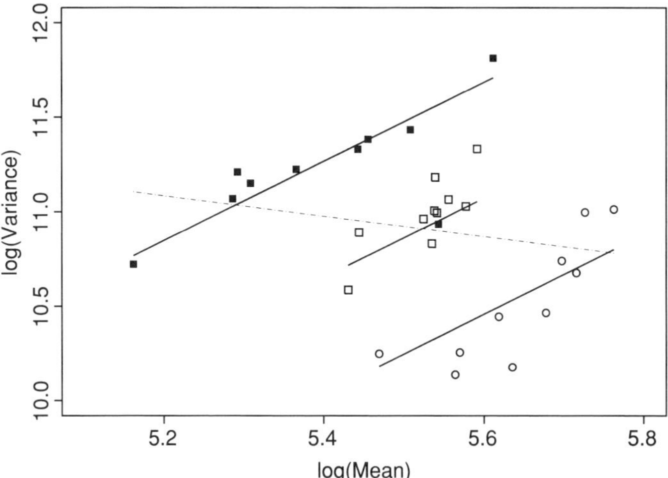

Figure 4 The slope of the ln-variance-ln-mean relationship (Taylor's Power Law) depends on the stochasticity in the carrying capacity. Within a group of tubes, the variance in the carrying capacity stays approximately the same, but the mean varies (depending on random differences between tubes). In this case, the slope of the power law should be 2.0 (Dennis and Costantino, 1988). Across groups of tubes, the variance in carrying capacity is bigger than the variance in means, and the slope of the log-mean-log variance relationship decreases (and can potentially take any value depending on specifics). Here the data are egg numbers for 7-day (circles), 14-day (squares), and 21-day variance (solid squares) regimes, with each symbol representing a tube. ANCOVA fits a common slope to each group (slope = 2.08 ± 0.37, not significantly different to 2.0), and the overall slope across all tubes is −0.54 ± 0.53.

replicate tubes (e.g., all 7-day variability tubes) we might expect the slope of the power law to be 2 as K may vary randomly between tubes; but as all tubes have the same environmental forcing, the CV would be approximately constant. Indeed this is what we find: within groups of tubes exposed to the same patterns of environmental forcing, the slope is not significantly different from 2 (Fig. 4). However, when all of the tubes are considered together, mean carrying capacities differ between tubes but so do the variances (7-,14- and 21-day forcing). In this case the slope of the relationship is not expected to be close to 2, and indeed becomes negative in our case (Fig. 4).

2. *Spectral Analysis of Population and Food Dynamics*

The match between the shape of the distributions of food and those of the populations is weak (Benton *et al.*, 2002). Spectral analysis decomposes the fluctuations in a time series into the contributions of different frequencies. If

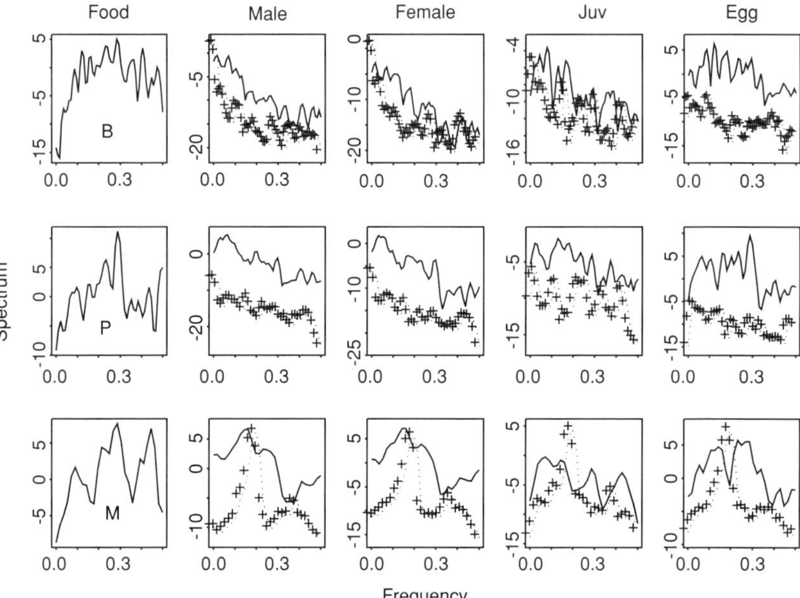

Figure 5 Smoothed Fourier spectra for the (logged) time series of food (1st column), males (2nd column), females (3rd column), juveniles (4th column), and eggs (5th column) for groups of five populations being fed balls of yeast (B, top row), or powdered yeast (P, middle row) or with lagged density dependent mortality being imposed (M, bottom row). Each combination of stage and treatment show spectra for mites receiving 14-day variability in food supply (solid lines) or the "near constant" food (crosses and dotted line). The census interval was four days, and the time series were 117 counts for B, 95 for P, and 56 for M. The time series of food supply was aggregated into 4-day bins for the purpose of this analysis. The mite spectra are the means of the 5 spectra for the five tubes, with the time series first being detrended by linear regression. The approximate 95% CI for the spectra are −4 to +8 units (decibels) on either side of the spectra, so any frequencies that have spectral values of >12 db are approximately significantly different. The mite time series are dominated by low frequency dynamics, whereas the food is dominated by higher frequency dynamics. A separate analysis of the near constant B and P time series, after detrending with a 2 df spline, indicates that the low-frequency behavior is different: the B cultures have a larger low-frequency component to their dynamics than the P cultures, as shown in the example time series in Fig. 1.

noise just creates variability in the existing population dynamics, the spectra for the mites fed on variable food and near constant food should be parallel, with the variable mites' spectra being absolutely larger. Conversely, if the variability in food creates matching variability in the mites, then the spectra for the mites fed variable food should be strongly correlated to the spectra for the food itself. In Fig. 5, we estimate the spectra for the food and for each life stage for mites in the near-constant conditions (crosses) and fed variable food (lines). For each of the 12 life-stage × treatment combinations in Fig. 5, we correlated the spectra to assess whether the mites fed variable food had spectra that were more similar to those from the variable food or the mites fed constant food. Overall, the correlation between the spectra from the variable time series and that of the food time series is −0.02, although this varies across the stages and the experiment (from −0.48 for males fed on balls to +0.61 for eggs on powder; Fig. 5), whereas the correlation between the spectra from the variable cultures and the constant cultures is much higher at 0.51 (and varies from 0.77 for males fed on balls to 0.23 for juveniles in the cycling cultures). Hence, the first conclusion is that the dynamics are modified by noise, but remain dominated by the deterministic patterns. This seems to arise due to low-frequency (i.e., long-term) variation in the deterministic dynamics not present in the food supply. Spectral analysis of the decaying oscillations shown in Fig. 1a indicates a period of ~70 days and such low-frequency components are clearly present in the mites fed both balls and powder (a period of ~70 days is equivalent to a frequency of 0.057 in Fig. 5, as the counts were made every 4 days).

A cursory examination of Fig. 5 indicates that the amount that the noisy dynamics are dominated by the deterministic patterns depends on the life history stage and the experimental dynamics. We analyzed the correlations between spectra with an analysis of variance (ANOVA) model with correlation between food and mite spectra as the dependent variable and life-history stage and dynamics as factors, and controlled for the correlation between the noise and the deterministic dynamics by adding it as a covariate. This model indicates that both stage and dynamics have significant influences on the synchrony between food supply and population dynamics (stage: $F_{3,5} = 5.4$, $p = 0.05$; dynamics: $F_{2,5} = 9.3$, $p = 0.02$). The model predicts that the correlations between the dynamic spectra and the food spectra (ignoring the life-history stage) increase by an average of 0.39 for the mites fed balls, 0.31 for mites fed powder, and −0.02 for the cycling mites; unsurprisingly, the strong deterministic patterns shown by the cycling mites prevent close synchrony between the food input and population response. The corresponding figures for life-history stage (ignoring the dynamics) are 0.09, 0.21, 0.19, and 0.42 for males, females, juveniles, and eggs, respectively. Thus, the time series of egg numbers is most sensitive to variation in food supply and, therefore, shows the greatest similarity to the food spectrum, and the greatest change from the "near constant" dynamics (e.g., see the spectrum from the eggs fed powder in Fig. 5). The other stages' spectra are modified by the addition of noise but in rather subtle ways.

The clearest interpretation of these spectra come from the cycling mites, where the cycle periodicity (frequency = 0.19, period = 4/0.19 = 21 days) surfaces as a sharp peak in all life stages under constant conditions. The addition of noise changes the spectrum differently in adults, juveniles, and eggs. In adults, the noise causes the sharp peak to become much broader. In juveniles, the noise abolishes the peak altogether. This occurs because the time an individual spends as a juvenile is very plastic. When food is constant, juvenile numbers cycle but with the cycle delayed by the time it takes to develop (juveniles hatch and grow to recruitment as adult numbers decline and competition is weak). Under noisy food, juvenile numbers no longer cycle in tune with adult numbers because the access to resources does not always correspond to periods of low competition from the adults. Thus, juveniles may stay juveniles for more than one adult cycle. Considering eggs, the noise causes the single peak (at 21 days) to become two separate peaks (one at 28 days and one at 14 days). The peak at 14 days matches a similar peak in the food spectrum (caused by the underlying periodicity of the fortnightly feeding regime). Theory suggests that periodic forcing on cycling populations should lead to resonant cycles that are the lowest common multiple of the forcing and cyclic periods (Henson, 2000; Greenman and Benton, 2004), which, in this case should be 6 weeks (periodic forcing of 2 weeks $\times$ cycles of 3 weeks), whereas we observe cycles of 4 weeks. This departure from theory might reflect the case that the forcing is noisy.

There has been some debate in the literature about the "color" of population dynamic spectra. The term 'color' is borrowed from optics, where light that is dominated by low frequencies is red and by high frequencies is blue. This debate has centered around a discussion of what factors are needed to turn the spectra from simple models, which are predominantly blue, to the predominately red spectra observed for real time series (Cohen, 1995; White *et al.,* 1996; Kaitala *et al.,* 1997b; Balmforth *et al.,* 1999; Akcakaya *et al.,* 2003). The addition of noise to a model may make a blue spectrum red or make a red spectrum blue, but the former effect perhaps predominates (Kaitala *et al.,* 1997b; Balmforth *et al.,* 1999; Laakso *et al.,* 2001; Akcakaya *et al.,* 2003; Greenman and Benton, 2003). A simple way to measure the color of a spectrum is to estimate the difference in area under the "red" end of the spectrum (frequencies 0 to 0.25) and the "blue" end of the spectrum (frequencies 0.25 to 50): if the difference is positive, the spectrum is red (White *et al.,* 1996). The three spectra for the food (Fig. 5) are all blue, but the spectra for the time series from the variable cultures are redder than the spectra from the constant cultures in 10 out of the 12 cases in Fig. 5 (binomial test, $p = 0.019$, the exceptions are mortality treatments, eggs and juveniles); thus, the addition of noise to the mite environment has reddened the dynamic spectrum. This reddening of population-dynamic spectra, even in the presence of blue noise, has also been shown recently in a microcosm system (Laakso *et al.,* 2003). The reddening can be explained by the biology of the mites. Perturbations in population density have long-term effects because a release in competition changes animals' growth rates, sizes and age at maturity, subsequent fecundity and, therefore, the competitive structure in the next generation (see the following section). These long-term responses to noise create signals at the red end of the spectrum (Figure 5).

3. Variability Between Replicates

It is clear from the experimental time series that there is considerable variability between the experimental cultures within a treatment group (Fig. 2 and 4), even when they are being fed the same regime of food. This variability between tubes means that the correlation in population dynamics is typically less than the correlation in the environmental forcing (Benton *et al.,* 2001a, 2002). Moran (1953) showed that in simple linear models the correlation between populations should be the same as the correlation in the noise. Recognition that many populations exhibit spatial synchrony (Kendall *et al.,* 2000), and that spatial synchrony is important in species management, has recently lead to considerable interest in the topic. As with the mites, Soay sheep living on adjacent islands are less correlated than they should be if Moran's theorem holds (Grenfell *et al.,* 1998, 2000). There are a number of likely explanations for this loss of synchrony. First, sampling error would tend to disrupt patterns of correlation. Second, "local noise" would reduce the impact of "global noise" (Ranta *et al.,* 1998). In our system, local noise would be inevitable with uncontrolled differences in each tube, including the quality of the plaster base and hence the humidity, and small variations in the food delivery. Third, density dependence ensures that population dynamics are non-linear, and the non-linearity tends to lead to asynchrony as populations at different densities respond to the same environmental state in different ways (Grenfell *et al.,* 1998, 2000; Greenman and Benton, 2001). For example, poor spring weather conditions have the most effect on red deer on the island of Rum when the population density is high (Albon *et al.,* 1987). When different ages or stages in the life history respond to different densities in different ways, this leads to complexity in the population response to a given environment, such that the response depends both on the stages and their population densities (Coulson *et al.,* 2001). We will return to this topic later in this chapter when we explore the mechanism by which the environmental state is translated into the life history and thus into population dynamics.

In summary, feeding the mites variable food affects their dynamics in ways that depend on the life-history stage and the deterministic dynamics (cycling or steady state). As variability increases, the average population size decreases and the variance increases. The mean-variance relationship across different populations depends on the relative way that the carrying capacity and the variance in the limiting resources change. The population dynamics, as measured by the spectrum, are dominated by the underlying deterministic dynamics, but nonetheless with important changes caused by noise.

C. Investigating the Mechanism: "Top-Down" Time Series Analysis

Having described, from a phenomenological point of view, the way that the properties of the dynamics are affected by environmental noise, we now move on to detecting and exploring the mechanisms by which the environment maps onto the dynamics. Understanding the mechanism becomes important if one wants to predict how the system will respond to environmental change or novel environments (Bradbury *et al.,* 2001). In applied ecology, there is often a paucity of information on the biology of the species in question, even if there are some time series data collected from censuses. In such cases, we may need to develop population models to aid management, and use the time series data to estimate unknown parameters in the model (Ellner *et al.,* 1997, 1998; Wood, 2001). In this section, our aim is to tackle the "inverse problem" of extracting biology from time series, because we are in a unique position to "test drive" the techniques of estimating biology from time series. We have the time series of the population counts, and perhaps most importantly, we also have the information on the environmental forcing that causes much of the population fluctuation. We can also compare the insights from such analyses to the knowledge we generate from the "bottom-up" approach of the following section. It is, of course, possible to ground truth time series analytical techniques using purely simulated data. However, the biological complexity inherent in the few systems living in variable environments that have been examined in depth indicates that "the devil is in the detail" and that different age classes and genders respond differently to environmental noise, density, and age structure (Clutton-Brock and Coulson, 2002). Hence, simulating data for modelling purposes risks creating a bias because the causation of the variability in the model is likely to be much more simple and "less messy" than real organisms living in a variable world.

The population dynamics of an organism are a complex interplay between density dependence and the environmental variation (broadly, density-independent factors). If the environment is very variable, the variability may hide the density dependence; conversely, if the environment is less variable, density-dependent processes become easier to detect. In the extreme, in a constant environment, the population may be at perfect equilibrium and, therefore, it becomes impossible to extract any biology as one cannot distinguish between, for example, no recruitment and survival of all organisms in a stage, or high recruitment and low survival. Reconstructing the biology from the time series, therefore, depends on a variety of factors. First, the amount of sampling error is clearly important. Second, whether or not there are parallel time series of different life stages is a great aid to inference. For example, if the number of adults was unchanged over time this could suggest either high adult survival or lower adult survival but with some recruitment from the juvenile stage. Inspection of the juvenile time series may allow these hypotheses to be differentiated. Third, the population size is likely to depend on the broader "environment," and knowledge of how influential environmental factors (the abiotic environment, or organisms in different trophic levels) vary over time can be instrumental in disentangling the complex causation of changes in population density over time.

Long-term population persistence is usually assumed to result from density-dependent factors regulating population size, although populations can potentially exist for surprisingly long times in stochastic density-independent environments (Orzack, 1993). The regulatory ability of density-dependent factors means that they are often considered more important. Solving the "inverse problem" involves extracting the biological information from time series, and is typically centred on identifying the way vital rates (e.g., survival and fecundity) change with population density, and then perhaps explaining residual variation in terms of other environmental factors. Density dependence can be direct, where the vital rates are a product of the current competitive environment, or it may be delayed. This typically arises because of the inevitable temporal separation of events in the life cycle. For example, delayed density dependence would occur if adult performance was a function of the competitive environment experienced during development, or if the number of predators was a function of population density at a previous time.

1. Time Series Modelling: The Process

Time series modelling is based on identifying the relationship between past population sizes and the current population size. The initial step involves estimating the order and delay of the population processes. The delay is the length of the time step or "lag" that is biologically most important (for example, if the generation time is 3 years, the delay might be 3, even if data were collected yearly). The order (also called the 'embedding dimension') is the number of previous time steps (lags) that need to be considered to estimate the current population size.

The approach to analyzing time series when there are multiple stages and non-linear relationships between density and biology involves fitting a non-linear model, of the correct order and dimension, that specifies some level of biological structure in the population (Ellner and Turchin, 1995; Ellner *et al.,* 1998; Kendall *et al.,* 1999; Smith *et al.,* 2000). The model-fitting process can involve parametric, semi-parametric, or non-parametric methods. Simulation studies repeatedly demonstrate that it should be statistically possible to estimate biology and even signals from forcing variables from noisy time series (Dennis and Taper, 1994; Lee, 1996; Ortega and Louis, 2000; Pascual and Ellner, 2000; Ranta *et al.*, 2000; Berryman and Turchin, 2001). Time series models have been powerful in explaining many aspects of a system's population dynamics, both in the laboratory (the Indian meal moth: Bjørnstad *et al.*, 1998; Briggs *et al.*, 2000; Bjørnstad *et al.*, 2001; blowflies: Smith *et al.*, 2000; Lingjaerde *et al.*, 2001; Moe *et al.*, 2002) and in the wild (cod: Stenseth *et al.*, 1999; Bjørnstad *et al.*, 1999a; Fromentin *et al.*, 2001; lynx: Royama 1992; Stenseth *et al.*, 1998; voles: Stenseth, 1999; Bjørnstad *et al.*, 1999b; Turchin and Ellner, 2000; Dungeness crab: Higgins *et al.*, 1997; leaf-eared mouse: Lima *et al.*, 1999; soay sheep: Coulson *et al.*, 2001).

2. *Time Series Analysis: Using Mite Data*

Our information on the stochastic forcing allows us to investigate not just the order, delay, and shape of the density dependence but also to ask how important our knowledge of the noise could be to explain the variability in the population data. The question "how much do we need to know about the biology and about the environment to make accurate predictions from models?" is not trivial, especially because identifying the relevant environmental factor driving the dynamics is often problematic (Ranta *et al.*, 2000; Lundberg *et al.*, 2002). Jonzen *et al.* (2002) suggest, through an extensive modelling exercise, that in the absence of environmental covariates, regression approaches are highly influenced by autocorrelated noise which can bias the estimates of autoregressive coefficients and estimates of the order of time series. Moreover, they found that noise in general undermines the repeatability of estimates of the coefficients. Therefore, knowledge of noise will help, but how much knowledge of the noise and the biology is necessary?

The data we analyze in the following stems from five replicate populations receiving 14-day variability in food, where the food supplies are correlated among the populations at 50%. For each replicate, we have counts from eggs, juveniles, and adults. Plotting the growth rate of each stage (ln $[N_{t+1}/N_t]$) against current (N_t) and past ($N_{t-1}, N_{t-2\ldots}$) numbers clearly reveals density dependence: there is a negative relationship between the rate of change of each stage and the numbers in that stage at the previous count, 4 days before (Fig. 6). This negative relationship is still discernable, but is weaker at lag 1 and generally has disappeared at lag 2. For the change in egg numbers, the relationship becomes positive at lag 3 (the slope of a simple linear regression is 0.0033 ± 0.0009). This time period, 12 to 16 days,

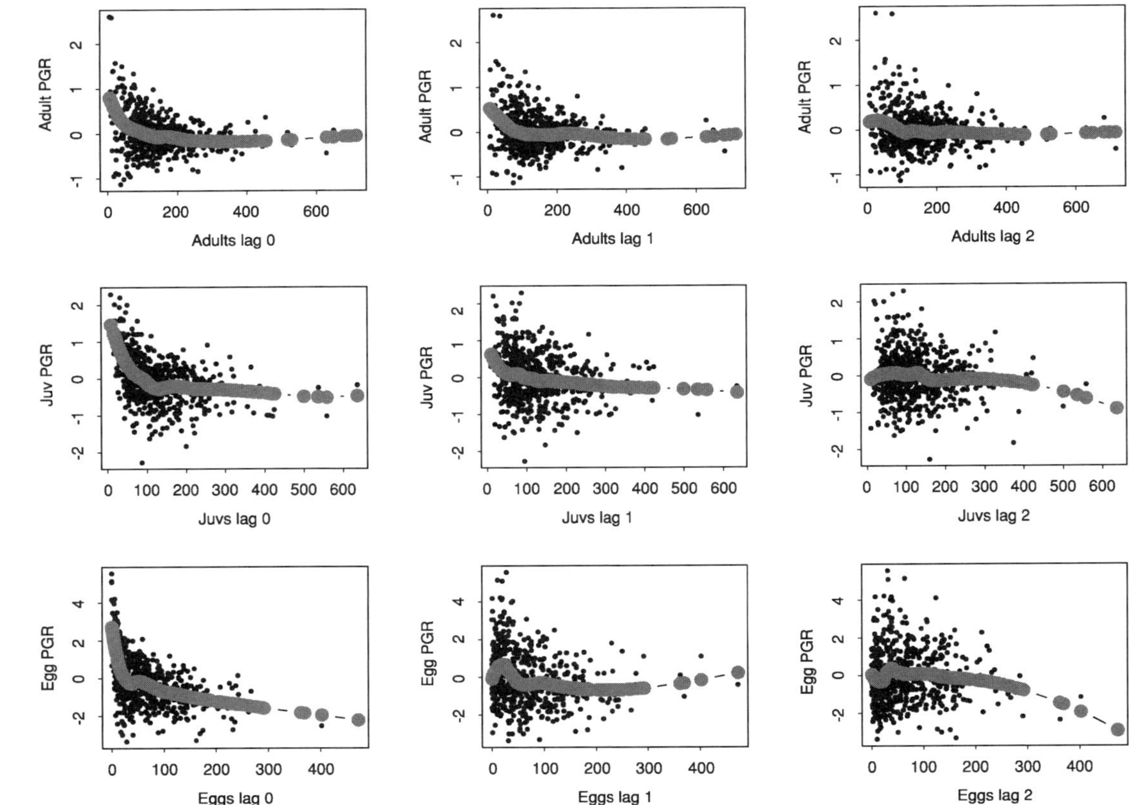

Figure 6 Population growth rates (PGR: $\ln[N_{t+1}/N_t]$) of each stage (Adult, Juv = Juvenile, Egg) as functions of the density of that stage at the three previous counts (lags 0,1 and 2). The data are plotted and a simple loess (span = 0.2) fitted to describe the shape (solid grey dots). Top row-adults, middle row-juveniles, bottom row-eggs.

approximates both the minimum generation time and the average time between feeds (2 weeks). Thus, positive rates of change in egg number between the current and next counts would result from a large number of eggs 12 days previously who have hatched, received food, recruited to adulthood, and laid eggs. Figure 7 shows the stage-specific growth rates plotted against the numbers of an additional life stage to that shown in the respective panels in Fig. 6. In each case, there is a three-dimensional relationship indicating that the additional life stage adds extra information and will explain some of the scatter around the relationships shown in Fig. 6.

We estimated the order and delay using a robust, non-parametric cross validation routine (S.P. Ellner, personal communication). The order-estimating algorithms suggest no more than 2 lags of order 2 (max 16 days) and often only 1 lag of 1 (4 days) are important in determining the dynamic properties, i.e., the recent past is the strongest determinant of dynamic behavior. We know that life histories may be influenced by the environments experienced by previous generations (via maternal effects; see the following text), but such historical effects seem not to affect the gross pattern of population dynamics. This dominance of recent history on dynamics is perhaps not surprising in a system with strong stochastic forcing.

The density-dependent relationship between per-capita fecundity (estimated as $Eggs_t/Female_{t-1}$) and female population density ($Females_{t-1}$) can be fitted by exponential or Ricker-type models of density dependence, although neither explains the scatter around the line (Fig. 8A). For ease of comparison with some data from life-history experiments, we will use the exponential model. To the simple model, we can add, as covariates, increasing information about the biologically important lags and also the environmental covariate (which we estimated as per capita food availability). It is clear from Fig. 8, that the building of a model that includes both information drawn from the order estimation and from our knowledge of the stochastic environment lead to dramatic increases in predictive ability and explanatory power in per-capita fecundity: our initial R^2 for the exponential model relating fecundity to female density is <20%. By the time we have added information on the history of food (lag-0 and lag-1), we have an R^2 of close to 50%. This filling-in of the scatter on the graph demonstrates that there is definite value in information on environmental variance, but simultaneously, it highlights that even knowing both female densities and the details of their food supply only explains about 50% of the variation in egg numbers over time.

The quality and amount of the environmental information in the model influences its explanatory power. To demonstrate this, we simulated the loss of 10% to 90% of the data and re-fitted the density-dependent model. The data on per capita food was lost, or lost and replaced (by cubic interpolation, replacing with a random sample of the data, or replaced by a random

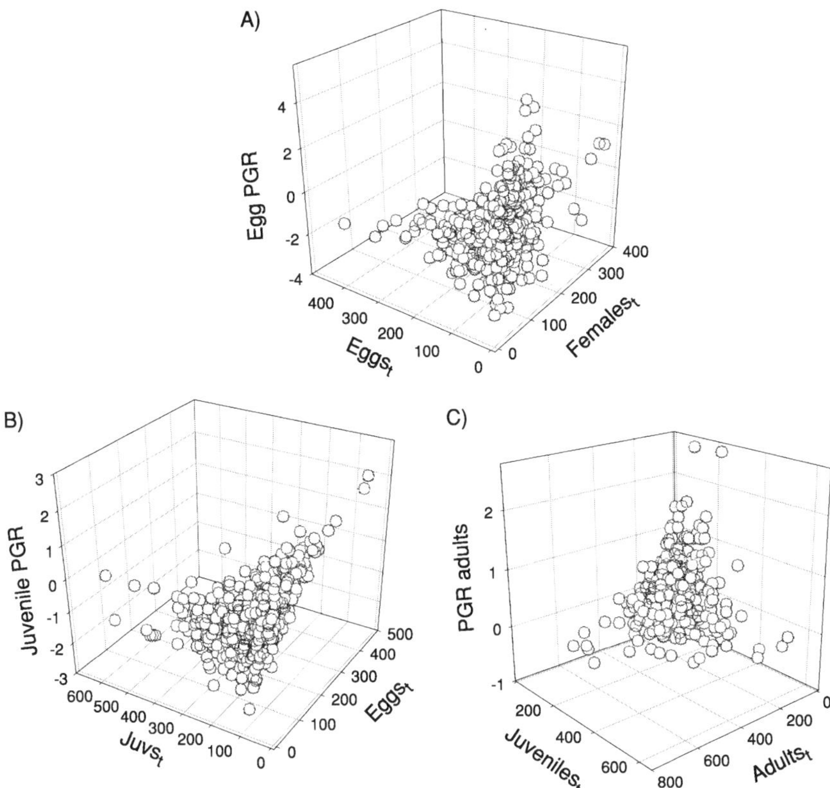

Figure 7 The data shown in the first column of Fig. 6 (the lag-0 plots) are replotted with the lag-0 counts of an additional relevant life stage. The presence of patterns due to both independent variables indicates that analysis of the time series would have greater explanatory power by considering the structure of the life cycle (e.g., future egg numbers are influenced by the number of current eggs and the numbers of females, via the vital rates of hatching and fecundity), as well as the lags. A) growth rate of the population of eggs as a function of the eggs at time t (impacts growth rate via hatching) and females at time t (influences growth rate via egg laying), B) juveniles' population growth rate as a function of juveniles at time t (impacts PGR via juvenile survival and loss of juveniles by recruitment) and eggs at time t (impacts PGR via hatching) and C) adults' population growth rate as a function of juveniles at time t (which impacts on PGR via recruitment to adulthood) and females at time t (which impacts via adult survival).

number drawn from a distribution with the observed mean and standard deviation). We generated 1,000 sets of each of these manipulations and then carried out the regression, estimating the R^2 for each model. Where the data was lost and not replaced, the sparse model, on average, had similar explanatory power as the model with the full data, but the range was much greater.

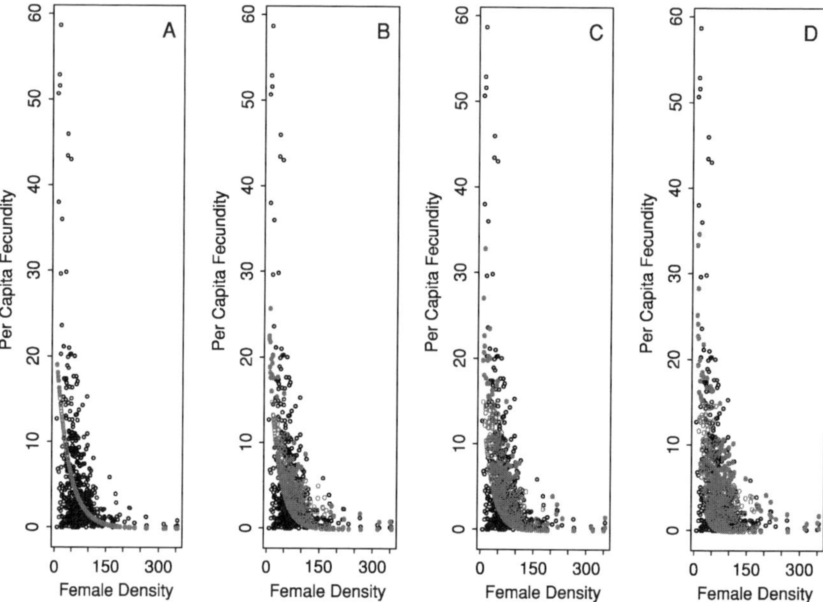

Figure 8 Explaining the patterns of per capita fecundity (PCF) involves knowledge of density and food supply. The data are plotted (black dots) and fitted values from models are grey. The fitted models are of the form $PCF_t = a \exp(-b^* \text{density}_t) + s(\text{food}_t)$, where per capita fecundity, $PCF_t = Eggs_{t+1}/Females_t$, $Food_t$ is per capita food (food delivered in period t to t+1 divided by population size at t), and s( ) is an n df spline. A) PCF_t as a function of $Females_t$, $R^2 = 22\%$; B) PCF_t as a function of $Females_t$ and $Food_t$, $R^2 = 29\%$; C) PCF_t as a function of $Females_t$ and $Food_{t-1}$, $R^2 = 38\%$; D) PCF_t as a function of $Females_t$ and $Food_t$ and $Food_{t-1}$, $R^2 = 48\%$.

If the lost data was particularly influential, the sparse model could explain much less variance than the complete model. The converse was also possible. Where the data was lost and replaced (whether by random sampling or interpolation), there was always a non-linear (a negative exponential) decrease in explanatory power as the amount of data lost and replaced increased. This suggests that a loss of relatively small amounts of environmental information causes a large decrease in the average R^2 of the model.

Data on the variance in the environment and on the order of density dependence in our time series allowed us to make precise estimates of per capita fecundity; but are they accurate? From experiments to investigate the life history, we can fit a negative exponential density-dependent model on real fecundity (eggs per female per day); this model can then be compared to the negative exponential model (controlling for food supply) estimated from the time series data (Fig. 9). The fitted models describe the same shapes of density dependence (Fig. 9), and the coefficients are not statistically

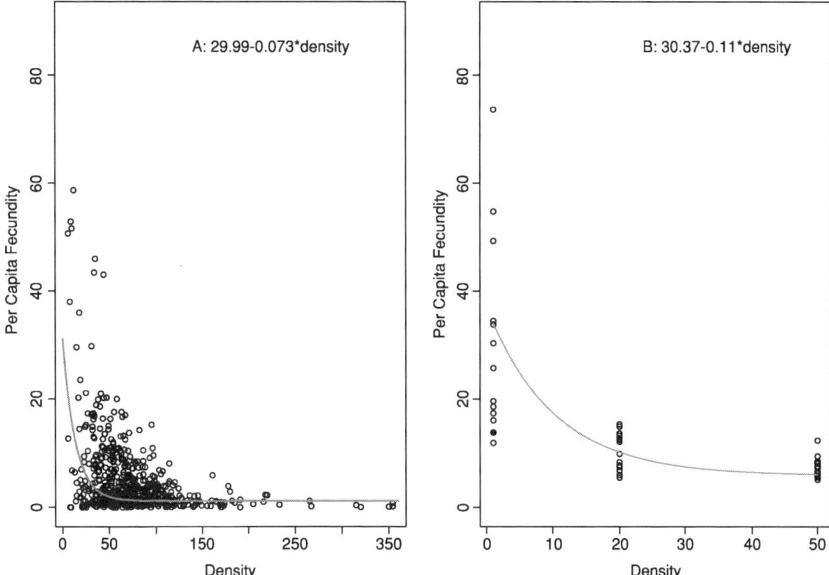

Figure 9 A comparison of the basic exponential models of density-dependent per capita fecundity from the time series (A) and the life-history data (B). Each model fit (grey) is generated by a model of the form $PCF_t = a^*\exp(-b^*density_t) + c^*Covariates$. In the time series data, the covariates correspond to estimates of per capita food availability at time t and t−1 (see Fig. 8). In the life-history data the covariates correspond to the treatments manipulating food availability at various stages in the life cycle, including their conditions experienced as a juvenile (Beckerman *et al.*, 2003). The fitted model is shown for the median values of the covariates. Intercepts are A: 29.99 ± 4.48(SE); B: 30.37 ± 4.43(SE). Slopes are A: −0.073 ± 0.0095(SE); B: −0.11 ± 0.067(SE).

different. This is encouraging as the considerable scatter around each line arises from different causes. The time series variance comes from a natural interaction between the biology of the organism and stochastic variation in food, while the independent life history data possesses variability from four independent manipulations of density, rearing conditions, food amount, and periods of starvation (Beckerman *et al.,* 2003). Estimating deterministic biology from noisy time series data is clearly possible.

As we noted in the introduction, from a biological perspective, the ε term (error) in models is effectively shorthand for "other factors we do not know about." Here, we have shown that it is possible to estimate components of the deterministic skeleton from time series data that is (deliberately) noisy, but also that knowing something about the environmental drivers can go a long way in explaining how different environmental states may cause predictable changes in the life histories of the organisms, which, in turn, may cause predictable changes in the population dynamics. Therefore, we can remove some of the "unknown" in ε by understanding the biological filter—lags in density and the shape of the function—and biologically relevant information about the environment. This approach is likely to have the greatest chance of success when the mechanism between forcing and biology is clear. For example, in a field system, "climate" may explain less variation in the dynamics than a direct measure of how climate affects the organism, such as through its food supply or influence on mortality (Hallett *et al.*, 2004).

3. *Age Structure and Population Responses to Perturbations*

A cursory examination of Fig. 1 indicates that there is considerable variability between different replicated populations subject to (approximately) the same pattern of environmental variation. This variability leads to lower synchrony between populations than would be expected from the correlation in the environment (Benton *et al.*, 2001a). What is the biological causation of different populations responding differently to the same environmental state?

We can address this question first by analyzing some time series. One set of five populations was fed on a periodic, non-stochastic, forcing regime (the same food as the 14-day variation regimes, but all food given on the first day in each period). Analysis of these time series indicates that the number of adults is dependent on the time since feeding, the population density at the time food was given, and also the population age structure (quasi-likelihood GLM, with log-link and variance equal to the mean, significance assessed by the bias corrected confidence limits estimated by bootstrapping: days-since-food^3 $p < 0.001$; log(population size $-A + J + E$ - at feeding), $p < 0.001$; proportion of adults at feeding $p < 0.001$; proportion of juveniles at feeding $p > 0.05$; interaction between proportion of adults and juveniles $0.01 < p < 0.05$; Fig. 10). In other words, populations of the same size respond to the same environmental state (a "feed") in different ways, depending on the relative proportions of adults and juveniles. The reason for this is, in part, because the life history inevitably takes time to complete and, therefore, adds lags into the population response. A population comprising of adults can increase in response to food more quickly than a population of recently hatched juveniles, simply because the time it takes for a female to lay eggs after a feed is less than the time it takes for a young juvenile to grow to adulthood. A similar lag is indicated in Fig. 10; the short-term increase in adult numbers in the first few days following a feed is due to older juveniles recruiting to adulthood.

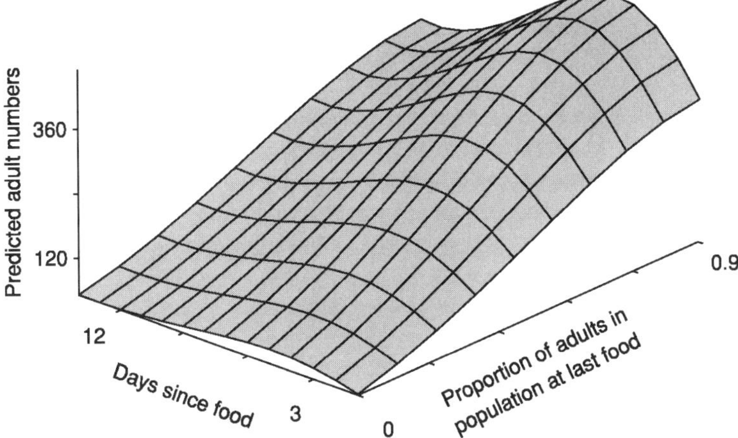

Figure 10 The population response to a given amount of food depends on both population density and age structure. Fitted surface for a model based on 5 replicated time series (each of 120 counts) of mites fed on a regular pulse of food every 14 days. This is equivalent to ~170 feeding events. The dependent variable was the number of adults, and the predictor variables were: tube, population size (eggs + juveniles + adults) at time of feeding (ln transformed), proportion of adults in population at feeding, proportion of juveniles, and days since feeding (as a cubic term). The model was a quasi-likelihood GLM, with log link and variance equal to the mean. Here we show the plane of fitted values for the number of adults versus time since feeding and proportion of adults, while controlling for population density (set at the mean value) and tube effects.

D. Investigating the Mechanism: "Bottom-Up" Experiments on Mite Biology

The sorts of time series analysis previously described have been instructive in indicating that aspects of the population biology depends on the population state and food supply over the previous one or two time periods. However, even the best model of Fig. 8 does not fully describe the scatter among the data points. To try to understand the biological processes that give rise to the time series, we now turn our focus to understanding what was gained from investigations on individuals, rather than populations.

Time series analysis identified that the response to food depends on age structure, and in part this is because life cycles take time to complete. A second reason the population response depends on age structure is because age structure, with population density, is a determinant of the competitive environment that an individual faces. As adults and juveniles are unequal consumers (the absolute intake rate of adults is some 3× higher than an average juvenile) and also unequal competitors (adults are better at excluding juveniles), the amount of food available per capita depends on both the population size and the stage structure. Therefore, individuals in different populations will receive different per capita food if the populations differ in structure, even if the population density and food supply are the same. Different resources will lead to different life-history responses and, therefore, different population effects.

In experiments on small groups of individuals, we find that there is strong density dependence in all aspects of the life history: growth, age, and size at maturity (see the following text), survival and fecundity and the shape of the density-dependence is modified by the amount of food that the tube receives (Beckerman *et al.*, 2003). Density dependence in a vital rate is an expression of plasticity in individuals' life histories, and plasticity in one trait may lead, through trade-offs, to plasticity in other traits. For example a predator may kill prey (causing a direct change in average survival), but the predator may also influence prey-foraging patterns, perhaps influencing future growth rates, survival, and fecundity. The way an individual responds to a given environment, therefore, is often contingent upon its history. This "historical contingency" can become self-propagating at the population level. As soon as two populations develop different age structures (for whatever reason), the competitive environment in the populations will differ, so individuals from the two populations will have different per capita food supplies. In turn, this will lead to differences in future growth, survival, and fecundity, affecting the competitive environments in the future. Hence, history in the life-history will tend to reinforce the effects of differences in initial conditions.

1. History in the Life History

Understanding trade-offs is central to our understanding of life-history evolution (Stearns, 2000; Roff, 2002). The fitness costs and benefits determined by the existence of trade-offs mean that different individuals, even if genetically identical but subject to different environments, may be positioned at a different place on a trade-off. The relationship between the expression of a trait, its phenotypic plasticity and the environment is known as a reaction norm.

To date, the interplay between life-history theory and population biology has largely been how dynamics affects evolution: from r- and K-selection (MacArthur and Wilson, 1967) to evolution in variable environments with density dependence (Kaitala *et al.*, 1997a; Benton and Grant, 1999; Diekmann *et al.*, 1999). However, recently there has been much interest about how the life history affects dynamics, either its evolution (Ebenman *et al.*, 1996; Abrams and Matsuda, 1997; Richards and Wilson, 2000) or its variability (Ginzburg and Taneyhill, 1994; Crone, 1997; Inchausti and Ginzburg, 1998; Ginzburg, 1998; Witting, 2000; Benton *et al.*, 2001b). This latter interest comes about because of the widespread recognition that maternal allocation decisions affect the offspring's quality, and these "maternal effects" can have profound influences on later life history (Mousseau and Fox, 1998). Coupled with this has been a recognition that an environment, especially early on in life, can itself lead to "delayed" life-history effects across the lifetime (Lindstrom, 1999; Beckerman *et al.*, 2003). For example, deprivation of food early in life is often followed by "compensatory growth" allowing an organism to mature at a "normal size" but paying a later cost in survival (Metcalfe and Monaghan, 2001), or maturation at the "normal time" but paying a cost in terms of smaller body size leading to lower reproductive success (Plaistow and Siva-Jothy, 1999). Such delayed life-history effects arise because the environment (whether directly through the weather, for example, or indirectly through maternal allocation decisions) causes movement along trade-offs leading to individuals making different life history decisions in the future. Delayed life history effects give rise to noticeable cohort variation which persists throughout life (Albon *et al.*, 1987; Langvatn *et al.*, 1996; Rose *et al.*, 1998; Forchhammer *et al.*, 2001; Grenouillet *et al.*, 2001; Reid *et al.*, 2003), and as the cohorts vary in growth, survival, and fecundity, cohort variation leads to population dynamic variation (Albon *et al.*, 1987; Saether, 1997; Lindstrom, 1999; Clutton-Brock and Coulson, 2002; Beckerman *et al.*, 2003).

We can illustrate the variation in mite life history that results from variation in per capita food (which, in culture, is determined by food supply and density). Individuals that grow at different rates mature at different times and at different ages. Reproductive success is related to body size, as in many invertebrates, and, as a result, the fastest growing individuals mature at a large size, perhaps even trading-off age in favor of size. Individuals with little food grow slowly and mature at the threshold size at whatever age it takes them to reach it. As a result, the reaction norm for age versus size in the mites, is like a "J" rotated +90° (Fig. 11) (Plaistow *et al.*, 2004).

Food during development alters the size and age at maturity and, therefore, patterns of reproductive allocation. Females in high-food environments mature at large sizes and have a high absolute reproductive investment. Typically, they will begin laying many small eggs and switch to fewer larger eggs as a terminal reproductive investment. Conversely, females in low-food environments mature smaller and later and have a lower reproductive investment. If food is not too limiting, females will lay fewer eggs, but invest more in each one (Fig. 12). These strategies are reminiscent of r- and K-selection: females with lots of resources tend to be in a low-density habitat and fitness depends on reproducing fast.

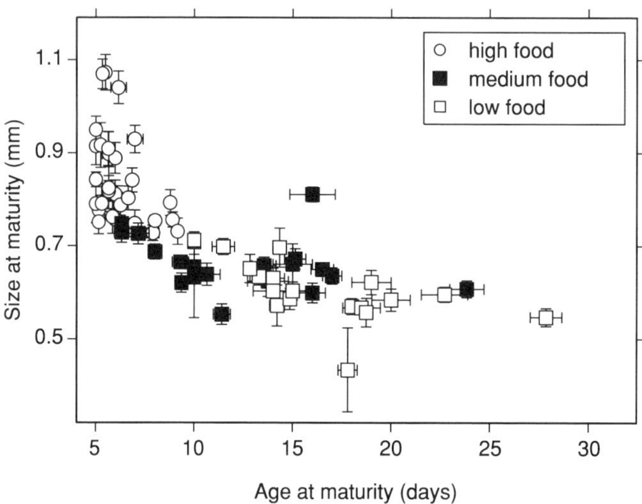

Figure 11 Age and size at maturity. Eggs were collected from females withdrawn from a common-garden stock culture, and isolated into groups of 20 in a tube. Each tube was randomly assigned to a treatment group receiving "high," "medium", or "low" food (filter paper circles soaked in yeast solutions of different concentration). The tubes were fed each day (the filter paper refreshed with fresh), and the adults were photographed and measured on the day of maturation. Each point is the mean per tube (average n = 8.8 ± SE0.2 females per tube matured, n = 70 tubes). The within-tube slope of the age-size relationship can be deduced from the error bars: positive slopes typically occur when Y errors are bigger than X errors, and negative slopes occur when X errors are bigger than Y errors. Fast growing females trade-off age against size, while slow growing females do the reverse.

Conversely, females who grew slowly tended to have faced more competition, and so fitness depends on competitive ability, enhanced by allocating more resources to each egg.

The reproductive strategy depends on both current food supply and also the food experienced during development (which affects adult size). Not surprisingly, the strategy undertaken also has future life-history effects. Female survival is correlated with reproductive output (Beckerman *et al.*, 2003), so females who developed with good food tend to die sooner. In addition, juvenile traits are linked to maternal investment in the eggs. For example, females whose mothers were reared on low food mature, on average, 61% faster than females whose mothers were reared on high food (LL: 14.5 ± SE 0.22 days, n = 13, cf HL: 23.9 ± 1.01 days, n = 18), but, as to be expected from the shape of the reaction norm, they matured at approximately the same size (LL: 0.58 ± 0.01 mm cf HL: 0.63 ± 0.02 mm) (S.J. Plaistow *et al.*, 2003, unpublished data). In turn, these juveniles' growth

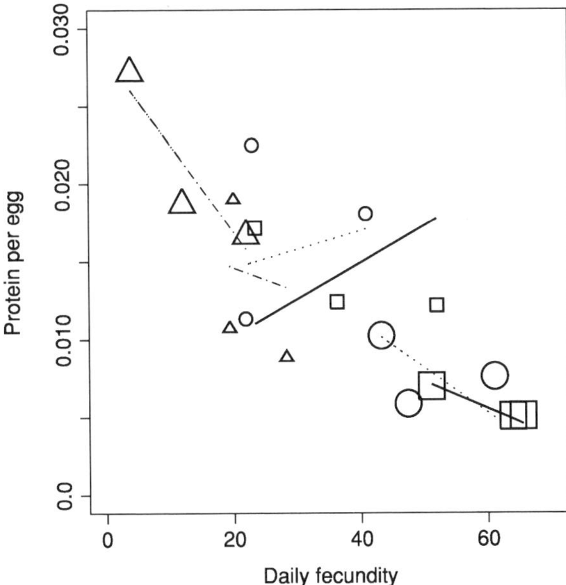

Figure 12 Reproductive allocation: egg quality versus fecundity. Eggs were collected from females withdrawn from a common-garden stock culture, and isolated into groups of 20 in a tube. They were randomly assigned to one of four food treatments (1= low, 4 = high). Upon maturity, the adults were set up in tubes of 10 pairs. Each treatment consisted of 3 replicate sets of 2 tubes. Daily fecundity (eggs laid per female per tube) was estimated. At three stages in their lives—days 2 to 4, 9 to 11, and 15 to 19—eggs were pooled across days and across the pairs of tubes and analyzed using the anthone assay to assess crude protein. The data points are mean female fecundity (across the paired sample) and the mean of three repeat measures of protein from the pooled egg samples. Only data for food = low (small symbols) and food = high (large symbols) are shown. Age is represented by squares = young, circles = middle age, triangles = old age. The lines are the fitted values from a GLM (on the whole dataset), with protein as the dependent variable and the following significant factors: fecundity $p = 0.002$, age $p = 0.03$, food level $p = 0.08$, food*fecundity $p = 0.009$, age*fecundity $p = 0.04$ (and tube, random effect, $p = 0.004$). Well-fed females produce larger numbers of small eggs than poorly fed females, until they get old, when the terminal reproductive investment switches to fewer but larger eggs.

rates, significantly influenced by the parental environment, affect their own reproductive allocation decisions, giving rise to weak grand maternal effects.

In the mites there can be significant historical effects on the life history, both from conditions experienced within an individual's lifetime and those of previous generations. However, the strength of these effects are context dependent. In good food environments the effects of the past tend to be erased by the high level of current resources. In medium-low food environments, history can matter greatly.

2. *From the Life History to Population Dynamics*

Evolutionary costs and benefits determine trade-offs which, in turn, determine how a phenotype is expressed in differing environments where both resources and competitors determine the environment. The mite life-history plasticity indicates that the allocation of a unit of resource to growth or maturation, or to large or small eggs, is strongly contingent on previous history and, in turn, influences the life histories of the next generation.

The affect on current conditions reaches into the future by two routes. First, is through maternal effects. How offspring perform partly depends on the investment in each egg. Second, plasticity in response to the current environment alters the number of competitors that offspring will face (by changing survival patterns and fecundity patterns). The change in fecundity is dynamically important: individuals who manage to grow fast, also grow very large and, as a consequence, can have a huge reproductive output relative to animals with less food. The interplay between maternal effects and the competitive effects created by a current environment can be influential for population dynamics. In one experiment initiated with cohorts of 250 "large" or "small" eggs, the population trajectories for the treatment groups differed for >3 generations (Benton *et al.*, in press). This difference arose because the large and small eggs hatched at different times, survived and grew at different rates and were consequently different sizes and ages at maturity. The adults varied in their fecundities and the size of the eggs they laid, leading to the next generation consisting of a high- or a low-density cohort. Individuals in the different-sized cohorts varied in the strength of competition they faced and, therefore, contrasted in their intake rates, and grew and matured at different sizes and times. Juveniles in the low-density cohort grew faster, matured together, and died close in time (leading to a noticeable cohort cycle in the adults). Juveniles in the high-density cohort grew more slowly and matured and subsequently died at different times (leading to population trajectories with less cohort structure).

Similar processes are likely to underlie the over-compensatory dynamics initially shown by cultures fed on balls relative to powder. Powdered food is spread so all individuals have access to the resources, but each one only accesses a small amount. Therefore, individuals grow relatively slowly, and because they are growing slowly, they mature at the minimum size and at different times. As females are small, they have a low fecundity. Conversely, aggregating food into balls allows some individuals to access large amounts of resources, therefore, grow fast, mature at a young age and large size and produce large numbers of eggs. This leads to a high-density, slow-growing cohort in the next generation. The cycles created by the delayed density-dependent mortality arise through a similar mechanism. Mortality decreases density, allowing a cohort of juveniles to grow fast and, therefore, mature together, creating an upturn in the cycle. At high densities, the adults monopolize food, so juveniles grow sufficiently slowly that few recruit into adulthood; mortality then outweighs recruitment and the population decreases.

The joint influence of environment (food supply and its aggregation) and competition (age structure, density, behavior) on life histories leads to population-level consequences. First, cohort effects will be common. Individuals born into "high growth conditions" will have very different life histories to those born under "low growth conditions." Individuals from the different cohorts, in turn, will respond differently to the same environmental state. The presence of strong cohort effects is likely to increase the variability in the dynamics, over and above the immediate effects of food supply and age structure, although under some circumstances it may reduce variability (Beckerman *et al.*, 2003). Second, populations which are initially close in age structure and density and are subject to the same sequence of environments are likely to diverge over time in their dynamics, as the effects of different histories interacts with and reinforces differences in age structure, leading to further differences in the way the populations respond to the noise. Third, predicting the population response to a given environment is likely to be difficult because it is so contingent. For example, the lag introduced by development time can vary by, in the extreme, an order of magnitude (minimum observed = 4 days, maximum = 95 days).

III. DISCUSSION

All organisms are exposed to temporal environmental variability in their natural habitats. To understand their ecology and evolution and make the best predictions of their future population dynamics for management, some insight is needed into the effects of noise. In modelling terms, noise creates stochasticity in parameters, which, in term changes the population dynamics, either quantitatively, by making the dynamics noisy, or qualitatively, by changing (for example, equilibrium behavior to cyclic behavior) (Bjørnstad and Grenfell, 2001). In biological terms, rather than a random fluctuation in the life history, there may be a relationship between the environmental variable (which may be predictable, periodic, or really stochastic), the way the organisms' life history responds and the resulting population dynamics. In principal, this causal chain is understandable. Only by incorporating mechanism into models will it be possible to predict how organisms may respond to changing environments, such as those brought about by, for example, climate change (Bradbury *et al.*, 2001).

Our studies of the population dynamics of the mites provide insight into the way that environmental variability affects the population dynamics from a phenomenological and mechanistic perspective. Phenomenologically, as the environmental noise increases, the mean population size decreases and the skew increases, as has been predicted from first principles (Dennis and Costantino, 1988). The data illustrate that the mean-variance relationship is not a species-specific character, as has been suggested (Taylor *et al.*, 1978), but instead depends on the way that the mean and variance of the key limiting resource varies across populations.

There has recently been considerable discussion about the relationship between the environmental noise (the "input noise") and the population dynamics (the "output"), with the organism's biology acting analogously to a "filter" transforming input to output (Pascual and Ellner, 2000; Ranta *et al.*, 2000; Laakso *et al.*, 2001). This discussion builds on an earlier one based on the observation that most biological time series have red spectra (i.e., are dominated by low-frequency fluctuations), whereas many population models have blue spectra (i.e., are dominated by high-frequency fluctuations) (Cohen, 1995). A major explanation for this phenomenon is that the addition of noise can turn a deterministically blue spectrum red (White *et al.*, 1996; Kaitala *et al.*, 1997b; Balmforth *et al.*, 1999). To add to this discussion we have shown that our near-constant cultures fed balls and powder conditions already exhibit red dynamics, and the addition of blue noise slightly reddens the dynamics further; however, the dynamics broadly retain the characteristics of the near-constant situation. This is similar to a recent study on microcosms (Laakso *et al.*, 2003). The cultures which show population cycles, however, are much more strongly affected by the addition of noise, with the single sharp peak in the near-constant spectrum being broadened considerably by noise (adults), abolished (juveniles), or split into two (eggs). Hence, the characteristics of the "filter" differ both between life stages and between the different experiments, representing manipulation of the deterministic density dependence.

Time series analysis allowed us to reconstruct elements of the deterministic density dependence which could be verified from other experimental data. The top-down approach, coupled with the confirmation from our "bottom-up" experiments, illustrates that time-series analysis can provide insight into the major biologically important factors (e.g., lags, density dependence, and so on). The ability to extract the biology from time series depends on the amount of prior knowledge that one has. In particular, existing knowledge of the life cycle can guide the search for appropriate order and dimension, and allow the fitting of semi-mechanistic models (Ellner *et al.*, 1998); time series from more than one stage can aid the investigation of how relevant vital rates change; inclusion of important environmental covariates can increase the explanatory power enormously. None of these results are new, but this is the first time that they have together been shown for a biological system where there is knowledge of the biology, forcing, and stage structure. As our time series were deliberately noisy, yet the estimated deterministic relationships (e.g., Fig. 9) still accurately matched those from non-population experiments, we can have some confidence that time-series analysis of real, unreplicated data can produce meaningful results.

The bottom-up approach illuminates the biological complexity of the system and provides considerable insight into the way individual decisions translate into population dynamics in a way that descriptions of the population dynamics (such as the spectra) or time series analysis do not. Individuals decide when to mature and at what size, depending on the their growth rate (Plaistow *et al.*, 2004). In turn, this depends on the environment the individual experiences (food supply, competition) as well as the allocation that the mother made to its egg (Benton *et al.*, in press). This decision then determines, at least partially, the individual's own reproductive strategy and, therefore, its survival and allocation to its eggs. Thus, the environment can create delayed effects in the life history over more than one generation, leading to lags in the density dependence which themselves are time-dependent. Delayed life-history effects, if they are synchronized across the population due to shared history, can lead to cohort effects and an increase in population variability (Beckerman *et al.*, 2002; Benton *et al.*, 2002; Beckerman *et al.*, 2003).

Plasticity in life-history traits and delayed life-history effects are common in many organisms (Stearns, 1992; Mousseau and Fox, 1998; Lindstrom, 1999; Beckerman *et al.*, 2002). It is likely that our experiences with the mite system, where we can analyze the population dynamics and conduct experiments at the individual level, will generalize to many real-life systems. One important lesson is that understanding the biology of the life-history is crucial to fully understanding the population dynamics. Simple models (whether they are analytically tractable, or time series models such as those previously discussed) do not incorporate enough of the biology to fully capture the effect of noise on the dynamics mainly because they treat all individuals within a stage or class as identical, when it is the differences that are important. Indeed, the historical dependencies we find can create much longer-term effects than are usually captured in simple models. In part, this explains why natural time series have a strong, low-frequency component in their variability (Bjørnstad *et al.*, 1999a) when simple models may predict shorter-term fluctuations (Cohen, 1995).

A challenge for the future will be to describe the full richness of the interaction between noise and phenomenological population models in terms of the biology. For example, the addition of noise to a model may cause a deterministically stable model to fluctuate widely (Higgins *et al.*, 1997; Blarer and Doebeli, 1999; Greenman and Benton, 2003, 2004). What is the biological mechanism that creates a bifurcation from equilibrium to cycling when a parameter is varied by a small amount? Although the mites' dynamics tend towards deterministic equilibrium behavior, we can imagine biological situations where there is excitation by noise. For example, the reaction norm for age and size at maturity provides the possibility of periodic over-compensatory dynamics. If the environment allows periods of excess food, then individuals may grow fast, trade-off age for size and have very high reproductive outputs, leading to subsequent periods of very high-density, low survival and, therefore, low competition. Details such as these, coupled with knowledge of the temporal distribution of limiting resources, may provide the biological mechanism behind the phenomenological complexity as exhibited by even the most simple dynamic model with added stochasticity.

ACKNOWLEDGMENTS

Much of the empirical work described here was undertaken at the University of Stirling: thanks to our colleagues for providing a stimulating working atmosphere. Craig Lapsley did the bulk of the work in counting the mites. NERC provided funding (grant GR3/12452 and NER/A/S/2001/00430). Sue Littlejohns analyzed the protein in the eggs. Stewart Plaistow oversaw the size and age at maturity experiment and commented on the manuscript. Two anonymous referees provided constructive feedback. Steve Ellner, Esa Ranta, Peter Hudson, and Bryan Grenfell supported this work at various times by discussion or encouragement. The first draft of this paper was written while on sabbatical leave at the NERC Centre for Population Biology, at Imperial College, Silwood Park. Thanks to Bob Desharnais and the other editors for inviting this contribution and steering it through publication.

REFERENCES

Abrams, P.A. and Matsuda, H. (1997) Prey adaptation as a cause of predator-prey cycles. *Evolution* **51**, 1742–1750.

Akcakaya, H.R., Halley, J.M. and Inchausti, P. (2003) Population-level mechanisms for reddened spectra in ecological time series. *J. Anim. Ecol.* **72**, 698–702.

Albon, S.D., Clutton-Brock, T.H. and Guinness, F.E. (1987) Early development and population-dynamics in red deer. Density-independent effects and cohort variation. *J. Anim. Ecol.* **56**, 69–81.

Anderson, R.M., Gordon, D.M., Crawley, M.J. and Hassell, M.P. (1982) Variability in the abundance of animal and plant-species. *Nature* **296**, 245–248.

Balmforth, N.J., Provenzale, A., Spiegel, E.A., Martens, M., Tresser, C. and Wu, C.W. (1999) Red spectra from white and blue noise. *Proc. R. Soc. Lond., B, Biol. Sci.* **266**, 311–314.

Beckerman, A., Benton, T.G., Ranta, E., Kaitala, V. and Lundberg, P. (2002) Population dynamic consequences of delayed life-history effects. *Trends Ecol. Evol.* **17**, 263–269.

Beckerman, A.P., Benton, T.G., Lapsley, C.T. and Koesters, N. (2002) Talking 'bout my generation: Environmental variability and cohort effects. *Am. Nat.* **162**, 754–767.

Benton, T.G. and Grant, A. (1999) Optimal reproductive effort in stochastic, density-dependent environments. *Evolution* **53**, 677–688.

Benton, T.G., Lapsley, C.T. and Beckerman, A.P. (2001a) Population synchrony and environmental variation: An experimental demonstration. *Ecol. Lett.* **4**, 236–243.

Benton, T.G., Lapsley, C.T. and Beckerman, A.P. (2002) The population response to environmental noise: Population size, variance and correlation in an experimental system. *J. Anim. Ecol.* **71**, 320–332.

Benton, T.G., Ranta, E., Kaitala, V. and Beckerman, A.P. (2001b) Maternal effects and the stability of population dynamics in noisy environments. *J. Anim. Ecol.* **70**, 590–599.

Berryman, A. and Turchin, P. (2001) Identifying the density-dependent structure underlying ecological time series. *Oikos* **92**, 265–270.

Bjørnstad, O.N., Begon, M., Stenseth, N.C., Falck, W., Sait, S.M. and Thompson, D.J. (1998) Population dynamics of the Indian meal moth: Demographic stochasticity and delayed regulatory mechanisms. *J. Anim. Ecol.* **67**, 110–126.

Bjørnstad, O.N., Fromentin, J.M., Stenseth, N.C. and Gjosaeter, J. (1999) Cycles and trends in cod populations. *Proc. Natl. Acad. Sci. USA* **96**, 5066–5071.

Bjørnstad, O.N. and Grenfell, B.T. (2001) Noisy clockwork: Time series analysis of population fluctuations in animals. *Science* **293**, 638–643.

Bjørnstad, O.N., Sait, S.M., Stenseth, N.C., Thompson, D.J. and Begon, M. (2001) The impact of specialized enemies on the dimensionality of host dynamics. *Nature* **409**, 1001–1006.

Bjørnstad, O.N., Stenseth, N.C. and Saitoh, T. (1999) Synchrony and scaling in dynamics of voles and mice in northern Japan. *Ecology* **80**, 622–637.

Blarer, A. and Doebeli, M. (1999) Resonance effects and outbreaks in ecological time series. *Ecol. Lett.* **2**, 167–177.

Bradbury, R.B., Payne, R.J.H., Wilson, J.D. and Krebs, J.R. (2001) Predicting population responses to resource management. *Trends Ecol. Evol.* **16**, 440–445.

Briggs, C.J., Sait, S.M., Begon, M., Thompson, D.J. and Godfray, H.C.J. (2000) What causes generation cycles in populations of stored-product moths? *J. Anim. Ecol.* **69**, 352–366.

Caswell, H. (2001) *Matrix Population Models*. Sinauer Associates Inc., Sunderland, MA.

Clutton-Brock, T.H. and Coulson, T. (2002) Comparative ungulate dynamics: The devil is in the detail. *Philos. Trans. R. Soc. Lond., B, Biol. Sci.* **357**, 1285–1298.

Cohen, J.E. (1995) Unexpected dominance of high-frequencies in chaotic nonlinear population-models. *Nature* **378**, 610–612.

Coulson, T., Catchpole, E.A., Albon, S.D., Morgan, B.J.T., Pemberton, J.M., Clutton-Brock, T.H., Crawley, M.J. and Grenfell, B.T. (2001) Age, sex, density, winter weather, and population crashes in Soay sheep. *Science* **292**, 1528–1531.

Crone, E.E. (1997) Parental environmental effects and cyclical dynamics in plant populations. *Am. Nat.* **150**, 708–729.

Cushing, J.M., Dennis, B., Desharnais, R.A. and Costantino, R.F. (1998) Moving toward an unstable equilibrium: Saddle nodes in population systems. *J. Anim. Ecol.* **67**, 298–306.

Dennis, B. and Costantino, R.F. (1988) Analysis of steady-state populations with the gamma-abundance model: Application to tribolium. *Ecology* **69**, 1200–1213.

Dennis, B. and Taper, M.L. (1994) Density dependence in time-series observations of natural-populations—estimation and testing. *Ecol. Monogr.* **64**, 205–224.

Diekmann, O., Mylius, S.D. and ten Donkelaar, J.R. (1999) Saumon a la Kaitala et Getz, sauce hollandaise. *Evol. Ecol. Res.* **1**, 261–275.

Ebenman, B., Johansson, A., Jonsson, T. and Wennergren, U. (1996) Evolution of stable population dynamics through natural selection. *Proc. R. Soc. Lond., B, Biol. Sci.* **263**, 1145–1151.

Ellner, S. and Turchin, P. (1995) Chaos in a noisy world—new methods and evidence from time-series analysis. *Am. Nat.* **145**, 343–375.

Ellner, S.P., Bailey, B.A., Bobashev, G.V., Gallant, A.R., Grenfell, B.T. and Nychka, D.W. (1998) Noise and nonlinearity in measles epidemics: Combining mechanistic and statistical approaches to population modeling. *Am. Nat.* **151**, 425–440.

Ellner, S.P., Kendall, B.E., Wood, S.N., McCauley, E. and Briggs, C.J. (1997) Inferring mechanism from time-series data: Delay-differential equations. *Physica D* **110**, 182–194.

Forchhammer, M.C., Clutton-Brock, T.H., Lindstrom, J. and Albon, S.D. (2001) Climate and population density induce long-term cohort variation in a northern ungulate. *J. Anim. Ecol.* **70**, 721–729.

Fox, G.A. and Kendall, B.E. (2002) Demographic stochasticity and the variance reduction effect. *Ecology* **83**, 1928–1934.

Fromentin, J.M., Myers, R.A., Bjørnstad, O.N., Stenseth, N.C., Gjosaeter, J. and Christie, H. (2001) Effects of density-dependent and stochastic processes on the regulation of cod populations. *Ecology* **82**, 567–579.

Ginzburg, L.R. (1998) Inertial growth—population dynamics based on maternal effects. In: *Maternal Effects as Adaptations* (Ed. by T.A. Mousseau and C.W. Fox), pp. 42–53. Oxford University Press, Oxford.

Ginzburg, L.R. and Taneyhill, D.E. (1994) Population cycles of forest Lepidoptera—a maternal effect hypothesis. *J. Anim. Ecol.* **63**, 79–92.

Greenman, J.V. and Benton, T.G. (2001) The impact of stochasticity on the behaviour of nonlinear population models: Synchrony and the Moran effect. *Oikos* **93**, 343–351.

Greenman, J.V. and Benton, T.G. (2003) The amplification of environmental noise in population models: Causes and consequences. *Am. Nat.* **161**, 225–239.

Greenman, J.V. and Benton, T.G. (2004) Large amplification in stage-structured models: Arnold's tongues revisited. *J. Math. Biol.* **48**, 647–671.

Grenfell, B.T., Finkenstadt, B.F., Wilson, K., Coulson, T.N. and Crawley, M.J. (2000) Ecology—nonlinearity and the Moran effect. *Nature* **406**, 847.

Grenfell, B.T., Wilson, K., Finkenstadt, B.F., Coulson, T.N., Murray, S., Albon, S. D., Pemberton, J.M., Clutton-Brock, T.H. and Crawley, M.J. (1998) Noise and determinism in synchronized sheep dynamics. *Nature* **394**, 674–677.

Grenouillet, G., Hugueny, B., Carrel, G.A., Olivier, J.M. and Pont, D. (2001) Large-scale synchrony and inter-annual variability in roach recruitment in the Rhone river: The relative role of climatic factors and density-dependent processes. *Freshw. Biol.* **46**, 11–26.

Hallett, T.B., Coulson, T., Pilkington, J.G., Clutton-Brock, T.H., Pemberton, J.M. and Grenfell, B.T. (2004) Why large-scale climate indices seem to predict ecological processes better than local weather. *Nature* **430**, 71–75.

Henson, S.M. (2000) Multiple attractors and resonance in periodically forced population models. *Physica D* **140**, 33–49.

Henson, S.M., King, A.A., Costantino, R.F., Cushing, J.M., Dennis, B. and Desharnais, R.A. (2003) Explaining and predicting patterns in stochastic population systems. *Proc. R. Soc. Lond., B, Biol. Sci.* **270**, 1549–1553.

Higgins, K., Hastings, A., Sarvela, J.N. and Botsford, L.W. (1997) Stochastic dynamics and deterministic skeletons: Population behavior of Dungeness crab. *Science* **276**, 1431–1435.

Inchausti, P. and Ginzburg, L.R. (1998) Small mammals cycles in northern Europe: Patterns and evidence for a maternal effect hypothesis. *J. Anim. Ecol.* **67**, 180–194.

Jonzen, N., Ripa, J. and Lundberg, P. (2002) A theory of stochastic harvesting in stochastic environments. *Am. Nat.* **159**, 427–437.

Kaitala, V., Mappes, T. and Ylonen, H. (1997a) Delayed female reproduction in equilibrium and chaotic populations. *Evol. Ecol.* **11**, 105–126.

Kaitala, V., Ylikarjula, J., Ranta, E. and Lundberg, P. (1997b) Population dynamics and the colour of environmental noise. *Proc. R. Soc. Lond., B, Biol. Sci.* **264**, 943–948.

Keeling, M.J. (2000) Simple stochastic models and their power-law type behavior. *Theor. Popul. Biol.* **58**, 21–31.

Kendall, B.E., Bjørnstad, O.N., Bascompte, J., Keitt, T.H. and Fagan, W.F. (2000) Dispersal, environmental correlation, and spatial synchrony in population dynamics. *Am. Nat.* **155**, 628–636.

Kendall, B.E., Briggs, C.J., Murdoch, W.W., Turchin, P., Ellner, S.P., McCauley, E., Nisbet, R.M. and Wood, S.N. (1999) Why do populations cycle? A synthesis of statistical and mechanistic modeling approaches. *Ecology* **80**, 1789–1805.

Kilpatrick, A.M. and Ives, A.R. (2003) Species interactions can explain Taylor's power law for ecological time series. *Nature* **422**, 65–68.

Laakso, J., Kaitala, V. and Ranta, E. (2001) How does environmental variation translate into biological processes? *Oikos* **92**, 119–122.

Laakso, J., Loytynoja, K. and Kaitala, V. (2003) Environmental noise and population dynamics of the ciliated protozoan *Tetrahymena thermophila* in aquatic microcosms. *Oikos* **102**, 663–671.

Langvatn, R., Albon, S.D., Burkey, T. and Clutton-Brock, T.H. (1996) Climate, plant phenology and variation in age of first reproduction in a temperate herbivore. *J. Anim. Ecol.* **65**, 653–670.

Lee, S.S. (1996) On a class of nonlinear time series models for biological population abundance data. *Appl. Stochastic Models and Data Analysis* **12**, 193–207.

Lima, M., Keymer, J.E. and Jaksic, F.M. (1999) El Nino-southern oscillation-driven rainfall variability and delayed density dependence cause rodent outbreaks in western South America: Linking demography and population dynamics. *Am. Nat.* **153**, 476–491.

Lindstrom, J. (1999) Early development and fitness in birds and mammals. *Trends Ecol. Evol.* **14**, 343–348.

Lingjaerde, O.C., Stenseth, N.C., Kristoffersen, A.B., Smith, R.H., Moe, S.J., Read, J.M., Daniels, S. and Simkiss, K. (2001) Exploring the density-dependent structure of blowfly populations by nonparametric additive modeling. *Ecology* **82**, 2645–2658.

Lundberg, P., Ripa, J., Kaitala, V. and Ranta, E. (2002) Visibility of demography-Modulating noise in population dynamics. *Oikos* **96**, 379–382.

MacArthur, R.H. and Wilson, E.O. (1967) *The Theory of Island Biogeography*. Princeton University Press, Princeton.

Metcalfe, N.B. and Monaghan, P. (2001) Compensation for a bad start: Grow now, pay later? *Trends Ecol. Evol.* **16**, 254–260.

Moe, S.J., Stenseth, N.C. and Smith, R.H. (2002) Density dependence in blowfly populations: Experimental evaluation of non-parametric time-series modelling. *Oikos* **98**, 523–533.

Moran, P.A.P. (1953) The statistical analysis of the Canadian lynx cycle. II. Synchronization and meteorology. *Aust. J. Zoo.* **1**, 291–298.

Mousseau, T.A. and Fox, C.W. (1998) *Maternal Effects as Adaptations*. Oxford University Press, Oxford.

Nisbet, R.M. and Gurney, W.S.C. (1976) A simple mechanism for population cycles. *Nature* **263**, 319–320.

Nisbet, R.M. and Onyiah, L.C. (1994) Population dynamic consequences of competition within and between age classes. *J. Math. Biol.* **32**, 329–344.

Ortega, G.J. and Louis, E. (2000) Using topological statistics to detect determinism in time series. *Physical Review E* **62**, 3419–3428.

Orzack, S.H. (1993) Life history evolution and population dynamics in variable environments: Some insights from stochastic demography. In: *Adaptation in Stochastic Environments* (Ed. by J. Yoshimura and C.W. Clark), pp. 63–104. Springer Verlag, Berlin.

Pascual, M. and Ellner, S.P. (2000) Linking ecological patterns to environmental forcing via nonlinear time series models. *Ecology* **81**, 2767–2780.

Perry, J.N. (1994) Chaotic dynamics can generate Taylors power-law. *Proc. R. Soc. Lond., B, Biol. Sci.* **257**, 221–226.

Plaistow, S. and Siva-Jothy, M.T. (1999) The ontogenetic switch between odonate life history stages: Effects on fitness when time and food are limited. *Anim. Behav.* **58**, 659–667.

Plaistow, S.J., Lapsley, C.T., Beckerman, A.P. and Benton, T.G. (2004) Age and size at maturity sex, environmental variability, and developmental thresholds. *Proc. R. Soc. Lond., B, Biol. Sci.*

Ranta, E., Kaitala, V. and Lundberg, P. (1998) Population variability in space and time: The dynamics of synchronous population fluctuations. *Oikos* **83**, 376–382.

Ranta, E., Lundberg, P., Kaitala, V. and Laakso, J. (2000) Visibility of the environmental noise modulating population dynamics. *Proc. R. Soc. Lond., B, Biol. Sci.* **267**, 1851–1856.

Reid, J.M., Bignal, E.M., Bignal, S., McCracken, D.I. and Monaghan, P. (2003) Environmental variability, life-history covariation and cohort effects in the red-billed chough Pyrrhocorax pyrrhocorax. *J. Anim. Ecol.* **72**, 36–46.

Richards, S.A. and Wilson, W.G. (2000) Adaptive feeding across environmental gradients and its effect on population dynamics. *Theor. Popul. Biol.* **57**, 377–390.

Roff, D.A. (2002) *Life History Evolution*. Sinauer Associates Inc., Sunderland, MA.

Rose, K.E., Clutton-Brock, T.H. and Guinness, F.E. (1998) Cohort variation in male survival and lifetime breeding success in red deer. *J. Anim. Ecol.* **67**, 979–986.

Royama, T. (1992) *Analytical Population Dynamics*. Chapman & Hall, London.

Saether, B.E. (1997) Environmental stochasticity and population dynamics of large herbivores: A search for mechanisms. *Trends Ecol. Evol.* **12**, 143–149.

Smith, R.H., Daniels, S., Simkiss, K., Bell, E.D., Ellner, S.P. and Forrest, M.B. (2000) Blowflies as a case study in non-linear population dynamics. In: *Chaos in Real Data: The Analysis of Non-Linear Dynamics From Short Ecological Time Series* (Ed. by J.N. Perry, RH. Smith, I.P. Woiwod and D.R. Morse), pp. 137–172. Kluwer Academic Publishers, Dordrecht.

Stearns, S.C. (1992) *The Evolution of Life Histories*. Oxford University Press, Oxford.

Stearns, S.C. (2000) Life history evolution: Successes, limitations, and prospects. *Naturwissenschaften* **87**, 476–486.

Stenseth, N.C. (1999) Population cycles in voles and lemmings: Density dependence and phase dependence in a stochastic world. *Oikos* **87**, 427–461.

Stenseth, N.C., Bjørnstad, O.N., Falck, W., Fromentin, J.M., Gjosaeter, J. and Gray, J.S. (1999) Dynamics of coastal cod populations: Intra- and intercohort density dependence and stochastic processes. *Proc. R. Soc. Lond., B, Biol. Sci.* **266**, 1645–1654.

Stenseth, N.C., Falck, W., Chan, K.S., Bjørnstad, O.N., O'Donoghue, M., Tong, H., Boonstra, R., Boutin, S., Krebs, C.J. and Yoccoz, N.G. (1998) From patterns to processes: Phase and density dependencies in the Canadian lynx cycle. *Proc. Nat. Acad. Sci. USA* **95**, 15430–15435.

Taylor, L.R. (1961) Aggregation, variance and the mean. *Nature* **189**, 732–735.

Taylor, L.R. and Woiwod, I.P. (1982) Comparative synoptic dynamics. 1. Relationships between interspecific and intraspecific spatial and temporal variance mean population parameters. *J. Anim. Ecol.* **51**, 879–906.

Taylor, L.R., Woiwod, I.P. and Perry, J.N. (1978) The density-dependence of spatial behaviour and the rarity of randomness. *J. Anim. Ecol.* **47**, 383–406.

Titmus, G. (1983) Are animal populations really aggregated? *Oikos* **40**, 64–68.

Turchin, P. and Ellner, S.P. (2000) Living on the edge of chaos: Population dynamics of Fennoscandian voles. *Ecology* **81**, 3099–3116.

White, A., Begon, M. and Bowers, R.G. (1996) Explaining the colour of power spectra in chaotic ecological models. *Proc. R. Soc. Lond., B, Biol. Sci.* **263**, 1731–1737.

Witting, L. (2000) Population cycles caused by selection by density dependent competitive interactions. *Bull. Math. Biol.* **62**, 1109–1136.

Wood, S.N. (2001) Partially specified ecological models. *Ecol. Monogr.* **71**, 1–25.

Global Persistence Despite Local Extinction in Acarine Predator-Prey Systems: Lessons From Experimental and Mathematical Exercises

MAURICE W. SABELIS, ARNE JANSSEN, ODO DIEKMANN, VINCENT A.A. JANSEN, ERIK VAN GOOL AND MINUS VAN BAALEN

I. SUMMARY

Empirical studies show that local populations of predator and prey mites on plants frequently become extinct, because of both intrinsic local dynamics and traveling waves of predators that overwhelm local prey populations. Despite frequent local extinctions, the predator-prey metapopulation persists and exhibits either large-amplitude cycles or small-amplitude aperiodic fluctuations. This chapter reviews how our understanding of predator and prey metapopulation dynamics in mites has been advanced by comparing laboratory experiments with both a "baseline" model and simulation models that break down the assumptions of the "baseline" model. The most critical

ADVANCES IN ECOLOGICAL RESEARCH VOL. 37 0065-2504/05 $35.00

DOI: 10.1016/S0065-2504(04)37006-6

assumptions are that: 1) habitat structure matters, but spatial configuration does not; and 2) local dynamics of prey and predator are deterministic, fully determined by the first colonization and always end in prey extinction after a single oscillation, followed by predator dispersal. The "baseline" model predicts multiple stable states, either with or without predators, when there are many patches that harbor only few prey. Stable predator-prey cycles surface, however, when the number of patches is low and can harbor many prey. Virtually all acarine predator-prey metapopulation experiments published to date have been done under the latter conditions and show metapopulation-level cycles that are either periodic with large amplitudes or aperiodic with small amplitudes. Simulations suggest the large amplitude cycles arise when survival in the dispersal phase is low, whereas the small amplitude cycles surface when survival in the dispersal phase is relatively higher, yet low enough to allow persistence of predators and prey at the metapopulation level. The "baseline" model is further analyzed to detect which factors cause persistence or extinction at the metapopulation level. Persistence was promoted by prey dispersal from predator-occupied patches, time spent by prey in the dispersal phase, interception of predators in predator-occupied patches, and heterogeneous colonization risks among patches. Extinction is promoted by predator dispersal after extermination of prey in a patch and time spent by predators in the dispersal phase. The relative importance of all of these factors is yet to be determined. Moreover, the relevance of spatial patterning of predator-prey interactions is much debated. Model-based simulations to analyze an experimentally established time series suggest that what matters to persistence is reduced discovery of prey patches, not the spatial configuration per se. However, the last word has not been said about this, since spatial predator-prey models suggest configuration to play an important role.

The "baseline" model is not evolutionary robust, in that its structure arises from individual traits that are not favored by natural selection. It is based on the assumption that first colonization is the single determinant of local predator or prey dynamics and that predator dispersal is delayed until after prey extermination. Delayed predator dispersal cannot evolve, however, when only the initial colonization of prey-only patches by a predator is taken into account. It also does not evolve in simulations with a model, parameterized for the acarine predator-prey system, in which not only the first colonization but also subsequent invasions are taken into account. However, observations on the behavior of field-collected predators show that delayed dispersal is the rule rather than the exception. We argue that delayed predator dispersal prevails in the field because local populations are subject to catastrophes which give predators little control over the exploitation of their local prey population. Hence, natural selection in the field will favor predators that exploit the local prey population as fast as possible and delay dispersal until after prey extermination.

The important lesson to be drawn from this review is that the insights came from comparing a suite of models that all differed in key assumptions from a comprehensive yet formal "baseline" model and from testing these models against laboratory experiments. Understanding the complex dynamics observed in acarine predator-prey metapopulations requires more than spotting an arbitrary caricatural model with a dynamic behavior that corresponds qualitatively to that of the biological system.

II. INTRODUCTION

Classical population ecology received its impetus from mathematical models that assumed individuals to be identical, populations to be well mixed, and interactions to obey mass action. In the field, populations may come close to these assumptions at a sufficiently small spatial scale, but at large spatial scales they exhibit patchy distributions where individuals interact strongly within patches, but weakly among patches. This led to the question of whether the patchy population structure and dispersal among patches add new properties to the dynamics of the overall population. The ensemble of local populations inhabiting patches is called the metapopulation (e.g., Hanski and Gilpin, 1997). Due to the spatial scale at which metapopulation structure becomes manifest in the field, their study poses logistic problems and it is even more difficult to carry out replicated experiments. Hence, there is a need to select model organisms small enough to exhibit a patchy structure at a manageable spatial scale and preferably in the laboratory. Apart from microbes (e.g., Holyoak and Lawler, 1996a,b), mites (Acari) and other small arthropods (e.g., Bonsall *et al.*, 2002) have become popular as model organisms for metapopulation studies in the laboratory.

Interestingly, the first metapopulation experiments were done at a time where classical models for unstructured populations were mainstream in ecological science. It was Carl Huffaker, who in 1958 published the results of his laboratory experiments with acarine predator-prey systems in a labyrinth of interconnected oranges (Huffaker, 1958; Huffaker *et al.*, 1963). These publications paved the way for what became known much later as metapopulation theory (e.g., Hanski and Gilpin, 1997; Keeling, 2002). In Huffaker's experiments, the habitat is given a patchy structure and patches may be colonized by prey and predators that have dispersed from other patches. Local populations may come close to the assumptions of classic predator-prey models, yet they are loosely connected to populations on patches elsewhere in the habitat. Thus, local predator-prey populations may exhibit strong coupling, but (except for dispersal-colonization events) their dynamics are largely decoupled from dynamic processes in other local populations. This may give rise to asynchrony among local population fluctuations and empty patches may be colonized by dispersers from some out-of-phase population elsewhere.

To conclude that asynchrony in the dynamics of local populations explains metapopulation persistence, however, is no more than a truism. If local populations are doomed to become extinct, for example, then asynchrony in these events is an alternative way to state that the ensemble of local populations persists. The crucial questions are where the asynchrony comes from in the first place and how it is maintained. Huffaker was ahead of his time. In his experiments with predatory mites and herbivorous mites on oranges, he noted that local populations frequently became extinct, either because herbivorous mites overexploit their local resource (an orange) or predatory mites overexploited the local population of herbivorous mites. Thus, patches once occupied by herbivores alone or by herbivores and predators became empty and then colonized again, first by a dispersing herbivore and later by a predator. This led to the so-called "hide-and-seek" hypothesis which explains global predator-prey persistence in the face of local extinction. Huffaker was well aware that to make this work special conditions on dispersal rates are required. Local outbreaks of predatory mites or herbivorous mites create an outburst of dispersers followed by interpatch movement, and tends to synchronize the phase of nearby local populations. Severe local outbreaks, therefore, may have the capacity to synchronize all local populations which may drive the predator-prey system as a whole to extinction. Extinction may also result when the local population size and the number of dispersers is too low to make colonizations balance extinctions. Thus, low connectedness between patches promotes "hide-and-seek," if local populations are sufficiently small, yet it is not a guarantee for escape of prey in time and space when local populations become too large. Huffaker clearly realized that global predator-prey persistence requires upper bounds to local populations and a degree of patch-to-patch connectedness which, given local population size, mediates dispersal rates that are neither too low nor too high. Thus, the conditions for metapopulation persistence seem to be restrictive; this prompted much research to identify these conditions experimentally and theoretically (Hanski and Gilpin, 1997).

Testing theories about metapopulation persistence, requires long-term experiments in systems with sufficient spatial heterogeneity. Huffaker *et al.* (1963) showed that, due to their small size and short generation time, plant-inhabiting mites are convenient for carrying out and for replicating such experiments under laboratory conditions. However, for the same reasons, the mites are subject to natural selection at the time scale of a metapopulation experiment. The environment created by the experimenter or feedback between mites and their environment may lead to selection for novel mite traits. In this way, experimental populations may change under the hands of the experimenter! This leads to questions about which individual traits relevant to metapopulation persistence are altered by natural selection and whether it is justified to assume that the time scale of (meta-)population dynamics is separated from the time scale of selection. A separation of time scales is an implicit assumption in most models and interpretations of experimental data (Mueller and Joshi, 2000). Therefore, it is pertinent to ask whether experimental outcomes and model predictions are robust against natural selection.

Here, we will show that to gain insight into metapopulation persistence a combination of theoretical and experimental approaches is required. The art of modeling is to include all relevant features and details of metapopulations, yet retain a formal mathematical representation. To do so we advocate a "baseline"-model approach guided by experimental observations. The "baseline" model serves to provide a mathematical representation of the full processes which may be at the expense of analytical tractability. The advantage of a comprehensive, mathematical description is that simplified models can be formally derived from the "baseline" model. In this way, the "baseline" model can then be tailored to answer specific questions. As we hope to show, it is in the way model simplifications are sought that experimental tests become meaningful.

In this chapter, we will first consider the main features of local dynamics in an acarine predator-prey system. Next, we will ask which of these features are crucial to the understanding of metapopulation dynamics. Does it suffice to represent local dynamics in terms of presence/absence of predators and prey, or is it crucial to incorporate the local dynamic processes? To answer this question, we will review the outcomes of a hierarchy of models of increasing complexity. Next, we review some of the predictions that have been experimentally tested, and identify some key predictions that still await rigorous tests.

Finally, we consider not only whether the models under consideration are robust against structural changes, but also whether they are evolutionarily robust; this means that the models maintain essentially the same structure when exposed to long-term natural selection (e.g., evolutionarily stable strategy [ESS]). We show how experimental tests play a crucial role in deciding how to comply with evolutionary robustness. The chapter closes with a review of the most pertinent lessons drawn from combining theory and experiment.

III. ZOOMING IN ON LOCAL PREDATOR-PREY DYNAMICS

Empirical observations may help identify the relevant features of local predator-prey dynamics. In this chapter, we focus on plant-dwelling mites as an example. Relative to the plants they inhabit, they are very small, approximately 0.5 mm in length, and have short generation times (approximately 1 to 2 weeks). The prey is the herbivorous mite, *Tetranychus urticae* (Acari: Tetranychidae) and the predator is also a mite, belonging to one of the following two species: *Phytoseiulus persimilis, Metaseiulus occidentalis* (Acari: Phytoseiidae). The herbivorous mite can feed on many different plant species, but the predatory mites only feed on herbivorous mites. Their life histories and interactions by predation have been extensively studied (Sabelis, 1986, 1990a, b, 1992; Janssen and Sabelis, 1992; Sabelis and Janssen, 1994).

As is the case for many plant-inhabiting arthropods, herbivorous mites exhibit a patchy distribution. As shown schematically in Fig. 1, a patch consists of a group of leaves that are infested by herbivorous mites and that are sufficiently connected to each other via petioles, stems, branches, or trunks, which makes the time scale of inter-leaf ambulatory dispersal smaller than the time scale of relevant intra-leaf processes (reproduction, development, predation). Thus, usually a patch represents the infested part of a single plant, but if leaves of neighboring plants touch each other (thereby facilitating ambulatory dispersal), a patch may cover several plants.

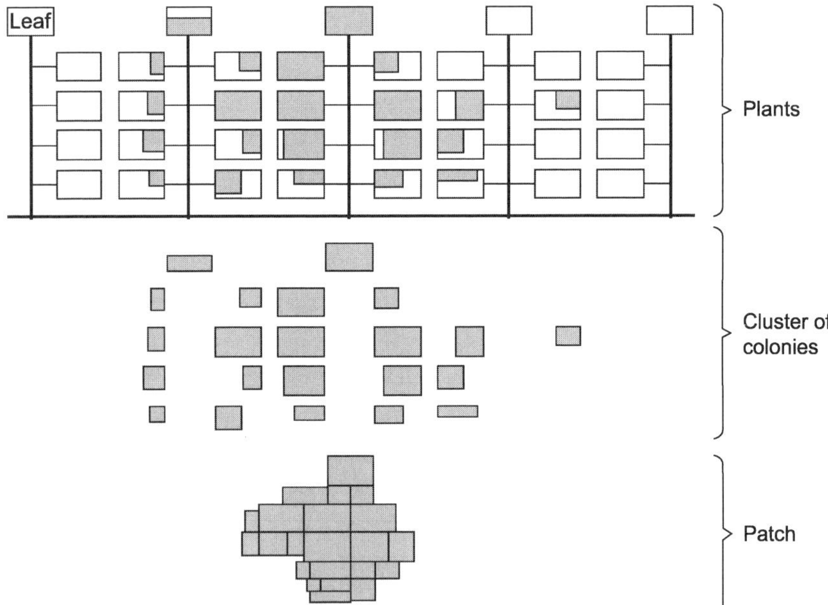

Figure 1 A patchy infestation of small herbivorous arthropods in a row of plants (top panel; shown from the side). It is inspired by observations on spider mites, but in essence applies to many other herbivorous arthropods. The infested leaf parts (shaded leaf areas) are excised (middle panel) and then put together as a jigsaw puzzle (bottom panel). The latter forms the patch or arena within which the interaction between predator and prey occurs (assuming negligible time spent in moving between infested leaves).

A local population on a patch is usually initiated by one or a few colonizing females. Since long-range dispersal is risky, the colonizers and their offspring tend to stay on the plant where they feed on leaf parenchym cells for growth, maintenance, or egg production. Their populations double in size every 2 to 4 days. Ultimately, the population of herbivorous mites-covers all leaves of the plant(s) and exhausts the plant as a food source. By this time, they take advantage of their small size, become airborne, and drift on air currents. In this way, they disperse passively over longer distances than they would cover by ambulatory dispersal. Apart from exhaustion of the food source, dispersal away from the local population can also be promoted by the presence of predatory mites (Pallini *et al.*, 1999).

Predatory mites have much of the same features. There are usually one or a few colonizing females that give rise to local populations which double in size every 2 to 4 days. Ultimately, the predator populations become so large and the predation pressure so intense, that the population of herbivorous mites on a patch is completely erased (Sabelis *et al.*, 1983). A detailed study of the local predator-prey dynamics showed that extermination of herbivorous mites is the general outcome for a large range of initial predator and prey numbers (Sabelis and van der Meer, 1986). This is surprising because these studies were carried out in wind tunnels providing the right conditions for aerial dispersal, and one would expect the predatory mites to disperse when the density of herbivorous mites declines. Instead, the predatory mites remained in the patch and did not initiate aerial dispersal until after the herbivorous mites were exterminated (Sabelis and van der Meer, 1986; Fig. 2a). Similar laboratory studies with field-collected strains of predatory mites yielded the same results (Fig. 2b) and, in addition, showed that after prey extermination some juvenile predators reach maturity by cannibalism, whereas others disperse possibly due to the threat of being cannibalized (Pels and Sabelis, 1999). The important lesson is that overexploitation of herbivores by predators, as well as overexploitation of the plant by the herbivores, are intrinsic properties of the acarine predator-prey system under consideration. Local extinctions, like those observed by Huffaker (1958) and Huffaker *et al.* (1963), need not necessarily result from local outbreaks followed by waves of dispersers synchronizing local dynamics on neighbor patches.

IV. A CARICATURE OF LOCAL PREDATOR-PREY DYNAMICS

Given the intrinsic tendency for local overexploitation and extinction, the question is how to model local predator-prey dynamics in the simplest and most essential way. Simulations with a parameterized computer model and experimental tests indicated that populations of herbivorous and

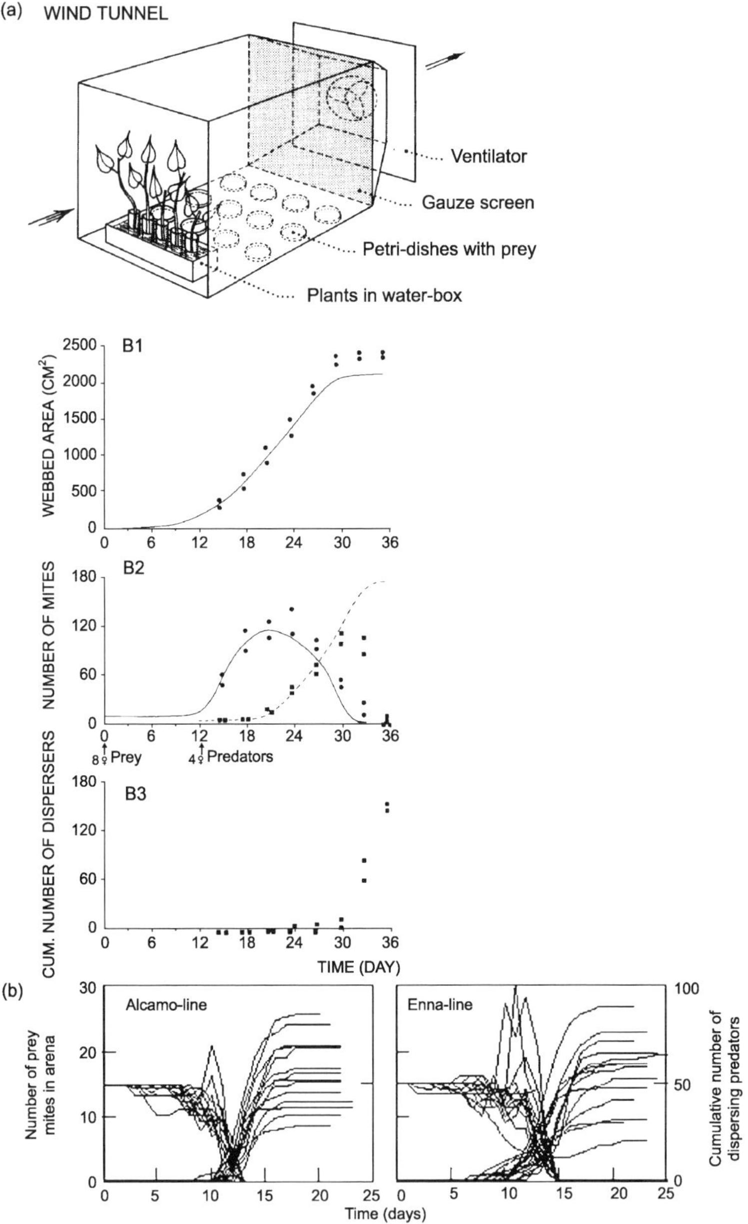

Figure 2 Extinction-prone, local predator-prey dynamics on a plant in a wind tunnel: (a) experiments of Sabelis and Van der Meer (1986), (b) experiments of Pels and Sabelis (1999). These experiments all show that predators delay dispersal until after local prey extermination.

predatory mites exhibit deterministic population growth and are strongly coupled (Sabelis and van der Meer, 1986). This is further strengthened by the experimental observation that predatory mites tend to space out over the leaf area infested by spider mites, because they avoid odors emanating from herbivores that are being attacked by other predatory mites (Janssen *et al.*, 1997a). The simulations also showed that predation rates are rather constant over much of the interaction period and only decline during a relatively short period in which the predator population peaks and the prey population crashes. This is because herbivorous mites create high densities per unit of infested leaf area, causing the functional response of the predator to prey density to settle in the plateau phase. The predation rate is then simply the product of the maximum predation rate and the number of predators in the patch. Under stable age distributions and age-independent predation risk, two linear differential equations suffice to describe local predator-prey dynamics (Metz and Diekmann, 1986; Diekmann *et al.*, 1988, 1989; Janssen and Sabelis, 1992; Sabelis, 1992):

$$\begin{aligned} \frac{dx}{dt} &= \alpha x - \beta y \\ \frac{dy}{dt} &= \gamma y. \end{aligned} \tag{1}$$

This model differs from a Lotka-Volterra model in that the predation and predator reproduction rates are constant (i.e., independent of prey density). Thus, the predation process resembles that of eating a pancake: each bite provides the same amount of food until suddenly the pancake is fully consumed. Hence, the model in equation (1) is called the "pancake predator" model. Here, t = the time elapsed since predator invasion, $x(t)$ = number of prey at time t, $y(t)$ = number of predators at time t, α = rate of prey population growth, β = maximum predation rate, and γ = rate of predator population growth. Integration yields:

$$\begin{aligned} x(t) &= x(0)e^{\alpha t} - y(0)\frac{\beta}{\gamma - \alpha}(e^{\gamma t} - e^{\alpha t}) \\ y(t) &= y(0)e^{\gamma t}. \end{aligned}$$

By the time the prey becomes extinct ($x(t = \tau) = 0$), the predator-prey interaction ends and all $y(\tau)$ predators are to disperse (hence cannibalism is ignored). Therefore, prey dispersal should occur before τ, but the term expressing this process is here ignored (or rather considered part of the net prey growth rate). The time from predator invasion to prey extinction (τ) is a function of parameters and initial numbers:

$$\tau = \frac{1}{\gamma - \alpha} \ln \left[1 + \frac{\gamma - \alpha}{\beta} \frac{x(0)}{y(0)} \right].$$

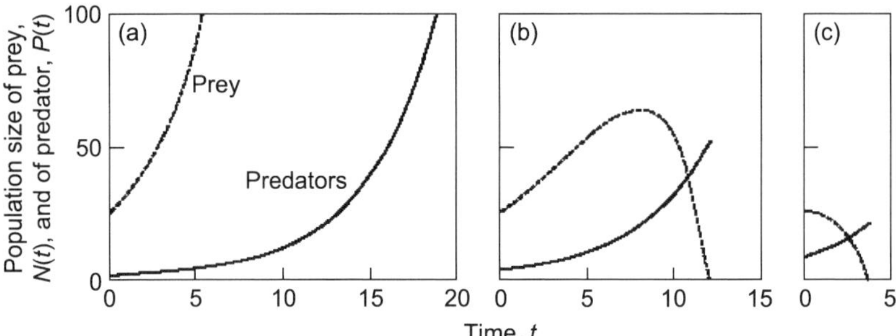

Figure 3 Three types of local predator-prey dynamics (solid line for predators and dashed line for prey) according to the pancake predation model (with parameters $\alpha = 0.3$, $\beta = 1$, $\gamma = 0.25$, $\mu = 0$, and $x(0) = 25$): (a) prey increase ($y(0) = 1$), (b) delayed prey decline ($y(0) = 3$), and (c) immediate prey decline ($y(0) = 8$). The general conditions for each of these types of dynamics are discussed in the text.

Three types of dynamic behavior of the prey population may occur (Figs. 3a, 3b, and 3c, respectively):

a. continuous increase (yet slower than α due to predation; predators surf on a prey wave), if τ is infinite and thus $y(0)/x(0) \leq (\alpha - \gamma)/\beta$;
b. initial increase and then (delayed) decline until prey extinction;
c. immediate decline until prey extinction, if $\alpha x(0) - \beta\, y(0) < 0$ or $y(0)/x(0) > \alpha/\beta$

In case (a), predators cannot suppress local prey outbreaks, but in cases (b) and (c) they eliminate the prey population and grow exponentially until τ when all prey are eaten and all predators disperse. In case (a), the prey population will inevitably reach a maximum (x_{max}) where the plant as a food source is exhausted. In the other two cases, this may also happen, but here this will be ignored, because it creates more complexity than insight in metapopulation models (e.g., Metz and Diekmann, 1986).

V. ZOOMING OUT TO THE METAPOPULATION LEVEL: THE "BASELINE" MODEL

For a mathematical description of metapopulation and local processes, some specific assumptions must be made. Newly founded prey populations start at $x = 1$ and newly founded predator populations at $y = 1$; all later arrivals are ignored, because their contribution to the fast-growing local population is negligible. Given a deterministic "pancake predator" model of local dynamics (equation 1), trajectories in the (x,y) plane are unique since prey-only patches start at (1,0) and predator-and-prey patches start at $(x,1)$. The modeling essentially involves bookkeeping of an ensemble of patches which are colonized at various times and describing the changes in the frequency of patches containing x prey and y predators, using the functions $n(t,x)$ to describe the frequency of prey-only patches and the function $m(t,x,y)$ for prey-and-predator patches. This is accomplished in a way analogous to modeling the transport of a heap of sand on a conveyor belt (Metz and Diekmann, 1986; Sabelis and Laane, 1986; Diekmann *et al.*, 1988, 1989).

Assuming mass action kinetics, the influx of new prey-only patches is in proportion to the product of the number of prey dispersers ($P(t)$) and empty patches ($n_0(t)$), with coefficient ζ. The balance at the invasion boundary $x = 1$ now yields:

$$\alpha n(t, 1) = \zeta P(t) n_o(t). \tag{2}$$

Given this boundary condition, changes in $n(t,x)$ result from changes due to prey growth and a shift from a prey-only to prey-and-predator state of the patch:

$$\frac{\partial n(t, x)}{\partial t} = -\frac{\partial(\alpha n(t, x))}{\partial x} - \eta(x) Q(t) n(t, x). \tag{3}$$

Assuming mass action kinetics, the outflux to a prey-and-predator state equals the product of the number of predator dispersers ($Q(t)$), the number of prey patches with x prey ($n(t,x)$) and a proportionality function describing the attractivity of a prey patch to predators in relation to the number of prey on a patch ($\eta(x)$). This outflux equals the influx of new prey-and-predator patches that start at the invasion boundary $y = 1$:

$$(\alpha x - \beta) m(t, x, 1) = \eta(x) Q(t) n(t, x). \tag{4}$$

Given this boundary condition, changes in the density function for predator-and-prey patches, $m(t,x,y)$, result from changes due to net prey growth and predator growth:

$$\frac{\partial m(t, x, y)}{\partial t} = -\frac{\partial((\alpha x - \beta y) m(t, x, y))}{\partial x} - \frac{\partial(\gamma y m(t, x, y))}{\partial y}. \tag{5}$$

The dynamics of the dispersers of prey and predators (P and Q) are governed by two differential equations, expressing four terms: (1) an increase due to dispersal from prey- or predator-occupied patches, (2) a decrease due to mortality in the dispersal phase, (3) a decrease due to successful colonization, and (4) a decrease due to interception. The term "interception" is used to express the removal of dispersers from the dispersal pool due to arrivals in already colonized patches. In the default version of the model, only term (1) and (2) are taken into account, thus assuming negligible loss due to colonization and interception, as in (3) and (4). The default version also assumes the number of free patches to be constant, whereas extended versions may include logistic growth of plants providing free patches.

VI. DERIVING A HIERARCHY OF MODELS FROM THE "BASELINE" MODEL

The "baseline" model described above (equations 2 to 5) is hard to analyze because it has an infinite number of dimensions and is parameter rich. However, it is a formal mathematical representation of the dynamic process. Hence, it can serve as a template for studying limiting cases and for applying time scale arguments (and consequently quasi–steady-state approximations), inspired by empirical knowledge of the biology of the acarine predator-prey system. This mathematical strategy may provide insight into the general case; it may guide computer experiments and perhaps show the route to a more comprehensive analysis of (close relatives of) the "baseline" model. In this way a hierarchy of models will result, increasing in complexity toward the top and increasing in tractability toward the bottom. This may help to understand how the global dynamic behavior of the "baseline" model is affected by its various ingredients and submodels.

As a first example of this modelling strategy, Diekmann *et al.* (1988) considered the case in which the time between predator invasion and prey extermination is negligible compared to the mean time spent in the dispersal phase and the mean length of the prey growth phase. Assuming further that prey dispersal from each prey-only patch is proportional to the number of prey (x) and prey patch attractivity to predators ($\eta(x)$) is a constant, the "baseline" model reduces to three differential equations, one for the prey patches, one for prey dispersers (P), and one for predator dispersers (Q). This limiting case has an asymptotically stable steady state. Assuming a short dispersal phase (i.e., processes increasing P and decreasing P are very fast), the equation for P can be replaced by a quasi–steady-state approximation which reduces the system to the familiar Lotka-Volterra equation with an unstable steady state (and neutrally stable cycles). Thus, a dispersal phase of non-negligible duration has a stabilizing effect on the global prey-predator interaction. This result is easy to understand because during the dispersal phase prey are safe as far as predation is concerned, despite other dangers specific for this phase. This result, the stabilizing effect of a prey dispersal phase acting as a refuge from predation, was also borne out from a modelling exercise in which local dynamics were ignored and replaced by patch-occupancies (Sabelis *et al.*, 1991; Jansen, 1995a).

Similarly, it can be shown that a predator dispersal phase of non-negligible duration has the opposite effect: it destabilizes the steady state of the global prey-predator interaction (Sabelis *et al.*, 1991; Jansen, 1995a). This result seems to contrast with later findings by Weisser *et al.* (1997) and Neubert *et al.* (2002), who concluded that dispersal delays have a stabilizing effect. However, these authors modeled individuals, not patch states, and took into account that predators were removed from the disperser pool due to arrivals. In the patch occupancy model of Sabelis *et al.* (1991) this loss of dispersers from the disperser pool was not taken into account in the default version of that model, but in an extended version with terms expressing disperser loss due to colonization and interception Sabelis *et al.* (1991) showed that interception in itself has a stabilizing effect on predator-prey metapopulation models of the patch occupancy type. Thus, the conclusions of Weisser *et al.* (1997) and Neubert *et al.* (2002) about the stabilizing effect of "dispersal" are actually due to interception, rather than due to the presence of a dispersal phase per se.

A second example of the modelling strategy concerns the case where the predator-prey interaction time (τ) and the overall yield of predators per patch are assumed constant and thus independent of the number of prey (x) at predator invasion in a patch (Diekmann *et al.*, 1988). If the prey dispersal rates from prey and predator-prey patches and the predator attractivity function are also assumed to be constant and host plants are never overexploited, the number of prey (x) per patch becomes irrelevant and it is only important to know the number of patches occupied by prey. If the processes determining the number of dispersers are very fast (short dispersal phases), the equations for P and Q can be replaced by quasi–steady-state approximations and the system reduces to two delayed differential equations. Analysis of this model revealed that the founding of new prey patches by prey dispersing from predator-and-prey patches has a stabilizing effect on the global predator-prey interaction (Diekmann *et al.*, 1988). If this model is modified to include uniform dispersal of predators over the interaction period instead of delaying dispersal until the end of the interaction period, the stability domain increases. Thus, the empirically established feature of delaying dispersal until all prey are wiped out decreases possibilities for a stable steady state in the global predator-prey interaction (but a parameter domain with stable steady states remains due to the stabilizing impact of prey dispersal from predator-and-prey patches). Similar conclusions were drawn from a patch-occupancy caricature of a predator-prey metapopulation model (Sabelis and Diekmann, 1988; Sabelis *et al.*, 1991).

As a third example, consider the case where the prey exhaust the plant resource very quickly compared to the time scale of prey dispersal (Diekmann *et al.*, 1989). In that case, the founding of a prey population leads almost instantaneously to the production of prey dispersers, or—if "meanwhile" invaded by predators that erase the prey—the production of predators. This verbal description translates into α, β, and γ approaching to infinity and requires prey-patch attractivity $\eta\xi$ to go to infinity while maintaining the same order of magnitude as α. Three special assumptions are made: (1) predators reproduce an order of magnitude faster than their prey (note that this prevents the plant from being overexploited!), (2) logistic growth of empty patches ($n_0(t)$ instead of constant n_0), and (3) prey patch attractivity $\eta(x)$ increases linearly with the number of prey in a patch (x). After scaling time and variables n_0, P, and Q to u, v, and w (for precise scaling relations for these new variables and the new parameters a, b, c, and d, see Diekmann *et al.*, 1989), the following system of equations results:

$$\begin{aligned}
\frac{du}{dt} &= bu\left(1-\frac{u}{c}\right) - uv;\\
\frac{dv}{dt} &= uve^{-w} - v;\\
\frac{dw}{dt} &= auv\left(1-e^{-w}+\rho\left(\frac{1-e^{-w}}{w}-1\right)\right) - dw;\\
&a, b, c, d > 0; u, v, w, > 0; 0 \le \rho < 1.
\end{aligned} \tag{6}$$

This system of equations (6) is transparent insofar as the flow of empty patches to prey-colonized patches (uv) is immediately split into a flow of prey patches escaping from predator colonization with probability e^{-w} (the zero-term of a Poisson distribution) and a flow of prey patches colonized by predators with probability $(1-e^{-w})$. The surprise comes from the second positive term in the third equation, which—for $\rho \neq 0$ (i.e. $0 < \rho < 1$)—has a shape that no one would anticipate when attempting to directly write down a caricature of the metapopulation model. Clearly, careful derivation of limiting cases from a "baseline" model can lead to counterintuitive results! This model (equation 6) also holds a surprise in that—unlike classic tritrophic models in the tradition of Hairston *et al.* (1960) and Oksanen *et al.* (1981)—it may exhibit a multiplicity of coexisting steady states, depending on the specific values of a, b, c, d, and ρ. For small a and d (i.e., representing a combination of low production of predators per prey patch and long predator dispersal phase in the scaled system) and some additional conditions on all five parameters, at least two stable steady states, one with plants and herbivores only and another with all three trophic levels, exists. For large a and d (i.e., high production of predators per prey patch and short predator dispersal phase in the scaled system) and the same set of additional conditions, there is one stable steady state with plants and herbivores as before and another unstable steady state where plants, herbivores, and predators coexist through a stable limit cycle. If the system happens to be in the plant-herbivore state, then introduction of a small number of predators may thus fail to bring the system to the plant-herbivore-predator state. Only by introducing a large enough number of predators will the system reach a state with all trophic levels present, and only by imposing large enough mortality on the predators (e.g., pesticides) will the system return from a state with all trophic levels to one with plants and herbivores only. Another important conclusion gleaned from analyzing this system is that the two steady states, one with two and the other with three trophic levels, are both stable when there are many patches that can harbor only few prey. If instead there are only a few patches that can harbor many prey, the steady state with three trophic levels becomes unstable and the system exhibits stable limit cycles (Diekmann *et al.*, 1989).

To complete the picture of stabilizing mechanisms, consider the role of heterogeneity in the system. We have already seen that heterogeneity in the period from predator invasion to prey extinction has a stabilizing influence (Diekmann *et al.*, 1988; Sabelis and Dieckmann, 1988). As this period (τ) depends on all parameters of the "pancake predator" model for local dynamics (equation 1), site-to-site variability in these parameters will also contribute to variability in τ and this in turn will have a stabilizing influence on the predator-prey interaction. In the extreme where local dynamics are governed by a stochastic birth-death process (Nachman, 2001), site-to-site variability will be high and, therefore, it is not surprising that this will promote persistence/stability. Since the local predator and prey dynamics in acarine systems are not dominated by stochastic effects, a deterministic representation of local dynamics is appropriate (Sabelis and van der Meer, 1986). If stochasticity plays a role, it will be in the colonization phase, when numbers per patch are low. There will also be significant heterogeneity in the probability of a prey patch to become colonized by predators. Patches differ in the number of herbivores (x) they harbor, the leaf area they occupy and, therefore, the chance that they will be detected by dispersing predators will also vary. Thus, the attractivity η is likely to be some function of x. Such heterogeneity in the ability to detect prey patches is known to have a stabilizing influence on the predator-prey equilibrium (Hassell and May, 1974; May, 1978; Murdoch and Oaten, 1989), albeit only shown in models where the distribution of predators over prey patches is of a given statistical type (e.g., the negative binomial distribution). A simple model that is free of such *a priori* assumptions and in which distributions arise from predator-prey feedback (Sabelis *et al.*, 1991), is analyzed in the Appendix, showing that *any* heterogeneity in patch-attack rates will tend to have a stabilizing effect.

The "baseline" model is based on the assumption that what matters is habitat structure, not spatial configuration. If this does not hold, which consequences can we expect? Suppose all local populations in the metapopulation fluctuate in unison. In that case, the metapopulation as a whole is as persistent as a local population. What would happen if the dynamics of a single local population is brought out of synchrony: will the metapopulation dynamics bring this patch back into synchrony or will this asynchrony spill over to other patches and thus cause asynchronous dynamics and thereby persistence? For two local predator-prey populations governed by

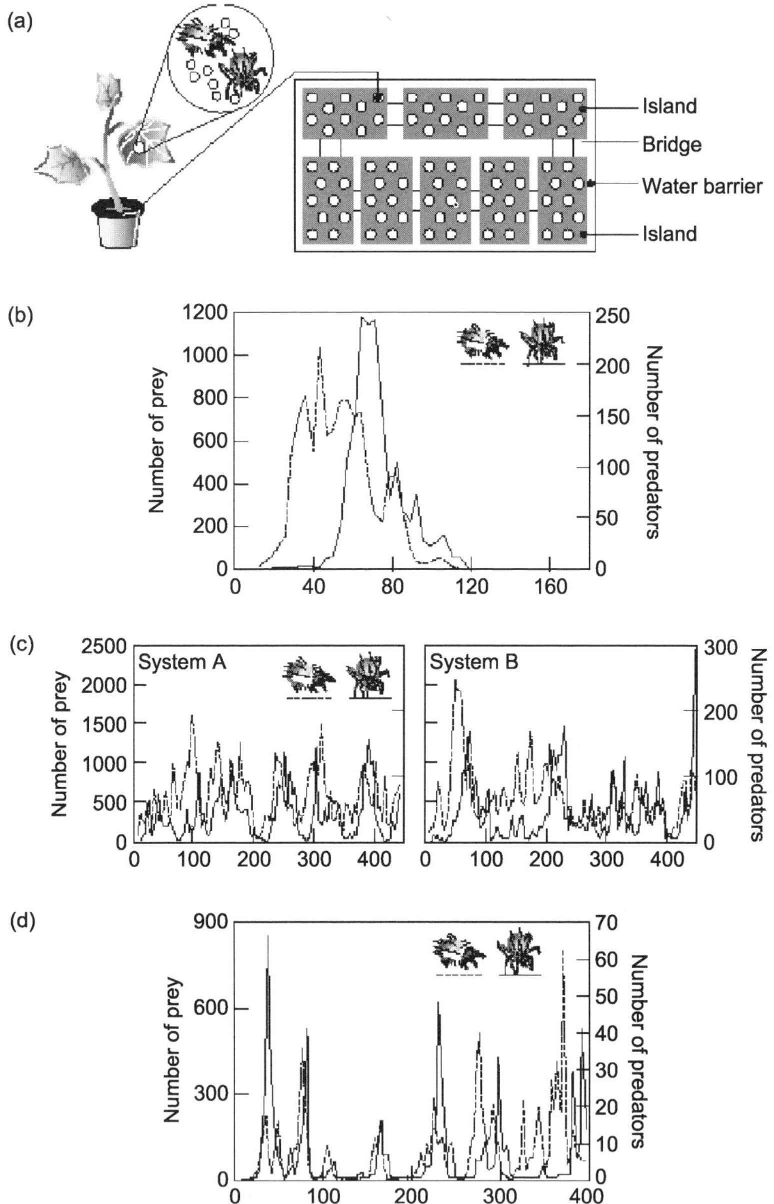

Figure 4 (Meta)population dynamics of predatory mites (*Phytoseiulus persimilis* Athias-Henriot; grey lines) and herbivorous mites (*Tetranychus urticae* Koch; black lines) at four structural levels in the habitat (Janssen *et al.*, 1997b). In (a), a schematic overview is presented of the circular system of eight interconnected islands (trays), each with 10 Lima-bean plants maintained in the two-leaf stage (by apex removal

Lotka-Volterra dynamics and weakly coupled by dispersal, models show that synchrony is maintained when the amplitude of the local population cycles is small. If large, however, a small perturbation will now grow and the dynamics of the two patches will become desynchronized for a while, until in the end the system settles again on a synchronized limit cycle, but one with a strongly reduced amplitude (Jansen, 1995b). This stabilizing effect of space is also found when local populations are governed by Rosenzweig-MacArthur dynamics (i.e., logistic prey growth and a Holling type II functional response) (Jansen, 1999, 2001). In this case, sustained asynchrony in local dynamics is possible when predator dispersal is not too high and not too weak. Often, this leads to a bistable state in which both the synchronized oscillations and asynchronous oscillations are resistant against perturbations.

VII. CONFRONTING THE "BASELINE" MODEL WITH EXPERIMENTAL TESTS

One way to test the model outlined above (Diekmann *et al.*, 1988, 1989) is to confront its predictions with empirical data. There are a number of published experiments showing cyclic dynamics. In the laboratory, Huffaker (1958) studied an acarine predator-prey system (the predatory mite, *Metaseiulus occidentalis*, and the herbivorous mite, *Eotetranychus sexmaculatus*) in a labyrinth of oranges loosely connected by bridges. This predator-prey system showed three cycles at the metapopulation level over a time span of one year, whereas increasing connectivity between oranges led to extinction after no more than one cycle of a few months. This experiment has become famous for demonstrating the potential for metapopulation persistence despite local extinction, but also infamous because it was unreplicated and some cycles depended on the survival of a single prey individual. In a later series of similar experiments, Huffaker *et al.* (1963) carried out two replicates and again found two or more cycles over a period of one year. Also in the laboratory, Janssen *et al.* (1997b) assessed the dynamics of another acarine predator-prey system on a circle of eight interconnected islands, each harboring 10 two-leaf bean plants (Fig. 4a). Within the islands, prey and predators could easily move from one plant to the other, but to reach a

and replacement of overexploited plants). In (b), a time series plot is shown of the extinction-prone predator-prey dynamics on one super-island (i.e., the size of eight islands together). In (c) two replicate experiments showing persistent predator-prey metapopulation dynamics on the eight-island system as shown in (a): one replicate showing large-amplitude oscillations and the other replicate showing small-amplitude fluctuations. In (d), extinction-prone predator-prey dynamics on one of the islands in the eight-island system is shown.

neighboring island was made more difficult (by positioning the island-to-island connection somewhat away from the rim of the islands, which are frequented by predators on the move). Janssen *et al.* (1997b) found frequent extinctions of island populations (Fig. 4d), but the overall population persisted for more than one year. There were six cycles per year in one replicate of the experiment, but aperiodic fluctuations in the other replicate of the experiment, even though both replicates were conducted simultaneously and under similar conditions (Fig. 4c). When all islands were fit together to form one superisland with 80 plants, first the prey and then the predators became extinct in approximately three months (Fig. 4b). The different results from the two replicates with the subdivided habitat may have arisen from stochastic effects in the initial phase or from the existence of a chaotic attractor. In a 0.5-year experiment using the same acarine predator-prey system on cucumber plants in a three-compartment greenhouse, Nachman (1981) found one or two cycles in three replicates. Using another acarine predator-prey system (*M. occidentalis* and *T. urticae*) on apple trees, van de Klashorst *et al.* (1992) studied local and global dynamics in a system of six benches, each with 48 trees, in a single greenhouse over a period of two years. This unreplicated long-term experiment resulted in a first year with four distinct large amplitude cycles at the metapopulation level, then followed by a second year of smaller-amplitude aperiodic fluctuations (Fig. 5a and 5b; Lingeman and van de Klashorst, 1992). The authors argued that there was no obvious externally imposed factor causing the marked change in dynamic pattern from year 1 to year 2, but unfortunately the management regime was not constant in all respects throughout the experimental period. Long-term persistence of acarine predator-prey systems in greenhouses has also been observed in practice. In two unheated plastic-covered glasshouses (southern Queensland, Australia), where the predatory mite *P. persimilis* was released to control the herbivorous mite *T. urticae* on rose hedges, there were no outbreaks of *T. urticae* and there was no need to reintroduce or redistribute the predatory mites over a period of three years (Gough, 1991)!

The laboratory and greenhouse experiments referred to in the previous text have shown that increasing habitat structure promotes persistence despite frequent local extinctions and creates a propensity for cyclic dynamics or aperiodic fluctuations at the metapopulation level. Strikingly, these two types of fluctuations were observed in two parallel replicates (Janssen *et al.*, 1997b; Fig. 4c) and in one and the same time series where large-amplitude periodic oscillations were followed by small-amplitude aperiodic oscillations (Lingeman and van de Klashorst, 1992; van de Klashorst *et al.*, 1992; Figs. 5a and 5b). This is indicative of a bi-stable system. The latter is consistent with predictions from the model by Diekmann *et al.* (1988, 1989); that is for the case of large a and d (high predator production per patch and short dispersal phase) and for the case of a small number of large (instead of many small) patches. Indeed, most of these experiments exhibit local extinction in spatial units larger than a single plant: groups of oranges in Huffaker *et al.* (1963); islands with 10 bean plants in Janssen *et al.* (1997b); groups of cucumber plants in a plant row in Nachman (1981) and Sabelis *et al.* (1983);

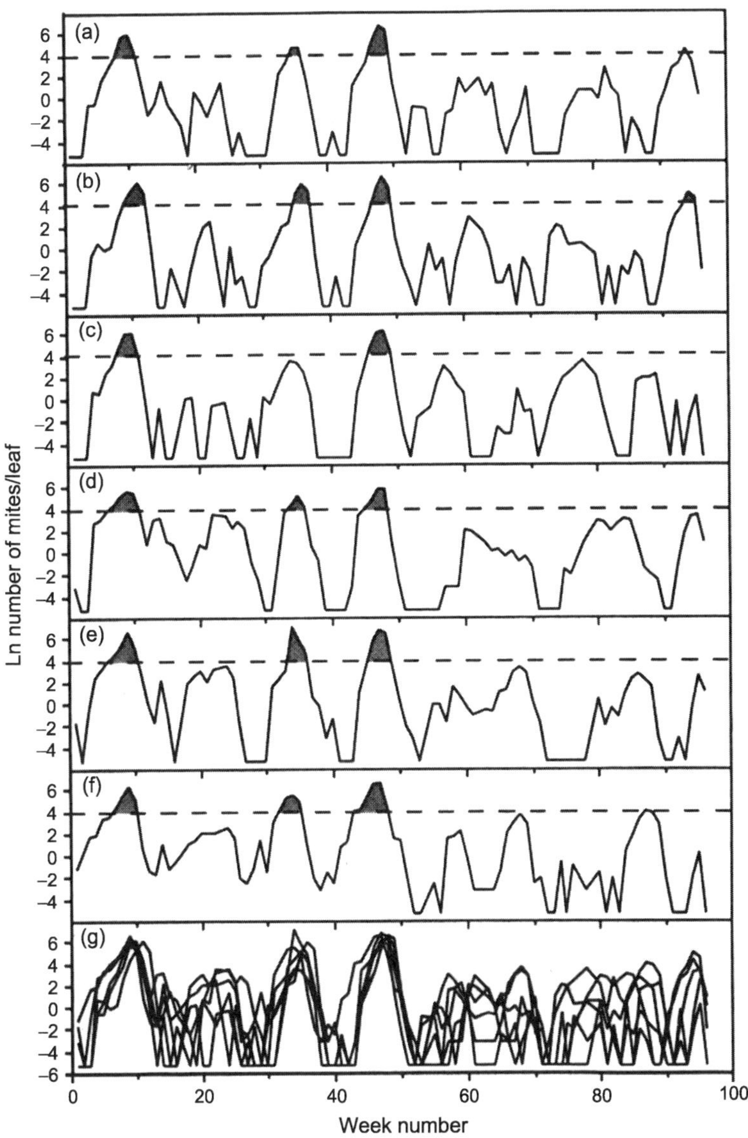

Figure 5 (*continued*)

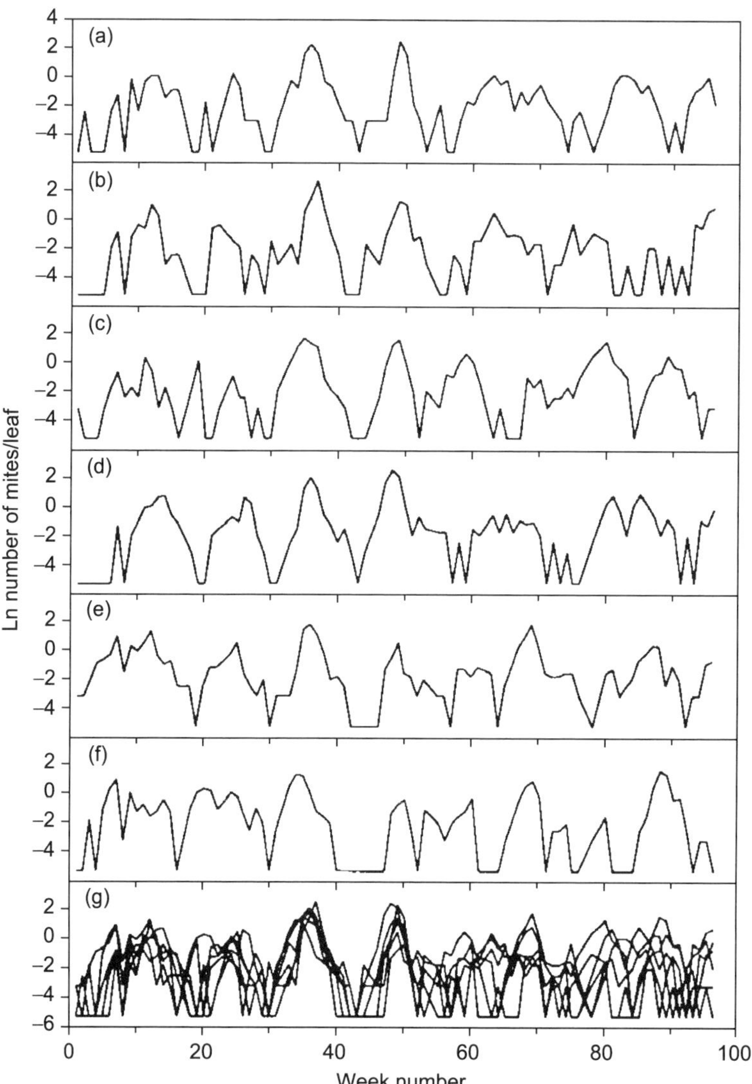

Figure 5 Dynamics of (a) herbivorous mites (*Tetranychus urticae* Koch; black lines) and (b) predatory mites (*Metaseiulus occidentalis*; grey lines) on each of six benches, each with 48 small apple trees in a greenhouse (Van de Klashorst *et al.*, 1992). Dynamics per bench is shown in the top six panels. The last (7th) panel is an overlay plot of all six panels above. It shows synchrony in the dynamics during the first year (large amplitude oscillations) and asynchrony in the dynamics during the second year (small-amplitude aperiodic fluctuations).

benches with 48 small apple trees in van de Klashorst *et al.* (1992). Moreover, the number of these spatial units was small: 8 islands in Janssen *et al.* (1997b) (Fig. 4a); 6 benches in van de Klashorst *et al.* (1992). Thus, there were effectively few large patches in these experiments, consistent with predictions of the Diekmann *et al.* (1989) model. The other prediction from this model about two co-existing stable states or a stable state co-existing with a stable limit cycle was not borne out by these experiments. However, none of the experiments were designed to detect bi-stability, thus leaving the question open for future experimental research.

VIII. TESTING THE ROBUSTNESS OF THE "BASELINE" MODEL BY EXTENDED SIMULATIONS

Another way to test the model is to perform computer simulations based on a model that takes small steps beyond the realm of assumptions underlying the "baseline" model of Diekmann *et al.* (1988, 1989). The parameterized computer simulation model by Pels *et al.* (2002) differed mainly from this "baseline" model in that it allows for stochasticity of the patch invasions, performs bookkeeping of not only first but also later invasions per patch and takes a carrying capacity for the prey into account. This model predicts persistence despite frequent local extinctions, but only if survival of predators in the dispersal phase is sufficiently low. As shown in Figs. 6a and 6c, survival during dispersal at a level low enough to be just within the persistence region gives rise to regular, large-amplitude oscillations. For relatively lower values of the survival during dispersal, however, small-amplitude, aperiodic oscillations appear (Figs. 6b and 6c). Both dynamic phenomena were observed in the long-term experiments by Janssen *et al.* (1997b) among replicate experiments (Fig. 4c) and van de Klashorst *et al.* (1992) among years in a single experiment (Figs. 5a and 5b).

Ellner *et al.* (2001) produced another parameterized simulation model with deterministic local dynamics and a stochastic description of the colonization process (but a very short dispersal phase!). It differs from Pels *et al.* (2002) mainly in taking age structure in the local populations into account by means of non-linear, stage-structured matrices, and in using regression models for the conditional probability of an available plant being colonized, given the current recent mite abundances on each plant, and the state of each plant. To identify significant "risk factors" affecting colonization, logistic regression with stepwise and backward variable selection was applied to the dataset of Janssen *et al.* (1997b) (McCauley *et al.*, 2000). To assess how spatial dynamics affect persistence in the metapopulation, two types of underlying mechanisms should be distinguished. The first mechanism is that spatial subdivision produces differing dispersal scales in predators and

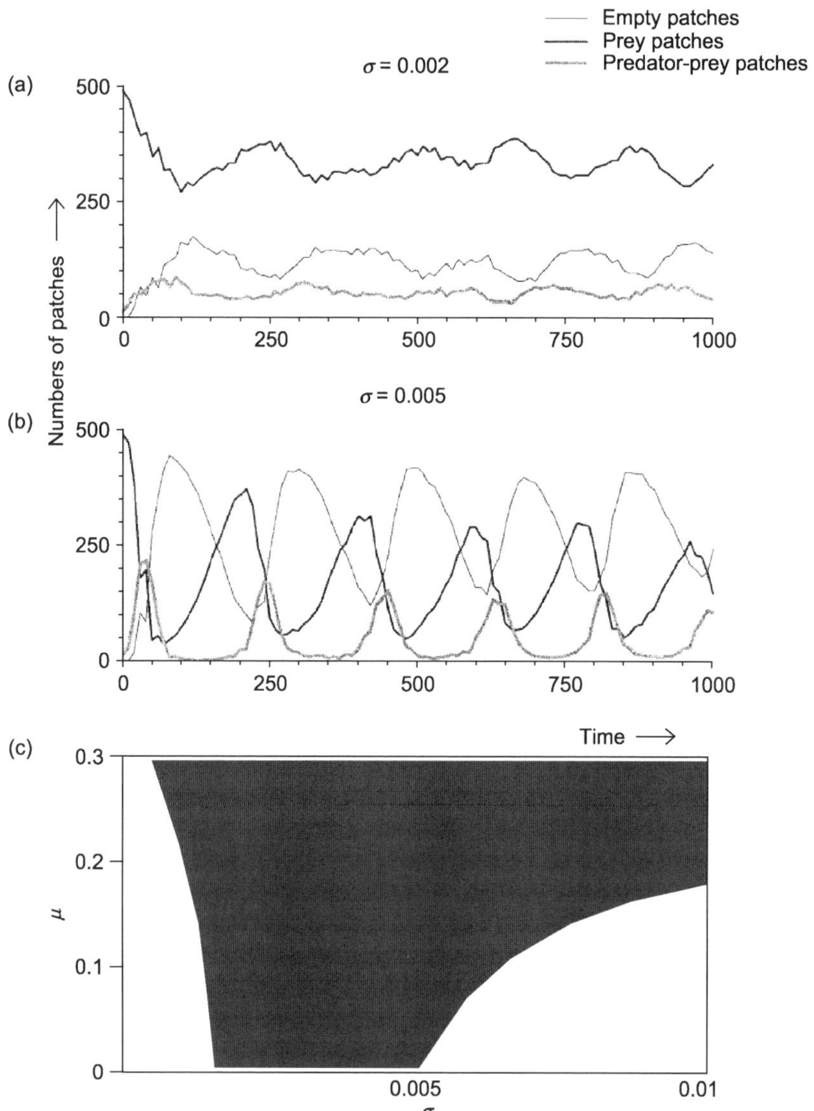

Figure 6 (A and B) Simulated metapopulation dynamics of the number of empty patches, prey patches, and predator-and-prey patches, for two low values of predator survival during dispersal ($\sigma = 0.002$ in Fig. 6a, and $\sigma = 0.005$ in Fig. 6b). For the lower of the two values, populations show large-amplitude cycles, whereas small-amplitude aperiodic fluctuations emerge for the other value. For higher values of predator survival during dispersal the metapopulations go extinct (Pels *et al.*, 2002). (C) Metapopulation persistence (shaded area) in parameter domain of survival during the dispersal phase (σ) and dispersal from a predator-and-prey patch (μ) (Pels *et al.*, 2002).

prey so that the exact position of an individual in space matters. The second mechanism is that spatial subdivision reduces the discovery rate of prey outbreaks by predators. This last mechanism has two components, neither involving spatial pattern: reduced average predator success in finding prey populations, and the resulting between-plant asynchrony, meaning that not all plants are simultaneously exploited by prey and not all prey outbreaks are simultaneously discovered and extinguished. Hence a non-spatial model for total numbers of plants occupied by each species can describe the system.

Under the first mechanism ("spatial pattern matters"), eliminating spatial pattern at the island and metapopulation levels should produce dynamics similar to those observed in the single-island system. Under the second mechanism ("reduced discovery matters"), decreasing local between-plant asynchrony should have this effect. These different aspects of spatial pattern were systematically eliminated in subsequent series of computer simulations, in which the densities of the mites were randomized using various schemes differing in the scale at which it was randomized. In "island shuffling" simulations, the sequence of islands in the ring (Fig. 4) was randomly shuffled each day, but plants remained on their home island. This preserves within-island correlation but eliminates pattern at the metapopulation level. In "plant scrambling" simulations, the location of all plants was randomized each day, and in "global dispersal" simulations, all colonization probabilities were calculated from system-wide average densities rather than local densities. These simulations eliminate spatial pattern at the island and metapopulation level, but retain between-plant asynchrony. In "plant blurring" simulations, the age-specific mite densities on each plant were replaced daily by the average density over all plants on the same island. This eliminates within-island asynchrony but preserves the pattern at the metapopulation level.

Island shuffling, plant scrambling, and global dispersal had only small effects on the dynamics. In contrast, plant blurring produced extinction similar to the single-island experiment, with predators finding all prey outbreaks because of the rapid within-island dispersal. These simulation results indicate that the first stabilizing mechanism, that of prey and predators having different dispersal scales, is not likely to be responsible for long-term persistence in the metapopulation system, but the second mechanism might be. If the second mechanism caused persistence, decreasing the predators' success at detecting prey outbreaks in the single-island should result in persistence. This prediction was confirmed in the single-island case by reducing the daily probability of predators finding each prey-only plant. Conversely, the metapopulation model persisted less often when predator colonization probabilities were increased. Thus, the dynamics of the systems result mainly from the average probability that prey mites find unoccupied plants, the average probability that predators find a prey outbreak, and the stochasticity of individual colonization events. The systems can be modelled successfully without regard to the spatial location of plants. A preliminary "plant scrambling" experiment was found to be consistent with this conclusion (supplementary information in Ellner *et al.*, 2001).

The approach by Ellner *et al.* (2001) showed that comparing alternative mechanistic models is a powerful approach for testing hypotheses regarding processes responsible for patterns in ecological dynamics. By comparing models with different assumptions about the effects of habitat subdivision, they showed that the one essential process allowing for persistence of the metapopulation was one of the earliest and simplest hypotheses about spatial dynamics: isolation by distance (e.g., de Roos *et al.*, 1991). Habitat subdivision increased the effective distance between plants from the viewpoint of the predators, giving prey a moving refuge from their enemies.

It can be concluded that there is at least some correspondence between the predictions from the "baseline" model and the extended versions used in the simulations (Ellner *et al.*, 2001; Pels *et al.*, 2002). In particular, the analysis by Ellner *et al.* (2001) provides strong support for the "baseline" assumption that habitat structure matters, not spatial configuration. Moreover, there is support for the "baseline" prediction that cyclic dynamics prevail when the dispersal phase is short and local production of predators is high (high values of the scaled parameters a and d) (Diekmann *et al.*, 1989) and receives some support from two parameterized simulation models, one by Pels *et al.* (2002) and the other by Ellner *et al.* (2001). These models have a short dispersal phase and produce either large amplitude cycles or small amplitude aperiodic fluctuations depending on survival during the dispersal phase. Most convincingly, these two types of dynamic behavior also are visible in population experiments currently published. In one experiment (van de Klashorst *et al.*, 1992), however, they showed up in one and the same experimental time series. Perhaps, the survival chances during dispersal changed in the course of the experiment; however, there are no data to substantiate this claim. Alternatively, it is the intensity of spatial coupling between local populations that under certain conditions can create bi-stability between synchrony and asynchrony in the dynamics of local populations (Jansen, 1999, 2001).

So far, there have been no explicit attempts to demonstrate the co-existence of stable states (one with herbivores and plants only, and one with predators)—another prediction of the "baseline" model (Diekmann *et al.*, 1989). This can be tested by imposing mortality events of different intensities on the predator population when the metapopulation persists with all three trophic levels present or by releasing increasingly more predators in a system where the metapopulation persists with plants and herbivores only. Such experiments need to be done, both using computer simulation models and "real-life" experiments.

IX. IS THE "BASELINE" MODEL EVOLUTIONARILY ROBUST?

So far, we have sought simplifications of metapopulation models to analyze processes underlying persistence and stability of metapopulations. One may ask, however, whether the simplifications represent key features that are robust against natural selection. In the model by Diekmann *et al.* (1988, 1989), such a key feature is that predators delay dispersal until after extermination of the local prey population. One may wonder whether this phenomenon would be favored by natural selection within the assumed structure of that model. For example, the model by Diekmann *et al.* (1988, 1989) is based on the assumption that the first predator colonizing a patch determines the local predator-prey dynamics. What would happen if there are later colonizers of the same patch and they have other strategies of exploiting the local prey population and other decisions with respect to the timing of dispersal? Should we expect the predators ultimately to delay dispersal until after local prey extermination or would natural selection favor other strategies? These questions are at the heart of a series of articles published by Van Baalen and Sabelis (1995a,b) and Sabelis *et al.* (1999a,b, 2002).

In principle, there can be many and varied dispersal strategies which has severe consequences for local predator-prey dynamics. To illustrate this point it is instructive to make use of the "pancake predator" model for local predator-herbivore dynamics (equation 1). Suppose the strategy set of the predator is determined by the per capita emigration rate of the predator. Of course, all predators will disperse when herbivores are exterminated, but they can also decide to move away *during* the interaction. Such increased emigration of the predator will relieve the prey of predation pressure, thereby causing the prey population to represent a larger future food source for the predators that stay put (Van Baalen and Sabelis, 1995b). Assume for simplicity that the per capita emigration rate is a constant μ. The effective per capita population growth rate of the predators is reduced by μ so that the effective predator growth rate now equals:

$$\frac{dy}{dt} = (\gamma - \mu)y$$

As illustrated in Fig. 7a, a small increase of μ from 0/day to 0.04/day, greatly alters the area under the prey curve—a measure of prey availability over the interaction period.

According to Metz and Gyllenberg (2001), an adequate "stand-in" fitness measure for individuals that occur in metapopulations is the number of dispersers produced per female. In this case it is the sum of the number

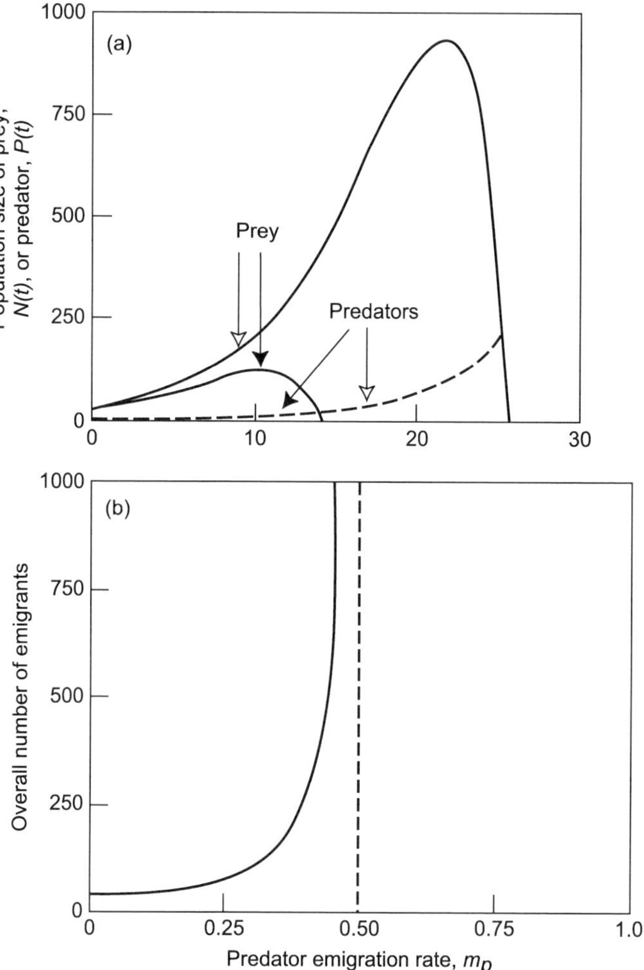

Figure 7 (A) Influence of emigration on local predator-prey dynamics according to the pancake model: predator emigration rate (μ) equals zero for bold predator curve and thin prey curve indicated by a drawn arrow, whereas it is 0.04 for the predator and prey curves indicated by a dashed arrow ($\mu = 0$ or 0.04, $\alpha = 0.3$, $\beta = 3$, $\gamma = 0.25$, $x(0) = 30$, $y(0) = 1$). (B) The relation between the overall production of predator dispersers per patch and the per capita rate of emigration (μ) during the predator-prey interaction period (from predator invasion to prey elimination) (Van Baalen and Sabelis, 1995b).

of dispersers produced during the interaction with prey, and those dispersing after prey extermination. As shown in Figure 7b, the production of dispersers increases disproportionally with μ and reaches an asymptote when the per capita emigration rate is so high that the predators cannot suppress the growth of the prey population any more; that is when $\mu = \gamma - \alpha + \beta\, y(0)/x(0)$. Thus, predators that suppress emigration during the interaction with prey reach their full capacity to reduce the local prey population, but by the time they eliminated the prey population they have produced the lowest number of dispersers per prey patch. This so-called "killer" strategy with a zero emigration rate does less well than the strategy of a "milker" which typically has a non-zero emigration rate during the predator-prey interaction period. When "killers" enter a prey patch with "milkers," "killers" will steal much of the prey the "milkers" had set aside for future use. Therefore, if there is a risk of invasions by "killers," it pays to anticipate such events; selection will favor exploitation strategies that are less "milker"-like and more "killer"-like.

The outcome of the above "milker-killer" dilemma is determined by a complex interplay of local competition between the exploitation strategies and global (=metapopulation) dynamics. It depends on the probability of multiple predator invasions in the same prey patches, on the resulting production of dispersers per prey patch, and on metapopulation dynamics, which in turn determine the probability of multiple invasions. The complexity of this ecological feedback is staggering; to keep track of the numbers of each strategy type when competing in local populations, dispersing into the global population and invading into local populations requires a massive bookkeeping procedure. Hence, it is necessary to simplify to gain some insight. For example, Van Baalen and Sabelis (1995b) assumed that all patches start with exactly the same number of predators and prey (which assumes metapopulation-wide equilibrium and ignores stochastic variation in the number of colonizers) and that the predators had enough time to reach their full production potential per prey patch (the assumption of sequential interaction rounds). In this setting, they considered a single mutant predator clone with a per capita emigration rate, μ_{mut}, and its success relative to that of the resident predator clone that has another per capita rate of emigration (μ_{res}). The question is whether an ESS value of μ_{res} for which it does not pay any mutant to deviate still exists. In particular, Van Baalen and Sabelis (1995b) calculated for which combinations of parameters ($y(0), x(0), \alpha, \beta, \gamma$) it does not pay to increase μ away from zero. The general outcome is that "killers" ($\mu = 0$) are favored by selection, except when the number of predator-foundresses is low and the number of prey foundresses is high (Fig. 8). In other words, "milkers" are favored as long as they have a sufficiently large share in the local populations to maintain control over the time to prey elimination (τ).

This analysis whets the appetite for more elaborate work taking into account: (1) asynchrony in local dynamics, (2) stochastic variation in predator and prey colonization rates (since these are probably low!), (3) an upper boundary to prey population size set by the local amount of food, and

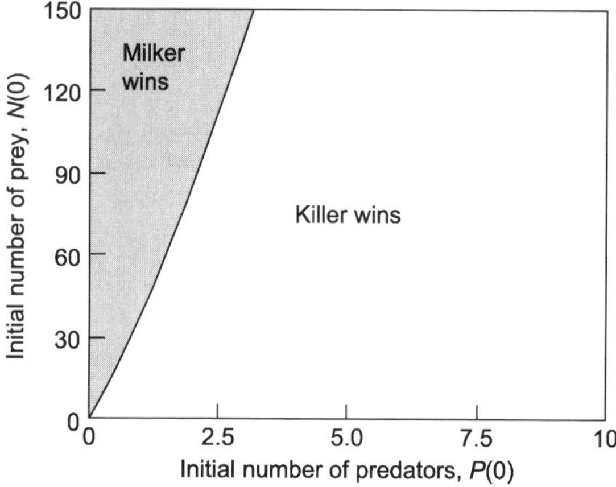

Figure 8 When does it pay to increase the per capita emigration rate of the predator (μ) away from zero? This diagram shows that milker strategies are only possible when $y(0)$ is low and $x(0)$ sufficiently high (modified after Van Baalen and Sabelis, 1995b).

(4) metapopulation dynamics. Such extensions are likely to show that "milkers" which achieve a longer interaction period are also exposed for a longer period to subsequent predator invasions (and thus, face competition with "killers" sooner or later), that stochastic rather than uniform invasions will help to isolate "milkers," which thereby gain full advantage of their exploitation strategy, and that limits to the amount of food available for the prey will decrease opportunities for a "milker," as it loses full control over the interaction period (τ). As these factors have opposite effects it is not immediately clear whether "killers" or "milkers" will win the battle or may even co-exist. Computer simulations of the metapopulation dynamics of "milkers" and "killers" using a model parameterized for phytoseiid mites (predators) and spider mites (prey) (Pels *et al.*, 2002) showed that the full ecological feedback gives rise to prey and predator densities where multiple predator invasions are sufficiently rare to make "milkers" the more successful strategy. The results of a large number of computer experiments to determine the average number of dispersers born from mutants with different emigration strategies released randomly in the metapopulation of the residents (but after 2,000 days to avoid the initial phase of transient metapopulation dynamics) is summarized in the pairwise invasibility plot shown in Fig. 9. This shows that metapopulation dynamics can force the predator-prey system into a state (i.e., number of predator and prey colonizing a patch, $N(0)$ and $P(0)$, in Fig. 8) where "milkers" are the winners of the competition.

Apart from the need for more theoretical work, experimental analysis of variation in exploitation strategies of predators seems a promising avenue for future research. Such an analysis, carried out for the case of the predatory mite *P. persimilis*, revealed that laboratory cultures exclusively harbor predators of the "killer"-type (Sabelis and van der Meer, 1986), whereas field-collected populations in the Mediterranean (Sicily) exhibit some variation in the onset of emigration before or after prey elimination (Pels and Sabelis, 1999). Interestingly, most populations collected along the coastal line where predators are more abundant (e.g., Alcamo-line in Fig. 2b), initiated emigration only after elimination of prey, whereas a line collected inland where local predator populations are very scarce and hence more isolated showed some emigration before prey elimination (e.g., Enna-line in Fig. 2b)! These results are in qualitative agreement with the analytic ESS analysis of Van Baalen and Sabelis (1995b) (Fig. 8), but they seem to contradict the results of the more "realistic" computer simulations that were not only parameterized for this particular mite system but also take stochastic colonization and the full ecological feedback into account (Pels *et al.*, 2002; Fig. 9). This discrepancy may well result from a variety of factors that cause the predators to loose control over the exploitation of the local prey population: environmental disasters (heavy rain, wind, or fire; plant consumption by large herbivores), overexploitation of plants by large herbivores, and exploitation competition with other predator species or herbivore diseases. The discrepancy between the simulations (Pels *et al.*, 2002; Fig. 9) and the analytic treatment (Van Baalen and Sabelis 1995b; Fig. 8) may emerge because: (1) the simulations were obviously only carried out for persisting resident populations whereas the analytic treatment implicitly assumed equilibrium (and thus persistence), (2) the simulated predator-prey feedback causes patches to be invaded by very low numbers of predators, whereas the analytic treatment presupposed equal invasion for all patches, and (3) the stochastic colonization process of the predators allows some patches to be invaded singly and gives the single invader full control over the exploitation of the local prey population, whereas the analytic treatment ignored stochastic variation in the number of colonizers.

Given the likelihood of local environmental disasters, we expect predator dispersal after local prey extermination to be the rule (Sabelis *et al.*, 1999a,b, 2002). This expectation would then provide a rationale for the "killer strategy" assumptions underlying the model of Diekmann *et al.* (1988, 1989). However, this is for reasons quite different from what is assumed in that model: a single invasion per patch, a case where selection favors "milkers," whereas selection favors "killers" under multiple invasion and loss of exploitation monopoly due to local disasters.

There may be a pitfall in the arguments and model choices explained in the preceding text because neither theoretical nor experimental analyses addressed the possibility of more flexible strategies, such as "milk when

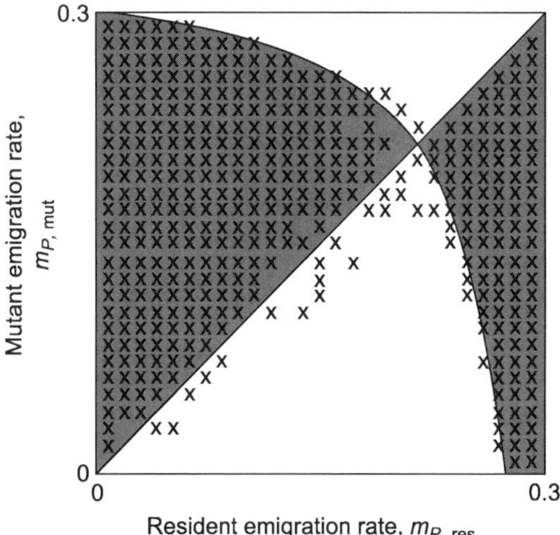

Figure 9 A pairwise invasibility plot for mutants that differ from the resident predators with respect to the emigration rate (μ). The resident's emigration rate is given at the horizontal axis and the mutant's emigration rate at the vertical axis. An "x" indicates a simulation where the mutant invades the resident metapopulation. The results shown are obtained for a value of the predator's survival rate that allows the predator-prey system to persist (see Figure 6). Natural selection moves the dispersal rate (μ) to values quite different from 0 and hence the simulations predict "milkers" to evolve (Pels *et al.*, 2002).

exploiting the prey patch alone, and kill when other (e.g., non-kin) predators have entered the same patch." This is not far fetched because predatory mites are known to perceive each other's presence (Janssen *et al.*, 1997a) and they can discriminate between kin and non-kin (Faraji *et al.*, 2000; Schausberger and Croft, 2001).

X. LESSONS FROM MATHEMATICAL AND EXPERIMENTAL EXERCISES

The approach discussed in this chapter has led to some general lessons on how to combine mathematical modelling and experimental approaches in analyzing metapopulation dynamics of predator-prey systems. These can be summarized, as follows:

1. *Deriving limiting cases from comprehensive ("baseline") mathematical models is a useful strategy to establish the array of relevant caricatural*

models. There is a danger in writing down the caricatures right from start. Careful derivation of caricatures from more comprehensive models seems a preferable modelling strategy.

2. *Mathematical formalisms may set limits to the complexity of a "baseline" model.* Whereas it is advocated to write down equations of the full process, the mathematical formalism may impose constraints on the complexity that can be taken into account. The model by Diekmann *et al.* (1988), discussed in this chapter, employs the formalism of physiologically structured population models (Metz and Diekmann, 1986). It is comprehensive and unique in taking local as well as global processes into account, but it had to be based on a simplification of the patch invasion process (local dynamics dominated by first invaders), because only in this way the formalism was sufficiently powerful to develop an appropriate bookkeeping procedure of predator and prey numbers on patches, on the move and in the metapopulation.
3. *Computer simulations are an important means to explore the consequences of relaxing assumptions that are inevitably made, also in the "baseline" model.* The "baseline" model discussed in this chapter is essentially non-spatial and is of the single-patch-invasion type. Simulations help to explore the consequences of altering key features of the "baseline" model, such as allowing for multiple invasions, creating habitat structure, and manipulating the degree of population mixing (e.g., computer experiments on plant blurring, plant scrambling, island shuffling, global dispersal).
4. *Population dynamic models may not be evolutionarily robust and, therefore, require careful interpretation when tested against biological data.* The key feature for constructing the "baseline" model of metapopulation dynamics was that predators delay dispersal until after local prey extermination. This was based on local population experiments with lab-reared and field-collected strains of predatory mites. However, under the model assumptions in Diekmann *et al.* (1988, 1989) (single invasions, no prey carrying capacity, no disasters), "milkers" are likely to evolve. The reasons why "killers" evolved to be the prevailing strategy in the field, are apparently not covered in the setting of the "baseline" model.
5. *Because of the complexity of the problem, model simplifications were necessary which make biological experiments fully relevant and useful.* The predictions from limiting cases of the "baseline" model were tested against more comprehensive simulation models and against long-term observations on metapopulation dynamics in laboratory settings and greenhouses. In particular, the prediction of large-amplitude metapopulation cycles is in agreement with part of the empirical observations. The existence of small-amplitude aperiodic fluctuations in the other part of the observations is in agreement with computer simulations for the

multiple invasion case, provided survival during the dispersal phase is not too low (leading to cycles) and not too high (leading to global extinction). As yet, there is no empirical evidence for the prediction of multiple states, a feature of fundamental importance for understanding tritrophic metapopulation systems.

In conclusion, mathematical models in combination with laboratory and greenhouse experiments with acarine predator-prey systems have been instrumental in demonstrating that habitat structure promotes metapopulation persistence despite local extinction (see also Bonsall *et al.*, 2002; Holyoak and Lawler, 1996a,b). Whether the observed and predicted phenomena carry over to the field remains to be tested. Whereas field populations of acarines are likely to be subject to a metapopulation structure (Charles and White, 1988; Walde, 1994; Strong *et al.*, 1999), the emergent dynamic phenomena (large or small amplitude limit cycles; multiple states) need critical assessment.

APPENDIX: DIFFERENTIAL PREY PATCH VULNERABILITY STABILIZES PREDATOR-PREY INTERACTION

O. Diekmann, J. Cheng-fu, J.A.J. Metz & M.W. Sabelis

Consider a patch-occupancy model for predator-prey metapopulation dynamics (Sabelis *et al.*, 1991). Assume the dispersal phase for predator and prey is infinitely short. Then there are two equations, one for the density of prey patches (N) and one for the density of predator patches (M). Suppose there are two types ($i = 1$ or 2) of prey patches that occur at densities N_1 and N_2 (with N being their sum). The prey patches differ only in their vulnerability to predator attack, as expressed by parameters b_1 and b_2. Prey patches of type i are "born" from other prey patches at a rate a times a fixed fraction k_i (so $0 \leq k_i \leq 1$ and $k_1 + k_2 = 1$). Due to prey growth in the prey patch and/or prey dispersal from a prey patch at rate h_i, prey patch 1 may change into prey patch 2 and the reverse. If a predator invades and attacks either of the two prey patches, it becomes a predator-and-prey patch. The parameter e scales the density of predator-and-prey patches into predator density equivalents M. Dispersal of prey from predator-and-prey patches is ignored since its effect is known (Sabelis and Diekmann, 1988) and its presence in the model would obscure the impact of differential prey vulnerability. For c and d expressing extinction rates of prey patches and predator patches respectively, we obtain the following system of equations:

$$\frac{dN_1}{dt} = a(N_1 + N_2)k_1 - h_1N_1 + h_2N_2 - b_1N_1M - cN_1;$$
$$\frac{dN_2}{dt} = a(N_1 + N_2)k_2 - h_2N_{21} + h_1N_1 - b_2N_2M - cN_2;$$
$$\frac{dM}{dt} = e(b_1N_1 + b_2N_2)M - dM;$$
$$a, b, c, d, e, h > 0; k_1 + k_2 = 1.$$

When $b_1 = b_2 = b$, we can rewrite this system of differential equations into a classic Lotka-Volterra analogue. Hence, there is a neutrally stable steady state and populations will oscillate indefinitely. So, when $b_1 = b_2$, we are precisely at the edge of the stable-unstable region. To analyze the above system for the case of differential attack rates, we first find expressions for the steady state and then we proceed to carry out a linearized stability analysis near the equilibrium.

Steady States

Under steady state conditions, the system equations yield two expressions;

$$N_1 + N_2 = \frac{1}{a-c}\frac{d}{e}M;$$
$$b_1N_1 + b_2N = \frac{d}{e}.$$

Combining these two results we can express N_1 and N_2 in terms of M

$$\begin{pmatrix} N_1 \\ N_2 \end{pmatrix} = \frac{d}{e}\frac{1}{b_2 - b_1}\begin{pmatrix} b_1M' - 1 \\ -b_2M' + 1 \end{pmatrix} \text{with} \quad M' = M/(a-c).$$

Substitution in the first system equation yields a quadratic function $F(M)$:

$$F(M){:} = b_1b_2M^2 - (a-c)(h_1 + h_2 + c) + \ldots$$
$$(h_1b_2 + h_2b_1 + cb_2 + ak_1(b_1 - b_2) - b_1(a-c))M = 0.$$

Since $F(0) < 0$ a unique positive solution M exists. The corresponding N_1 and N_2 are positive as well since sign $F((a-c)/b_2)) = \text{sign}(b_1 - b_2) = -\text{sign}((a-c)/b_1))$. It is now easy to see that under steady state conditions N_1 and N_2 may be related in various ways (e.g., $N_1 = m\,N_2$) provided that a, c, b_i, h_i and k_i satisfy the relation given above. Thus, even when $b_1 \neq b_2$ were to be an additional requirement, there are still many degrees of freedom left to find a suitable combination of parameters allowing for N_1 to bear a particular relation to N_2.

Stability Analysis

The Jacobi matrix corresponding to linearization at the steady state is given by

$$\begin{pmatrix} ak_1 - h_1 - b_1M - c & ak_1 + h_2 & -b_1N_1 \\ ak_2 + h_1 & ak_2 - h_2 - b_2M - c & -b_2N_2 \\ eMb_1 & eMb_2 & 0 \end{pmatrix}.$$

This yields the characteristic equation with coefficients A_i:

$$\begin{aligned} &\lambda^3 + A_1\lambda^2 + A_2\lambda + A_3 = 0; \\ &A_1 = (ak_1 + h_2)\frac{N_2}{N_1} + (ak_2 + h_1)\frac{N_1}{N_2}; \\ &A_2 = eM(b_1^2N_1 + b_2^2N_2); \\ &A_3 = (ak_1 + h_2)eMN_2(b_2^2\frac{N_2}{N_1} + b_1b_2) + (ak_2 + h_1)eMN_1(b_1^2\frac{N_1}{N_2} + b_1b_2). \end{aligned}$$

The Routh-Hurwitz criteria $A_1 > 0$ and $A_2 > 0$ are always fulfilled. So it remains to verify the third criterium $A_1A_2 - A_3 > 0$ or equivalently (after division by $eM\,(ak_2 + h_1)b_2N_2$):

$$(b_1 - b_2)\left[\frac{b_1}{b_2}\theta - 1\right] > 0, \text{where } \theta = \frac{(ak_1 + h_2)N_2}{(ak_2 + h_1)N_1}.$$

Note first that when $b_1 = b_2$ the Routh-Hurwitz condition is *not* fulfilled and that the system is on the verge of being *unstable*.

It can be shown that:

$$\theta = \frac{ak_2 + h_1 + \phi}{ak_2 + h_1}, \text{with } \phi = \frac{e}{d}(a - c)(b_1 - b_2)N_2.$$

Clearly when $b_1 > b_2$, then ϕ is positive, $\theta > 1$ and consequently $A_1A_2 - A_3 > 0$. Alternatively, when $b_1 < b_2$, ϕ is negative, $\theta < 1$ and $A_1A_2 - A_3 > 0$. Hence $A_1A_2 - A_3 > 0$ for $b_1 \neq b_2$. We conclude that the steady state of system (2.2) is *stable* for $b_1 \neq b_2$.

Our main conclusion is that, starting from the familiar Volterra-Lotka system with its neutrally stable oscillations, *any* variability in predation risk among prey (patch) types leads to a model with a unique positive steady state that is at least locally stable. The idea, that heterogeneity in predation risk confers stability to predator-prey systems, does not only survive our scrutiny while studying continuous time models, but, in fact, this idea is reinforced as it applies to both continuous time models and discrete generation models (Bailey *et al.*, 1962; Hassell and May, 1973, 1974; Hassell, 1978, 1984, 2000; May, 1978; Chesson and Murdoch, 1986; Murdoch and Oaten, 1989). However, all of these models assume some statistical distribution of predators over patches (usually the negative binomial distribution). Hence, they do not incorporate that the predation process itself changes the prevailing prey distribution. We emphasize that we did not make a-priori assumptions on prey distributions over the state space. In fact, our model can be interpreted as representing the dynamics in densities of two patch types that do not differ in the number of prey they harbor, but only in the risk of being detected by predators. Thus, stability of the steady state does not necessarily hinge on a clumped spatial prey distribution alone.

REFERENCES

Bailey, V.A., Nicholson, A.J. and Williams, E.J. (1962) Interactions between hosts and parasites when some host individuals are more difficult to find than others. *J. Theor. Biol.* **3**, 1–18.

Bonsall, M.B., French, D.R. and Hassell, M.P. (2002) Metapopulation structures affect persistence of predator–prey interactions. *J. Anim. Ecol.* **71**, 1075–1084.

Charles, J.G. and White, V. (1988) Airborne dispersal of *Phytoseiulus persimilis* (Acarina: Phytoseiidae) from a raspberry garden in New Zealand. *Exp. Appl. Acarol.* **5**, 47–54.

Chesson, P.L. and Murdoch, W.W. (1986) Aggregation of risk: Relationships among host-parasitoid models. *Am. Nat.* **127**, 696–715.

de Roos, A.M., McCauley, E. and Wilson, W. (1991) Mobility versus density limited predator-prey dynamics on different spatial scales. *Proc. R. Soc. Lond. B* **246**, 117–122.

Diekmann, O., Metz, J.A.J. and Sabelis, M.W. (1988) Mathematical models of predator-prey-plant interactions in a patchy environment. *Exp. Appl. Acarol.* **5**, 319–342.

Diekmann, O., Metz, J.A.J. and Sabelis, M.W. (1989) Reflections and calculations on a prey-predator-patch problem. *Acta Appl. Math.* **14**, 23–35.

Ellner, S.P., McCauley, E., Kendall, B.E., Briggs, C.J., Hosseini, P., Wood, S., Janssen, A., Sabelis, M.W., Turchin, P., Nisbet, R.M. and Murdoch, W.W. (2001) Habitat structure and population persistence in an experimental community. *Nature* **412**, 538–543.

Faraji, F., Janssen, A., van Rijn, P.C.J. and Sabelis, M.W. (2000) Kin recognition by the predatory mite *Iphiseius degenerans*: Discrimination among own, conspecific and heterospecific eggs. *Ecol. Entomol.* **25**, 147–155.

Gough, N. (1991) Long-term stability in the interaction between *Tetranychus-urticae* and *Phytoseiulus-persimilis* producing successful integrated control on roses in Southeast Queensland. *Exp. Appl. Acarol.* **12**, 83–101.

Hairston, N.G., Smith, F.E. and Slobodkin, L.B. (1960) Community structure, population control and competition. *Am. Nat.* **94**, 421–424.

Hanski, I.A. and Gilpin, M.E. (1997) *Metapopulation Biology: Ecology, Genetics and Evolution*. Academic Press, London.

Hassell, M.P. (1978) Dynamics of Arthropod Predator-Prey Systems. *Monographs in Population Biology*. Princeton University Press, Princeton, New Jersey.

Hassell, M.P. and May, R.M. (1974) Aggregation of predators and insect parasites and its effect on stability. *J. Anim. Ecol.* **43**, 567–594.

Hassell, M.P. and May, R.M. (1973) Stability in insect host-parasite models. *J. Anim. Ecol.* **42**, 693–726.

Hassell, M.P. (2000) *The Spatial and Temporal Dynamics of Host-Parasitoid Interactions*. Oxford University Press, Oxford.

Hassell, M.P. (1984) Parasitism in patchy environments: Inverse density dependence can be stabilizing. *IMA J. Math. Appl. Med. Biol.* **1**, 123–133.

Holyoak, M. and Lawler, S.P. (1996a) Persistence of an extinction-prone predator-prey interaction through metapopulation dynamics. *Ecology* **77**, 1867–1879.

Holyoak, M. and Lawler, S.P. (1996b) The role of dispersal in predator-prey metapopulation dynamics. *J. Anim. Ecol.* **65**, 640–652.

Huffaker, C.B. (1958) Experimental studies predation: Dispersion factors and predator-prey oscillations. *Hilgardia* **27**, 343–383.

Huffaker, C.B., Shea, K.P. and Herman, S.G. (1963) Experimental studies on predation: Complex dispersion and levels of food in acarine predator-prey interaction. *Hilgardia* **34**, 305–330.

Jansen, V.A.A. (1995a) Effects of dispersal in a tri-trophic metapopulation model. *J. Math. Biol.* **34**, 195–224.

Jansen, V.A.A. (1995b) Regulation of predator-prey systems through spatial interactions: A possible solution to the paradox of enrichment. *Oikos* **74**, 384–390.

Jansen, V.A.A. (1999) Phase locking: Another cause of synchronicity in predator-prey systems. *Trends Ecol. Evol.* **14**, 278–279.

Jansen, V.A.A. (2001) The dynamics of two diffusively coupled predator-prey populations. *Theor. Popul. Biol.* **59**, 119–131.

Janssen, A. and Sabelis, M.W. (1992) Phytoseiid life-histories, local predator-prey dynamics and strategies for control of tetranychid mites. *Exp. Appl. Acarol.* **14**, 233–250.

Janssen, A., van Gool, E., Lingeman, R., Jacas, J. and van de Klashorst, G. (1997a) Metapopulation dynamics of a persisting predator-prey system in the laboratory: Time series analysis. *Exp. Appl. Acarol.* **21**, 415–430.

Janssen, A., Bruin, J., Schraag, R., Jacobs, J. and Sabelis, M.W. (1997b) Predators use volatiles to avoid patches with conspecifics. *J. Anim. Ecol.* **66**, 223–232.

Keeling, M. (2002) Using individual-based simulations to test the Levins metapopulation paradigm. *J. Anim. Ecol.* **71**, 270–279.

Lingeman, R. and van de Klashorst, G. (1992) Local and global cycles in an acarine predator-prey system: A frequency domain analysis. *Exp. Appl. Acarol.* **14**, 201–214.

May, R.M. (1978) Host-parasitoid systems in patchy environments: A phenomenological model. *J. Anim. Ecol.* **47**, 833–844.

McCauley, E., Kendall, B.E., Janssen, A., Wood, S., Murdoch, W.W., Hosseini, P., Briggs, C.J., Ellner, S.P., Nisbet, R.M., Sabelis, M.W. and Turchin, P. (2000) Inferring colonization processes from population dynamics in spatially-structured predator-prey systems. *Ecology* **81**, 3350–3361.

Metz, J.A.J. and Diekmann, O. (1986) Dynamics of Physiologically Structured Populations. *Lecture Notes in Biomathematics 68*. Springer-Verlag, Berlin, Germany.

Metz, J.A.J. and Gyllenberg, M. (2001) How should we define fitness in structured metapopulation models? Including an application to the calculation of evolutionary stable dispersal strategies. *Proc. R. Soc., Lond. B* **268**, 499–508.

Mueller, L.D. and Joshi, A. (2000) *Stability in Model Populations*. Princeton University Press, Princeton, New Jersey.

Murdoch, W.W. and Oaten, A. (1989) Aggregation by parasitoids and predators: Effects on equilibrium and stability. *Am. Nat.* **134**, 288–310.

Nachman, G. (2001) Predator-prey interactions in a nonequilibrium context: The metapopulation approach to modeling "hide-and-seek" dynamics in a spatially explicit tri-trophic system. *Oikos* **94**, 72–88.

Nachman, G. (1981) Temporal and spatial dynamics of an acarine predator-prey system. *J. Anim. Ecol.* **50**, 435–451.

Neubert, M.G., Klepac, P. and van den Driessche, P. (2002) Stabilizing dispersal delays in predator–prey metapopulation models. *Theor. Popul. Biol.* **61**, 339–347.

Oksanen, L., Fretwell, S.D., Arruda, J. and Niemala, P. (1981) Exploitation ecosystems in gradients of primary productivity. *Am. Nat.* **118**, 240–261.

Pallini, A., Janssen, A. and Sabelis, M.W. (1999) Spider mites avoid plants with predators. *Exp. Appl. Ecol.* **23**, 803–815.

Pels, B. and Sabelis, M.W. (1999) Local dynamics, overexploitation and predator dispersal in an acarine predator-prey system. *Oikos* **86**, 573–583.

Pels, B., de Roos, A.M. and Sabelis, M.W. (2002) Evolutionary dynamics of prey exploitation in a metapopulation of predators. *Am. Nat.* **159**, 172–189.

Sabelis, M.W. (1986) The functional response of predatory mites to the density of two-spotted spider mites. In: *Dynamics of Physiologically Structured Populations* (Ed. by A.J. Metz and O. Diekmann), pp. 298–321. Lecture Notes in Biomathematics 68. Springer-Verlag, Berlin, Germany.

Sabelis, M.W. (1990a) How to analyse prey preference when prey density varies? A new method to discriminate between effects of gut fullness and prey type composition. *Oecologia* **82**, 289–298.

Sabelis, M.W. (1990b) Life history evolution in spider mites. In: *The Acari: Reproduction, Development and Life-History Strategies* (Ed. by R. Schuster and P.W. Murphy), pp. 23–50. Chapman and Hall, New York.

Sabelis, M.W. (1992) Arthropod predators. In: *Natural Enemies, The Population Biology of Predators, Parasites and Diseases* (Ed. by M.J. Crawley), pp. 225–264. Blackwell, Oxford.

Sabelis, M.W. and Janssen, A. (1994) Evolution of life-history patterns in the Phytoseiidae. In: *Mites. Ecological and Evolutionary Analyses of Life-History Patterns* (Ed. by M.A. Houck), pp. 70–98. Chapman and Hall, New York.

Sabelis, M.W. and Diekmann, O. (1988) Overall population stability despite local extinction: The stabilizing influence of prey dispersal from predator-invaded patches. *Theor. Popul. Biol.* **34**, 169–176.

Sabelis, M.W. and van der Meer, J. (1986) Local dynamics of the interaction between predatory mites and two-spotted spider mites. In: *Dynamics of Physiologically Structured Populations* (Ed. by J.A.J. Metz and O. Diekmann), pp. 322–344. Lecture Notes in Biomathematics 68. Springer-Verlag, Berlin, Germany.

Sabelis, M.W. and Laane, W.E.M. (1986) Regional dynamics of spider-mite populations that become extinct locally because of food source depletion and predation by phytoseiid mites (Acarina: Tetranychidae, Phytoseiidae). In: *Dynamics of Physiologically Structured Populations* (Ed. by J.A.J. Metz and O. Diekmann), pp. 345–375. Lecture Notes in Biomathematics 68. Springer-Verlag, Berlin, Germany.

Sabelis, M.W., van Alebeek, F., Bal, A., van Bilsen, J., van Heijningen, T., Kaizer, P., Kramer, G., Snellen, H., Veenebos, R. and Vogelezang, J. (1983) Experimental validation of a simulation model of the interaction between *Phytoseiulus persimilis* and *Tetranychus urticae* on cucumber. *OILB-Bulletin SROP/WPRS* **6**, 207–229.

Sabelis, M.W., van Baalen, M., Pels, B., Egas, M. and Janssen, A. (2002) Evolution of exploitation and defense in plant-herbivore-predator interactions. In: *The Adaptive Dynamics of Infectious Diseases: In Pursuit of Virulence Management* (Ed. by

U. Dieckmann, J.A.J. Metz, M.W. Sabelis and K. Sigmund), pp. 297–321. Cambridge University Press, Cambridge, United Kingdom.

Sabelis, M.W., Diekmann, O. and Jansen, V.A.A. (1991) Metapopulation persistence despite local extinction: Predator-prey patch models of the Lotka-Volterra type. *Biol. J. Linn. Soc.* **42**, 267–283.

Sabelis, M.W., van Baalen, M., Bakker, F.M., Bruin, J., Drukker, B., Egas, M., Janssen, A., Lesna, I., Pels, B., van Rijn, P.C.J. and Scutareanu, P. (1999a) Evolution of direct and indirect plant defence against herbivorous arthropods. In: *Herbivores: Between Plants and Predators* (Ed. by H. Olff, V.K. Brown and R. H. Drent), pp. 109–166. Blackwell Science, Oxford.

Sabelis, M.W., van Baalen, M., Bruin, J., Egas, M., Jansen, V.A.A., Janssen, A. and Pels, B. (1999b) The evolution of overexploitation and mutualism in plant-herbivore-predator interactions and its impact on population dynamics. In: *Theoretical Approaches to Biological Control* (Ed. by B.A. Hawkins and H.V. Cornell), pp. 259–282. Cambridge University Press, Cambridge.

Schausberger, P. and Croft, B.A. (2001) Kin recognition and larval cannibalism by adult females in specialist predaceous mites. *Anim. Behav.* **61**, 459–464.

Strong, W.B., Slone, D.H. and Croft, B.A. (1999) Hops as a metapopulation landscape for tetranychid-phytoseiid interactions: Perspectives for intra- and interplant dispersal. *Exp. Appl. Acarol.* **23**, 581–597.

van de Klashorst, G., Readshaw, J.L., Sabelis, M.W. and Lingeman, R. (1992) A demonstration of asynchronous local cycles in an acarine predator-prey system. *Exp. Appl. Acarol.* **14**, 185–199.

Van Baalen, M. and Sabelis, M.W. (1995a) The dynamics of multiple infection and the evolution of virulence. *Am. Nat.* **146**, 881–910.

Van Baalen, M. and Sabelis, M.W. (1995b) The milker-killer dilemma in spatially structured predator-prey interactions. *Oikos* **74**, 391–413.

Walde, S.J. (1994) Immigration and the dynamics of a predator-prey interaction in biological control. *J. Anim. Ecol.* **63**, 337–346.

Weisser, W.W., Jansen, V.A.A. and Hassell, M.P. (1997) The effects of a pool of dispersers on host–parasitoid systems. *J. Theor. Biol.* **189**, 413–425.

FURTHER READING

Walde, S.J. and Nachman, G. (1999) Dynamics of spatially structured spider mite populations. In: *Theoretical Approaches to Biological Control* (Ed. by B.A. Hawkins and H.V. Cornell), pp. 163–189. Cambridge University Press, Cambridge.

Ecological and Evolutionary Dynamics of Experimental Plankton Communities

GREGOR F. FUSSMANN, STEPHEN P. ELLNER, NELSON G. HAIRSTON, JR., LAURA E. JONES, KYLE W. SHERTZER AND TAKEHITO YOSHIDA

I. SUMMARY

We used microcosm systems to test whether simple mathematical models can be valid descriptions of population and community dynamics. Our conclusion is that *a priori* mathematical formulations of interacting populations are unlikely to produce completely satisfying predictions because they tend to ignore important biological mechanisms. We employed the feedback between microcosm experiments and increasingly refined dynamic models to identify the critical biology governing population cycles in our system, a predator-prey (rotifer-algal) chemostat. Here we summarize the results from five years of work with this system, which confronted us with unexpected yet resolvable biological complexities. First, we found that age structure of the predator population is necessary to generate qualitatively correct predictions of population dynamics (stability versus cycles). Second, we identified rapid evolution of both the algae and rotifers (in separate instances) as critical processes occurring on the same time scale as the ecological dynamics. We have learned that microcosms may not just serve as a means to check model assumptions, but that the results of microcosm studies can lead to novel insights into the functioning of biological communities.

ADVANCES IN ECOLOGICAL RESEARCH VOL. 37 0065-2504/05 $35.00

DOI: 10.1016/S0065-2504(04)37007-8

II. INTRODUCTION

Natural communities are complex associations of organisms that interact with each other and with the environment. Communities are intrinsically dynamic: both the number of populations comprising a community and the number of individuals within populations can vary over time (e.g., Tilman, 1996; Kendall *et al.*, 1999). One central goal of ecology is to understand the temporal fluctuations in population numbers that occur in communities—the "abundance" half of one standard definition of ecology (Krebs, 2002). Although there are only a handful of dynamic states that populations may exhibit (stable equilibrium, deterministic extinction, stable population oscillations, and irregular fluctuations), there is a long list of factors that may interact to determine the community dynamics (e.g., competition, predation, parasitism, mutualism; age, stage and genetic structure, and evolution of populations; spatial structure of the habitat, climate, physical and chemical parameters). Thus, a key step in analyzing the community dynamics is to untangle this mixture of interacting factors and to identify those essential for the observed dynamics.

In this chapter we describe our (eventually successful) attempt to reconcile the dynamics of a simple laboratory community consisting of algae and rotifers with the predictions of an even simpler mechanistic model. Our initial goals were to establish a predator-prey system (planktonic algae and rotifers) in laboratory culture and to develop a tractable model that would adequately describe the range of dynamic behaviors. Following these "preparatory steps" our plan was to run the laboratory system under conditions producing sustained, regular predator-prey oscillations and then "force" it by periodic enrichment with nutrients. In theory (Kot *et al.*, 1992), this treatment should drive the population dynamics into deterministic chaos (Fig. 1), a dynamic state frequently predicted by population models (May and Oster, 1976) but rarely observed in the field or the laboratory (Ellner and Turchin, 1995; Costantino *et al.*, 1997). By doing so, we hoped to gain insight into the conflict between the apparent rarity of chaos in nature, despite its prevalence in models and the demonstration (in the *Tribolium* model system) that chaos is a biological possibility (Costantino *et al.*, 1997). The ability to induce chaos "on demand" in a real predator-prey culture would allow us to investigate open questions about nonlinear population dynamics, such as "are noise and chaos distinguishable in population dynamics?" and "do organisms experiencing chaotic dynamics evolve life-history characteristics that reduce the likelihood of chaos?" However, our "simple" experimental system yielded some biological surprises that soon led our research into different but no less interesting avenues.

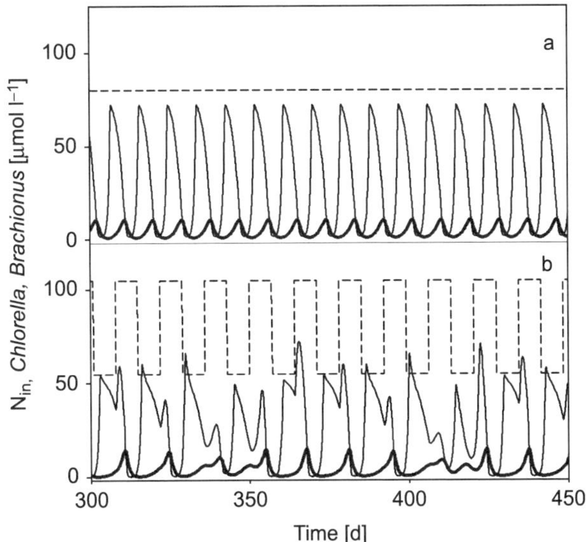

Figure 1 Double Monod model (1) to (4) simulations with (a) constant and (b) pulsed nutrient inflow into the chemostat. Parameter values (see Fussmann *et al.*, 2000): $\delta = 1.2\ \mathrm{d}^{-1}$; $b_C = 3.3\ \mathrm{d}^{-1}$; $b_B = 2.25\ \mathrm{d}^{-1}$; $K_C = 4.3\ \mu\mathrm{mol\ l}^{-1}$; $K_B = 15\ \mu\mathrm{mol\ l}^{-1}$; $\epsilon = 0.25$. In (a) $N_{in} = 80\ \mu\mathrm{mol\ l}^{-1}$, in (b) $N_{in} = 80 \pm 25\ \mu\mathrm{mol\ l}^{-1}$, with a pulsing period of 7 d. Thin solid line: *Chlorella*; thick solid line: *Brachionus*; dashed line: nitrogen inflow N_{in}. *Chlorella* and *Brachionus* concentrations are plotted in model units μmol nitrogen l^{-1}.

A quote attributed to Albert Einstein could serve as the motto for our theoretical and experimental explorations: "Everything should be made as simple as possible, but not simpler." Following this guideline and starting with a model system made tractable by allowing only for trophic interactions among populations, cumulative experimental evidence (often surprising to ourselves) led us to re-evaluate repeatedly the set of mechanisms we considered essential for explaining the dynamics observed in the laboratory cultures. This is not to say that our study was a string of failures with a happy ending. We would rather understand it as a demonstration of how hypothesis-driven, experimental research can progressively advance ecological understanding (Turchin, 2003). Mathematical modeling enabled us to formulate hypotheses about the expected community dynamics. Experimentation sent us back to the drawing board to refine our hypotheses and to identify the most promising among the competing hypotheses. Repeating this cycle resulted in our identifying evolution as a mechanism essential for explaining the experimental dynamics. The most important lesson we learned is that even in an extremely simplified and heavily controlled model system a multitude of essential biological processes can operate together, precluding the success of overly naïve modeling approaches. However, purposeful investigation can nonetheless identify the essential biological factors, which can lead to models with more predictive power and better understanding of the mechanisms underlying population dynamics.

We present this work in (nearly) historical order; beginning with Section III, our initial model formulation and the minor adjustments needed to obtain qualitative agreement with the dynamic patterns revealed by the experimental data. Section IV introduces theoretical evidence that the cyclical algal-rotifer dynamics we observed could not result from classical predator-prey interaction alone but that some additional biology was needed to explain the observed population oscillations. We mathematically tested several competing hypotheses about that missing biological mechanism and identified evolution of the algae as the most likely explanation. Section V describes a successful experimental test of this hypothesis showing that evolution (as cyclical clonal selection) does indeed play the role suggested by the models. Finally, Section VI provides evidence that rapid evolution of rotifers can also influence the population dynamics.

III. PREDATOR AND PREY IN THE CHEMOSTAT—A SIMPLE STORY?

Chemostats are laboratory flow-through microcosms used for the continuous culture of planktonic organisms (e.g., Boraas, 1980; Walz, 1993). Chemostats are very suitable for studying population dynamics because all basic processes occurring in the system can be easily translated into differential equations. We cultured two model organisms in the chemostats, the rotifer *Brachionus calyciflorus* and the unicellular green alga *Chlorella vulgaris*, to explore the dynamics of populations interacting in a predator-prey relationship. While the growth rate of *Brachionus* depends on the concentration of *Chlorella*, the growth rate of *Chlorella* depends on the availability of nitrogen, the limiting nutrient in our experiments. The "double Monod model" (DMM) (Jost *et al.*, 1973a; Nisbet *et al.*, 1983) is a simple mathematical formulation of this chemostat system. "Double Monod" refers to the fact that the uptake and per-capita recruitment rate of *Chlorella* (on nitrogen) and *Brachionus* (on *Chlorella*) are monotonically increasing but saturating (=Monod) functions (F_B and F_C in (4)) of the nutrient and prey concentrations, respectively (Monod, 1950; Holling, 1959), a more realistic assumption than the overly simplistic linear uptake functions of very early models.

$$\frac{dN}{dt} = \delta(N_{in} - N) - F_C(N)C \tag{1}$$

$$\frac{dC}{dt} = F_C(N)C - \frac{1}{\epsilon}F_B(C)B - \delta C \tag{2}$$

$$\frac{dB}{dt} = F_B(C)B - \delta C \tag{3}$$

$$\text{with} \quad F_C(N) = \frac{b_C N}{K_C + N} \quad \text{and} \quad F_B(C) = \frac{b_B C}{K_B + C} \tag{4}$$

In this chemostat DMM, N is the concentration of nitrogen, C the concentration of *Chlorella,* and B the concentration of *Brachionus.* ϵ is the assimilation efficiency of *Brachionus.* Nitrogen is continuously added to the system at the dilution rate δ, all components are removed from the system at the same δ. N_{in} is the nitrogen concentration in the inflow medium, b_C and b_B are the maximum birth rates of *Chlorella* and *Brachionus.* K_C and K_B are the half saturation constants of *Chlorella* and *Brachionus.* The model equations are balanced, i.e., nitrogen removed from one trophic level is transferred to the next level up, only decremented by the conversion efficiency. Thus, δ and N_{in} are the parameters available for experimental manipulation.

The DMM predicts washout of only *Brachionus* or *Brachionus* and *Chlorella* at high δ. Lowering δ results in co-existence of algae and rotifers at either an equilibrium or at sustained, regular oscillations (stable limit cycles (Fig. 1a), whose amplitudes increase with decreased δ. In the model, periodic input of nutrients at the right intensity and frequency forces otherwise regular oscillations into erratic, chaotic fluctuations (Fig. 1b). Chaos is likely to occur when the frequencies of intrinsic and extrinsic fluctuations are incommensurate, generating "dynamic friction" between the two oscillators. It is possible to imagine an experiment in which we attempt to produce chaos in an oscillating chemostat culture by pulsing the nutrient input. In order to minimize unintended side effects, we wanted to do this with the smallest possible amplitude of variation in the nutrient supply rate. We then asked several experts on nonlinear oscillations what ratio between the forcing and natural cycle frequencies would be most effective. We were surprised to find that our panel of experts gave very different answers. They had worked on different kinds of models, and the answer to our question appeared to be highly system-specific. So as a first step towards a forcing experiment, which was still our long-term goal, we needed a trustworthy model for our system.

A look at the results from our system (Figs. 2 and 3; Fussmann *et al.*, 2000) reveals that the experimental dynamics are in partial agreement with

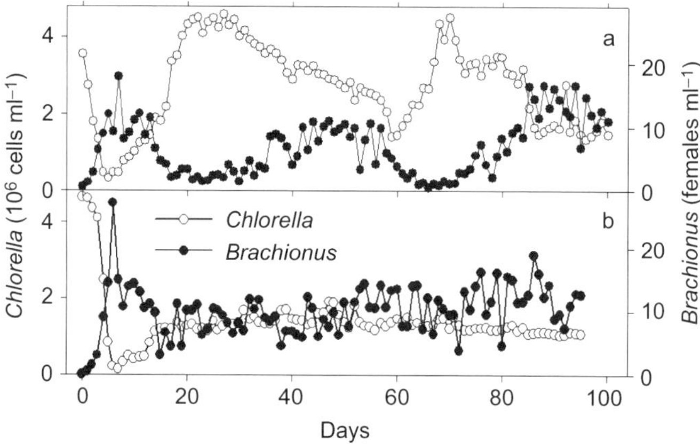

Figure 2 Two example experimental results of equilibria and oscillations in algal-rotifer chemostats. Nitrogen inflow concentration $N_{in} = 80\ \mu\text{mol l}^{-1}$. See Fussmann *et al.* (2000) and Yoshida *et al.* (2003) for more examples. (a) Long-period, out-of-phase cycles at dilution rate $\delta = 0.64\ \text{d}^{-1}$. (b) Equilibria (plus noise) at $\delta = 0.12\ \text{d}^{-1}$. Reproduced with permission from Nature Publishing Group.

our model predictions. We found co-existence at equilibria when δ was high, and decreasing δ destabilized the system to oscillations (Fig. 3c and f). However, we also identified three major discrepancies with model predictions.

1. Decreasing δ to very low values did not produce high-amplitude oscillations, as the DMM predicts (Fig. 3a and d). Rather, the system re-stabilized to equilibrium dynamics (Fig. 3c and f).
2. Cycle periods were irregular and too long compared to prediction. The predicted period was <10 days; those observed were >20 days.
3. The predator and prey cycles were nearly exactly out of phase, meaning that maxima of algae and minima of rotifers (and vice versa) occurred at almost the same time.

These discrepancies between observed and predicted dynamics prevented us from pursuing our original plan to force the chemostat system into chaotic population dynamics. First, we needed to develop a better understanding of the dynamic behavior of this deceptively simple system.

One obviously unsatisfying assumption of the DMM is that the predators experience no losses other than washout from the chemostat vessel. At very low dilution rates (and, therefore, slow washout), this leads to extreme, sustained oscillations because even a miniscule nutrient input suffices to maintain positive net growth of algae and rotifers. Introducing a mortality term for the rotifers into the DMM produces stable dynamics at low dilution rates (Nisbet *et al.*, 1983), a more realistic behavior that has also

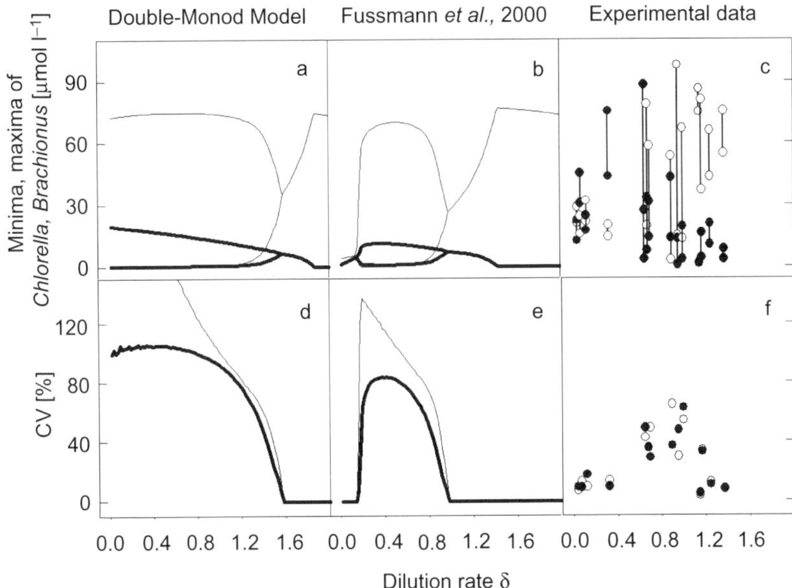

Figure 3 Stable and unstable dynamics in model simulations and experimental predator-prey cultures as a function of chemostat dilution rate δ. Model predictions (a, b, d, e; t = 1500 to 2000 days) and experimental data (c, f) for minima, maxima (a to c) and coefficient of variation (CV = standard deviation/mean [%]; d to f) of time series. Thin lines or empty symbols: *Chlorella.* Thick lines or filled symbols: *Brachionus*. Headings refer to the model used for computation. Experimental data computed for 14 chemostat trials. Lines in (c, f) connect minima and maxima of one chemostat trial. Large differences between minima and maxima and high CV indicate unstable, oscillating dynamics, low differences and CV (noisy) equilibria. Because minima/maxima in long experimental time series are usually outliers, minima/maxima were computed from smoothed data (local polynomial regression). See Fussmann *et al.* (2000) for transformation from individual to molar concentrations in (c) and for computation of CV in (f). $\lambda = 0.4\ \mathrm{d}^{-1}$, $m = 0.055\ \mathrm{d}^{-1}$, other parameters as in Fig. 1. Reproduced with permission from AAAS/Science.

been observed in other multispecies chemostat studies (Jost *et al.*, 1973b; Dent *et al.*, 1976). For our specific system it became evident (from counts of dead rotifers and eggs in the cultures) that the rotifers suffered two different kinds of loss in addition to washout: mortality in the chemostat and loss of fecundity as they senesce. To accommodate these two factors we partitioned (3) into two equations, one for the total concentration of *Brachionus* B and one for the concentration of reproducing *Brachionus* R (Fussmann *et al.*, 2000):

$$\frac{dN}{dt} = \delta(N_{in} - N) - F_C(N)C \tag{5}$$

$$\frac{dC}{dt} = F_C(N)C - \frac{1}{\epsilon}F_B(C)B - \delta C \tag{6}$$

$$\frac{dR}{dt} = F_B(C)R - (\delta + m + \lambda)R \tag{7}$$

$$\frac{dB}{dt} = F_B(C)R - (\delta + m)B, \tag{8}$$

with $F_C(N)$ and $F_B(C)$ defined in (4), and m and λ as the rotifer instantaneous mortality and senescence rates, respectively. We used counts of dead rotifers from chemostat cultures to determine m, counts of subitaneous eggs per rotifer provided estimates for fecundity and λ.

With the addition of rotifer mortality and senescence, the model correctly predicted the transitions between different qualitative dynamic behaviors observed in the chemostat system: extinction, equilibria, and population cycles (Fig. 3b, c, e and f). Its successes demonstrated that a simple mechanistic model could capture many of the major features of complex multispecies dynamics. However, the model was still unable to account for the observed cycle period and phase relations. These failures indicated that the chemostat system held ecological complexity yet to be discovered.

IV. TESTING HYPOTHESES OF MECHANISM

Because the Fussmann *et al.* (2000) model was unable to account for some key features in the observed cycles, we inferred that the model neglected at least one biological mechanism crucial to the system—but which one?

A closer look at the predator's per-capita recruitment rates provided a clue. To begin, we smoothed the observed time series of prey and predator densities with penalized regression splines (Simonoff, 1996). Such splines are continuously differentiable and thereby provide smoothed measures of population growth rates over time. To estimate the predator's per-capita recruitment rates over time, we divided smoothed population growth rates by smoothed densities and adjusted for dilution and mortality rates. Plots of those recruitment rates against smoothed prey densities show a revealing pattern.

For any level of prey density, the Monod function (4) predicts a single level of predator recruitment rate. As populations cycle, theory would have the recruitment rate slide back and forth along the Monod curve. That did not happen. Instead, as our experimental populations cycled, recruitment rate itself cycled (e.g., Fig. 4), with more than one level of predator recruitment rate for any level of prey density. Furthermore, these rates were consistently lower after peaks in predator density and higher after troughs.

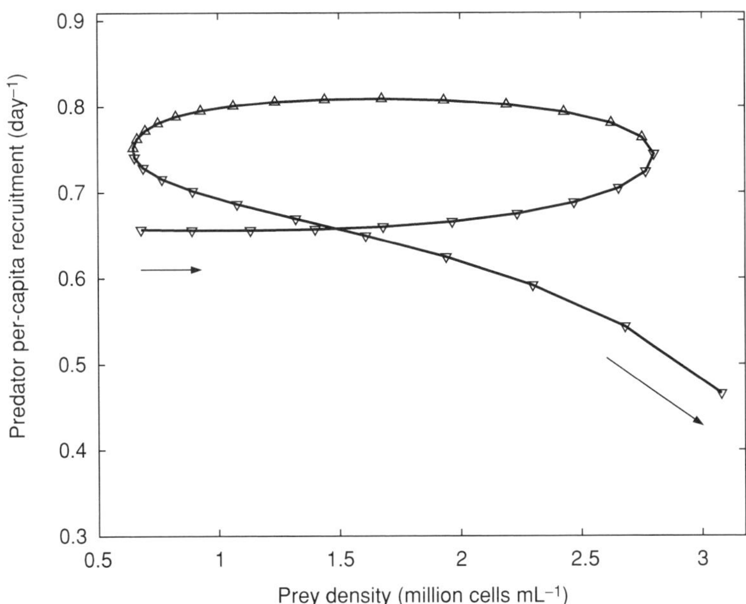

Figure 4 Estimated predator (rotifers) per-capita recruitment rates plotted as a function of prey (algae) density. Data were from an experiment where populations cycled (in this example, dilution rate = 0.69 d^{-1}, but the pattern was pervasive). For any level of prey density, per-capita recruitment rates were lower when predator density was decreasing (downward triangles), and higher when increasing (upward triangles). Arrows indicate the progression of time. Reproduced with permission from Blackwell Publishing Group.

We initially suspected predator interference. However, predator interference alone could not account for the pattern, because there were many pairs of times (t_1, t_2) when population densities were approximately equal [$C(t_1) = C(t_2)$, $B(t_1) = B(t_2)$], but recruitment rates were still quite different [$F_B(t_1) \neq F_B(t_2)$] (Shertzer *et al.*, 2002). Predator per-capita recruitment rates were consistently low after peaks in predator density and high after troughs. This suggested an explanation drawing increased attention in planktonic communities: organism quality (Nelson *et al.*, 2001; Scheuerell *et al.*, 2002).

Our first hypothesis was that algal size might be changing through the course of the population cycles, hence algal cell counts would not properly reflect the actual amount of food available to rotifers. To investigate this we used an automated particle counter to track changes in the algal size distribution through the prey cycle. We quickly discovered that there were no systematic trends in algal size that could account for the trends in rotifer recruitment seen in Fig. 4.

We next undertook a more systematic search for the missing mechanism. We developed four simple mechanistic models, each an extension to the original Fussmann *et al.* (2000) model, incorporating a single additional mechanism for which some empirical support existed (Shertzer *et al.*, 2002). The first model posited that viability of rotifer eggs increases with food availability; the second that the nutritional value of algae increases with their cellular carbon:nitrogen content; the third that rotifer waste products limit population growth; and the fourth model assumed that the algae could evolve rapidly in response to predation. Abrams and Matsuda (1997) had shown that a model including prey evolution in response to varying predation pressure could produce out-of-phase predator-prey cycles qualitatively similar to those observed in our system, and our model (described in the following text) was similar to theirs in assuming a trade-off in prey between intrinsic growth rate and vulnerability to predation. The new models were treated as hypotheses to be tested. Each was fit to the time-series data using two separate methods: trajectory matching and probe matching (Kendall *et al.*, 1999). The models were evaluated for the ability to fit prey and predator densities and to explain the key features in observed cycles that the original model could not. Those models (hypotheses) inconsistent with the chemostat data could be discarded.

After all four models were evaluated against the data, only one hypothesis was left standing—rapid prey evolution (Shertzer *et al.*, 2002). In our model for that hypothesis, like the Abrams and Matsuda (1997) model, prey are able to evolve defense against predation, but only with a trade-off in their maximum population growth rate. Our model differed in assuming that the trade-off is mediated by a quantitative physiological trait (z), that determines both maximum growth rate and the vulnerability to predation and is under stabilizing selection in the absence of predation. The dynamics of z were assumed to depend on the partial gradient of the algal per-capita growth rate with respect to the trait value of an invading mutant, following the standard quantitative-trait model (Lynch and Walsh, 1998):

$$\frac{dz}{dt} = V\frac{\partial}{\partial z_i}\left(\frac{1}{C_i}\frac{dC_i}{dt}\right), \tag{9}$$

where V is the additive genetic variance, z_i, C_i are the trait value and density of a rare invading algal clone and the right-hand side is evaluated at $z_i = z$, the current trait mean in the population. As the value of z increases, the model dictates that the vulnerability to predation decreases, but so does the population growth rate in the absence of predators. Thus, the quantitative-trait model explicitly assumes a nonlinear trade-off between algal vulnerability to predation and maximum population growth rate. Only the prey evolution model reproduced the long, out-of-phase predator-prey cycles we observed in our chemostats. Our success with the quantitative trait model provided more than just a satisfying explanation of the observed cycles: it identified a plausible mechanism that could be tested directly with fresh experiments.

V. RAPID EVOLUTION: A CLONAL APPROACH

Given the success of the prey evolution model in producing a qualitative fit to our empirical observations, we then returned to the organisms themselves to test its core predictions: (1) a trade-off in algal phenotypes between defense against rotifer grazing and the cost of that defense, (2) the existence of multiple algal genotypes that differ in their position on the trade-off curve, and (3) prey evolution in the chemostats that produces long, out-of-phase predator-prey cycles. We found that chemostats containing clonally variable algae had dynamics that match closely to a model modified (see the following) to permit clonal evolution, whereas chemostats containing a single algal genotype had dynamics matching those of standard non-evolutionary predator-prey models.

To test for the existence of different *Chlorella* genotypes along the trade-off curve, algae from our stock cultures were exposed to either of two different selection conditions (rotifers present or rotifers absent), and then they were used to determine the effects of this selection under conditions common to both groups (Yoshida *et al.*, 2004). We measured "algal food value" as a proxy for the strength of algal defense against predation. Algal food value was estimated as rotifer population growth rate when fed algae at a sufficiently high concentration. Algae cultivated under constant and intense rotifer grazing pressure became lower in food value relative to algae grown in the absence of rotifers but with a comparable mortality rate imposed by an elevated chemostat dilution rate (Table 1). The low-food-value algae were also slightly smaller, denser in terms of C and N content, but lacked obvious structural differences under transmission electron microscopy (T. Yoshida, 2003, unpublished data). This heritable response to rotifer predation shows that the selected low food-value algal genotypes are better able to survive rotifer grazing, although the mechanism for this remains to be determined. However, in contrast to assumptions made in our quantitative trait model and the Abrams-Matsuda (1997) model (Section IV), no significant difference was found in algal maximum population growth rate under nutrient rich conditions (Table 1). Instead, algae selected in the presence of rotifers were competitively inferior in the sense that their growth rate was lower in nutrient *deficient* conditions, compared to algae selected in the absence of rotifers (Table 1). Thus, the cost of algal defense only becomes apparent when nutrients are scarce. In essence, the

Table 1 Evolutionary trade-off between algal food value and competitive ability for *Chlorella vulgaris*

Treatment	Algal food value (d^{-1})	Algal growth rate (d^{-1})	
		80 μmol nitrate l^{-1}	1 μmol nitrate l^{-1}
Rotifers present	1.07 ± 0.05	2.00 ± 0.05	1.25 ± 0.02
Rotifers absent	1.58 ± 0.05	2.02 ± 0.02	1.73 ± 0.03

Values are means ± S.E.M. Algal food value was measured by rotifer population growth rate when feeding on algae at sufficiently high concentration. Difference in algal growth rate at a nutrient-deficient condition (1 μmol nitrate l^{-1}) but not at a nutrient-rich condition (80 μmol nitrate l^{-1}) indicates that algae selected under rotifer grazing pressure are poorer competitors for scarce nutrients but have the same maximum population growth rate when nutrients are plentiful.

low-food-value algae have a higher half-saturation constant for growth response to N in Monod growth kinetics. Contrary to the assumptions of the quantitative-trait model, the results demonstrated the existence of different genotypes in the algal population and a trade-off between algal food value and competitive ability among clones.

These findings required us to change the evolutionary model and re-examine whether it could still account for observed cycle properties. At the same time we abandoned the simplifying assumption of quantitative trait dynamics, which had been imposed so that the four contending models would have similar numbers of parameters, in favor of explicitly modeling clonal selection. Clonal selection is appropriate because *C. vulgaris* is obligately asexual (Pickett-Heaps, 1975), and because we demonstrated the presence of multiple clones in our laboratory populations (Table 1; Yoshida *et al.*, 2004). Just as importantly, a clonal model allows genetic variation to be a dynamic variable in our model whereas variation is fixed in the quantitative trait model. It turns out that allowing clonal variation to respond to selection yields a much better fit to the dynamics we observe. Thus, we replaced the single algal variable with a set of competing clones specified in terms of their food value, p, and related competitive ability. Here, p represents relative value as food for rotifers, and ranges between 1 ("good") and ≈0 ("poor"), with lower p clones having a reduced predation risk. Thus, low food value comes at the cost of reduced fitness when nutrients are scarce. The relationship between food value and fitness is conveniently defined by a trade-off curve. In our experiments (noted above), the cost of defense was only evident when nutrients are scarce, so we assumed a trade-off between algal food value, p, and the half-saturation constant for uptake of the limiting nutrient, K_c. A two-parameter trade-off curve (shape parameter α_1, and scale or "cost" parameter α_2) was used:

$$K_c(p) = K_{c,\min} + \alpha_2(1 - p^{\alpha_1})^{1/\alpha_1}. \tag{10}$$

Note that K_c attains its minimum value at $p = 1$, so reducing a clone's food value also effectively reduces its ability to compete for nutrients. Examples of trade-off curves are found in Fig. 5a and c. In the simulations we set a minimum food value, $p_{\min}$, above which food values were randomly chosen, since a food value too close to 0 would cause rotifer populations to collapse immediately (Table 1), contrary to our observations.

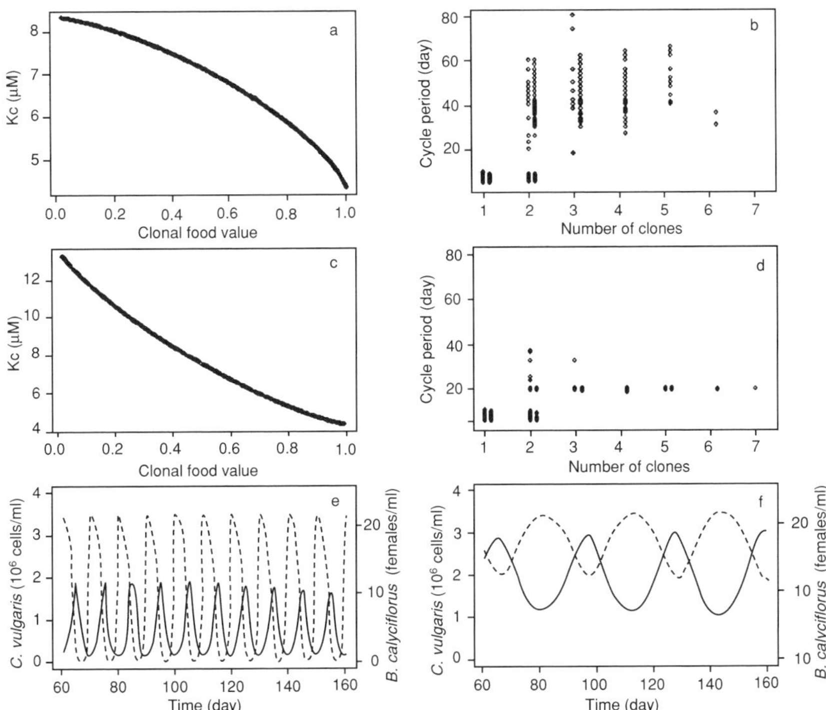

Figure 5 Model-predicted effects of prey clonal diversity on population dynamics (from Yoshida *et al.*, 2003). (a, c) Trade-off curves between algal food value and competitive ability (where competitive fitness is represented by half-saturation, K_c, for nutrient uptake by algae), with $\alpha_1 > 1$ (a) and $\alpha_1 < 1$ (c). (b, d) Effects of clonal diversity on predator-prey cycle length for the two trade-off curves between given $\alpha_1 > 1$ (b) and $\alpha_1 < 1$ (d). Cycle periods in (b) and (d) result from 700 simulation runs each for one, two, three, five, or seven clones initially present. Cycle periods are plotted against the number of clones surviving after the model is run long enough for competitively inferior clones to be gone extinct. 'Degenerate' cases (in which some of the clones initially present were eliminated) are plotted immediately to the right of non-degenerate cases in panels (b) and (d). (e, f) Cycles predicted by the model for the *Brachionus* predator (solid line) and *Chlorella* prey (dotted line): (e) single-clone system; (f) multiple-clone system.

Where the original model (5) to (8) assumes that rotifer feeding rate is a function of total algal density, our clonal model allows rotifers to respond to the relative abundance of clones with different food values. The equation for rotifer feeding rate is based on the "clearance rate," which is the volume of water per unit time that a rotifer clears of prey. Feeding trials with our study organisms (T. Yoshida, 2003, unpublished data) showed that clearance rate increases as algal density falls, but eventually reaches a maximum value at some critical algal density C^*. Further decreases in algal density below C^* are not compensated by increases in clearance rate. Above C^*, the total feeding rate is described well by a Monod equation, as in (4). This led us to use the following equation for the rotifer clearance rate:

$$g(\tilde{C}) = G/(K_b + \max(\tilde{C}, C^*)), \quad \text{where} \quad \tilde{C} = \sum_i p_i C_i \tag{11}$$

Here G is a parameter determining the maximum clearance rate, and p_{i} is the food value of the ith clone; therefore, $\tilde{C}$ is the total food value of the algal population. An alternate form of (11) was also tried, in which clearance rate depended on total algal density irrespective of food value, but our experimental data from cycling populations could be fitted better using (11). The rate at which rotifers consume algae with food value p_{i} is therefore

$$F_{b,i} = C_i p_i G/(K_b + \max(\tilde{C}, C^*)) \tag{12}$$

and consequently the total feeding rate on the entire *Chlorella* population is

$$F_b(\tilde{C}) = \tilde{C}G/(K_b + \max(\tilde{C}, C^*)) \tag{13}$$

In addition, we expressed the clonal model in terms of cell counts rather than nitrogen concentration, because this made it easier to do parameter estimation by calibrating the model against experimental cell-count data. The model then becomes (Yoshida *et al.*, 2003, online supplement):

$$\begin{aligned}
\frac{dn}{dt} &= \delta(VN_{in} - n) - \sum_{i=1}^{q} F_{c,i}(n/V)c_i \\
\frac{dc_i}{dt} &= \chi_c F_{c,i}(n/V)c_i - F_{b,i}b - \delta c_i, \quad i = 1, 2, \ldots, q \\
\frac{db}{dt} &= \chi_b F_b(\tilde{C})r - (\delta + m)b \\
\frac{dr}{dt} &= \chi_b F_b(\tilde{C})r - (\delta + m + \lambda)r
\end{aligned} \tag{14}$$

Here V is the chemostat volume, $n = NV$ is the concentration of nitrogen (μmol per chemostat), c_i is the cell count for *Chlorella* clone i (10^9 cells per chemostat), q is the number of clones, and b and r are the total population and the fertile population of *Brachionus* (individuals per chemostat), respectively. The χ parameters are conversions between consumption and recruitment and were not present in (5) to (8), because all state variables there were expressed in terms of total nitrogen.

Bringing the evolutionary model into line with our subsequent experiments did not change its fundamental prediction: evolution of algal food value substantially alters population cycles. Model simulations with randomly generated sets of 1, 2, 3, 5, or 7 clones show that when multiple clones are present, long cycles are possible (Fig. 5b and d) irrespective of the trade-off curve specified between food value and competitive ability (Fig. 5a and c). If the trade-off curve is concave *down* (Fig. 5a), it is more likely that the surviving co-existing clones are two extremes in food value. However, where the trade-off curve is concave *up* (Fig. 5c), intermediate (in terms of food value) algal types are most likely to persist.

Cycle lengths for two or more co-existing clones varied in period. Short cycles could occur if the algal population only contains two very similar clones, but with more than two clones short cycles become rare. In contrast, a single-clone system always produces relatively short cycles with the classic phase relation: peaks in predator abundance follow peaks in prey abundance by a quarter cycle (Fig. 5e). However, multi-clone cycles with two or more phenotypically divergent clones have phase relations resembling those observed, with algae and rotifers almost exactly out of phase (Fig. 5f), as a result of ongoing changes in clonal frequency driven by rotifer predation and competition for nutrients.

These cycle properties result from low food-value clones becoming dominant while the algal population is being reduced by grazing. Then, once the rotifer population has crashed due to food shortage, the algal density again increases but the rotifers do not respond immediately. Rotifer populations cannot increase until the algae return to high density, at which point the higher food-value clones (which are better competitors) increase in abundance. The rotifers can then increase, eventually becoming abundant enough to start grazing the algae down and start another round of the cycle.

To test these model predictions we returned to the laboratory, manipulating prey evolution by altering clonal diversity (i.e., genetic variability) in the prey population. We initiated replicated chemostat trials either with a single clone as an evolutionarily "stagnant" population or with multiple clones as an evolutionarily "active" one, at dilution rates and nutrient concentrations that would produce population cycles. We then compared cycle periods and phase relations between the treatments (Yoshida *et al.*, 2003).

Exactly as predicted, the population cycles in the single-clone, evolutionarily stagnant treatment were much shorter and of smaller variance than those in the multiple-clone treatment. In addition, the single-clone treatments had a much shorter phase delay, just slightly longer than the classical quarter cycle (Figs. 6 and 7). Furthermore, the clonal model was able to

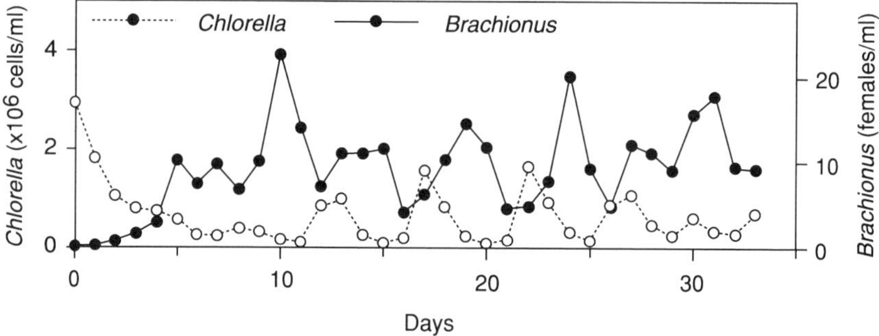

Figure 6 Example experimental result showing the short-period, classic-phase population cycles of rotifer-algal systems where the algal population consisted of a single clone. Dilution rate $\delta = 0.65\,d^{-1}$. Note that population cycles were long and out-of-phase when using multiple-clone algal population (Figure 2, also see more examples in Yoshida *et al.*, 2003). Reproduced with permission from Nature Publishing Group.

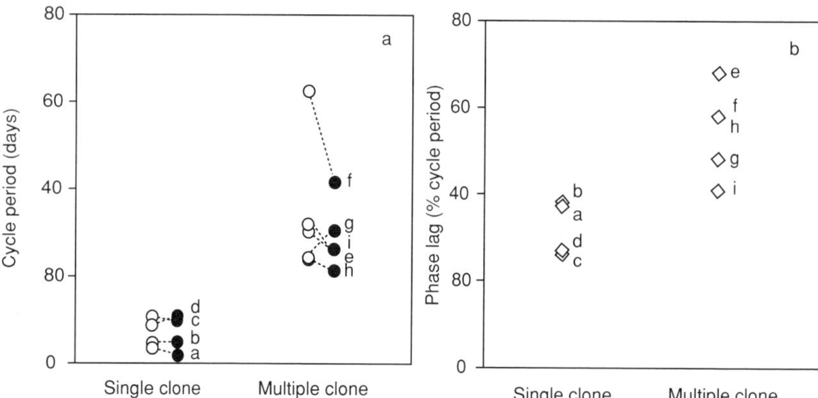

Figure 7 Periods (a) and phase lags (b) of population cycles of *B. calyciflorus* (filled circles in left panel) and *C. vulgaris* (open circles in left panel). Lines connect period estimates from the same experimental trial, and the labels correspond to the different trials. The large difference between prey and predator period estimates by spectral analysis in experiment f resulted from the strongly non-sinusoidal shape of the cycles. From Yoshida *et al.*, 2003. Reproduced with permission from Nature Publishing Group.

match the experimentally determined bifurcation diagram of the system (the transition points between stable and cyclic dynamics as the dilution rate is varied) more accurately than the non-evolutionary model (Fig. 8).

Because the algal taxon that we used, *C. vulgaris*, is obligately asexual, it is perhaps necessary to explain why we consider the process of cyclical clonal replacement to be algal evolution rather than algal community

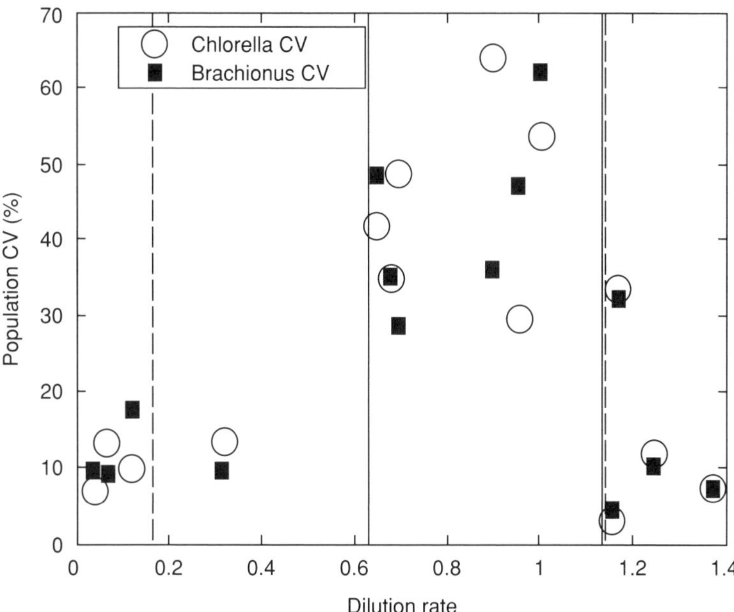

Figure 8 Bifurcation diagram comparing experimental results with model predictions for stability versus cycles in the chemostat system, as a function of the dilution rate δ at $N_{\text{in}} = 80\ \mu\text{mol}$ nitrate 1^{-1}. Vertical lines show the predicted locations of Hopf bifurcations, yielding stability at low and high dilution rates and cycles at intermediate rates. Solid lines are the predictions of the clonal selection model, dashed lines the predictions of the single-clone model. Plotted values (circles and squares) are the coefficient of variation (CV) of predator and prey populations, corrected for effects of sampling error; these are re-drawn using the same values as in Fig. 3 of Fussmann *et al.* (2000). Two of the seven high-CV experiments plotted in this figure had only one cycle of data, and therefore were not used for estimates of cycle period and phase lag (Fig. 7). Reproduced with permission from AAAS/Science.

dynamics. We do not wish to review here the long-standing debate over what constitutes a species, particularly under conditions of asexual reproduction. Rather, we simply state that we are satisfied that the dynamics are driven by a change in genotypes through time and that this satisfies fundamental definitions of evolution that refer to *heritable* change in form over many generations. To get the pattern we observe empirically, our model tells us that the genetic mean and variance of our algal population must have changed over the course of many cell divisions through the process of natural selection and response. Additionally, it is worth noting that the documented impact of rapid selection response on predator-prey dynamics does not depend upon the clonal nature of the prey in our system. Both our quantitative trait model (Shertzer *et al.*, 2002) and that of Abrams and

Matsuda (1997) produce qualitatively similar results without specifying the nature of inheritance. A similar argument applies to our interpretation of rotifer evolution in the next section.

VI. RAPIDLY EVOLVING ROTIFERS

Predators can also evolve rapidly, either separately or through co-evolution with their prey (Abrams, 2000; Bohannan and Lenski, 2000). We have not yet investigated whether the predators in our experimental system display evolutionary changes in response to the cyclically varying food value of their prey, but it is significant that the experimental results previously described made sense in the light of prey evolution alone. We conclude this review of our work on algal-rotifer chemostats with an example demonstrating the importance of predator evolution in a non-trophic context (Fussmann *et al.*, 2003).

Monogonont rotifers like *Brachionus* exhibit cyclical parthenogenesis, that is, a population persists through time with phases of amictic (ameiotic, asexual) and mictic (sexual) reproduction (e.g., Gilbert, 2003). In *B. calyciflorus*, crowding induces amictic females to produce mictic daughters (Gilbert, 1963) but there is significant clonal variation in the propensity to produce mictic females in response to an (unknown) crowding stimulus (Gilbert, 2002). We found that this genetic variance, which is usually present in naturally occurring rotifer populations, may be subject to rapid clonal selection, with serious implications for the population dynamics of the algal-rotifer system (Fussmann *et al.*, 2003).

We ran replicated chemostat experiments that differed in two important ways from the experiments described in the previous sections. (1) We started our cultures with rotifers directly sampled from the field, that is, our inoculum presumably comprised a multitude of clones. (2) We ran the experiments at higher nutrient concentrations of 514 μmol nitrate l^{-1} which, due to enrichment, resulted in the extinction of the rotifer population after one predator-prey cycle (Fig. 9). What we observed in all four replicate chemostats were not the simple "boom-and-crash dynamics" that our standard model (Fussmann *et al.*, 2000) predicted. Instead, the experimental rotifer dynamics showed two distinct local maxima of abundance before going extinct (Fig. 9). Again we found ourselves in a situation where our basic mathematical model was just a bit too simple and where we needed to invoke some additional biology to reconcile experimental observation with mathematical prediction. Daily inspection of our cultures gave us a strong indication of what that additional biology might consist: our cultures were rife with mictic rotifer females and, consequently, males (Fussmann *et al.*, 2003) (historically, the experiments described in this section were the earliest we

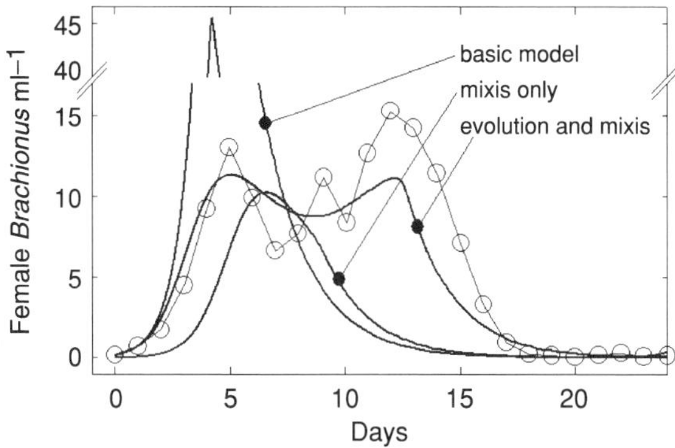

Figure 9 Predator dynamics of *Brachionus* subject to evolution against mictic reproduction. Comparison of predictions by three different model versions with real, experimental data (empty circles). "Basic model" indicates the time series predicted by model (5) to (8) (Fussmann *et al.*, 2000) which does neither account for mixis nor selection against mixis. The best fit of a model version including "mixis only" provides a quantitatively better prediction but fails to match the bimodal abundance pattern of the experimental data that only a model version including selection is able to produce. Corresponding data of the prey (*Chlorella*) are not shown.

performed; the propensity to reproduce sexually faded quickly away under permanent chemostat culture and did not interfere with experimentation reported in Sections III to V). It is important to realize that initiating sexuality in the chemostats could not result in *reproduction*, because experimental conditions (constant bubbling) prevented mating, and even if not, sexually produced eggs undergo diapause and would thus wash out of the chemostat vessel long before hatching. Therefore, investing in reproduction in the chemostat means, for a rotifer, investing in an evolutionary dead-end. Given the enormous number of sexually produced animals, we concluded that mixis had to be the key to explaining the peculiar dynamics we observed.

Using successive amendments to the basic model (Fussmann *et al.*, 2000) we found that both mixis and evolution against mixis were necessary to produce the bimodal pattern of abundance of *Brachionus* (Fig. 9). Evolution was incorporated into the mathematical model as directional selection for the quantitative trait "propensity to reproduce sexually" (which evolves toward asexual reproduction; see Fussmann *et al.*, 2003, for details). Incidentally, invoking the evolutionary process is not just a mathematical gimmick that does the trick but is corroborated by biological facts: (1) our initial sample of rotifers was very likely to be clonally diverse and, therefore, provided the genetic variance on which selection could act; and (2) in long-term chemostat cultures of *Brachionus* we observed an exponential decay of the propensity to reproduce sexually (see Fig. 1 in Fussmann *et al.*, 2003).

VII. CONCLUSIONS

Interaction and feedback between theory and experiment allowed us to learn much about our system, especially from conflicts between models and data: stability at low dilution rate rather than large-amplitude cycles, out-of-phase cycles in persistent populations, and the double-hump in rotifer abundance during the unstable boom-crash cycle under highly enriched conditions. Working with short-lifespan organisms in microcosms allowed rapid feedback between theory and experiment, with our experiments providing real challenges to theory due to tight experimental controls and good data. Instead of being free to explain the discrepancies by invoking unmeasured or uncontrolled variables (or misinformation from our allies), we had to clarify them by erecting and then testing new hypotheses and the corresponding new models.

Some ecologists have complained that microcosms are not real ecological systems, so that microcosm studies, at best, have limited relevance (Carpenter, 1996; see Huston, 1999, for a different opinion). However, framing the issue that way is a red herring: the real question is whether microcosms help us to understand real ecological systems. One does not ask if a novel is real, one asks if it provides real insights. One important way that microcosms can enhance our insight, illustrated by our studies, is that the necessity to fully explain our data forces us to invent new hypotheses which may carry over to the field. Our studies showed that evolutionary change could be an important "engine" driving ecological dynamics, even though the evolutionary trade-offs are so subtle that experiments under extreme conditions (Table 1) were required to reveal the costs associated with the benefit of reduced predation risk. The implication for the field is clear: even cryptic evolutionary changes may have large impacts on ecological interactions and dynamics. Another way microcosm studies can enhance our insight into real systems is by providing "quality control" on ecological theory, so that the models and hypotheses we use as the basis for field experiments or for interpreting data on macroscopic ecosystems have survived rigorous scrutiny using real organisms.

We believe that these potential payoffs are amply validated by past successes. Approaches to stage-structured population modeling that were originally developed to explain contrasting patterns of cyclicity in laboratory insect populations (Gurney and Nisbet, 1985) are now providing important insights into natural population regulation and cycles (Murdoch *et al.*, 2002, 2003), and chemostat studies provided the first validation for the R^* rule, now one of the basic principles of resource competition theory (Tilman, 1982). We hope to emulate these examples, initially with planktonic systems in small temperate lakes, where large-amplitude oscillations in phytoplankton and zooplankton abundance are prominent and regular features of seasonal succession (e.g., the annual "clear water phase"). In addition, the pervasive importance of rapid evolution in our system brings us back full circle to the question that originally motivated our study: if we readily find complex dynamics in models and the laboratory, why do we mostly see cycles and stability in nature (Kendall *et al.*, 1999)? Theoretical work to date including our own (Schliekelman and Ellner, 2001; Shertzer and Ellner, 2002) has assumed "slow" evolution in which organisms' life histories are fixed adaptations to long-term patterns of variability, and has failed to resolve the issue of why evolution does not emerge as a strong or consistent force for population stability (Mueller and Joshi, 2000). So instead of testing current models directly (as we had planned), we have done so indirectly by calling into doubt their basic premise about evolutionary dynamics, thus setting the stage for fresh theoretical approaches to the macroecology of population stability.

ACKNOWLEDGMENTS

This research was supported by a SGER grant from the National Science Foundation and a grant from the Andrew W. Mellon Foundation. We thank M. Armsby, R. Babcock, K. Brewer, K. Check, C. Cline, L. Davias, S. Hammer, A. Holmes, M. Hung, P. Kalika, M. Kalvestrand, A. Katholos, C. Kearns, K. Keller, and J. Meyer for assistance with the experiments. M. Boraas provided the initial rotifer culture and advice on chemostat setup. T.Y. was partly supported by a JSPS Postdoctoral Fellowship for Research Abroad.

REFERENCES

Abrams, P.A. (2000) The evolution of predator-prey interactions: Theory and evidence. *Annu. Rev. Ecol. Syst.* **31**, 79–105.

Abrams, P.A. and Matsuda, H. (1997) Prey evolution as a cause of predator-prey cycles. *Evolution* **51**, 1740–1748.

Bohannan, B.J. and Lenski, R.E. (2000) Linking genetic change to community evolution: Insights from studies of bacteria and bacteriophage. *Ecol. Lett.* **3**, 362–377.

Boraas, M.E. (1980) A chemostat system for the study of rotifer-algal-nitrate interactions. *Am. Soc. Limnol. Oceanogr. Spec. Symp.* **3**, 173–182.

Carpenter, S.R. (1996) Microcosm experiments have limited relevance for community and ecosystem ecology. *Ecology* **77**, 677–680.

Costantino, R.F., Desharnais, R.A., Cushing, J.M. and Dennis, B. (1997) Chaotic dynamics in an insect population. *Science* **275**, 389–391.

Dent, V.E., Bazin, M.J. and Saunders, P.T. (1976) Behaviour of *Dictyostelium discoideum* amoebae and *Escherichia coli* grown together in chemostat culture. *Arch. Microbiol.* **109**, 187–194.

Ellner, S. and Turchin, P. (1995) Chaos in a noisy world: New methods and evidence from time-series analysis. *Am. Nat.* **145**, 343–375.

Fussmann, G.F., Ellner, S.P., Shertzer, K.W. and Hairston, N.G., Jr. (2000) Crossing the Hopf bifurcation in a live predator-prey system. *Science* **290**, 1358–1360.

Fussmann, G.F., Ellner, S.P. and Hairston, N.G., Jr. (2003) Evolution as a critical component of plankton dynamics. *Proc. R. Soc. Lond. B* **270**, 1015–1022.

Gilbert, J.J. (1963) Mictic female production in the rotifer *Brachionus calyciflorus*. *J. Exp. Zool.* **153**, 113–124.

Gilbert, J.J. (2002) Endogenous regulation of environmentally induced sexuality in a rotifer: A multigenerational parental effect induced by fertilisation. *Freshw. Biol.* **47**, 1633–1641.

Gilbert, J.J. (2003) Environmental and endogenous control of sexuality in a rotifer life cycle: Development and population biology. *Evol. Dev.* **5**, 19–24.

Gurney, W.S.C. and Nisbet, R.M. (1985) Fluctuation periodicity, generation separation, and the expression of larval competition. *Theor. Pop. Biol.* **28**, 150–180.

Holling, C.S. (1959) The components of predation as revealed by a study of small-mammal predation of the European Pine Sawfly. *The Canadian Entomologist* **91**, 293–320.

Huston, M.A. (1999) Microcosm experiments have limited relevance for community and ecosystem ecology: Synthesis of comments. *Ecology* **80**, 1088–1089.

Jost, J.L., Drake, J.F., Tsuchiya, H.M. and Frederickson, A.G. (1973a) Microbial food chains and food webs. *J. Theor. Biol.* **41**, 461–484.

Jost, J.L., Drake, J.F., Frederickson, A.G. and Tsuchiya, H.M. (1973b) Interactions of *Tetrahymena pyriformis, Escherichia coli, Azotobacter vinelandii*, and glucose in a minimal medium. *J. Bacteriol.* **113**, 834–840.

Kendall, B.E., Briggs, C.J., Murdoch, W.W., Turchin, P., Ellner, S.P., McCauley, E., Nisbet, R.M. and Wood, S.N. (1999) Why do populations cycle? A synthesis of statistical and mechanistic modeling approaches. *Ecology* **80**, 1789–1805.

Kot, M., Sayler, G.S. and Schultz, T.W. (1992) Complex dynamics in a model microbial system. *Bull. Math. Biol.* **54**, 619–648.

Krebs, C.J. (2002) *Ecology: The Experimental Analysis of Distribution and Abundance*, 5th ed. Benjamin Cummings.

Lynch, M. and Walsh, J.B. (1998) *Genetics and Analysis of Quantitative Traits*. Sinauer Associates, Inc., Sunderland, MA.

May, R.M. and Oster, G.F. (1976) Bifurcations and dynamic complexity in simple ecological models. *Am. Nat.* **110**, 573–599.

Monod, J. (1950) La technique de culture continue. Théorie et applications. *Ann. Inst. Pasteur* **79**, 390–410.

Mueller, L.D. and Joshi, A. (2000) *Stability in Model Populations*. Princeton University Press, Princeton, NJ.

Murdoch, W.W., Kendall, B.E., Nisbet, R.M., Briggs, C.J., McCauley, E. and Bolser, R. (2002) Single-species models for many-species food webs. *Nature* **417**, 541–543.

Murdoch, W.W., Briggs, C.J. and Nisbet, R.M. (2003) *Consumer-Resource Dynamics*. Princeton University Press, Princeton, NJ.

Nelson, W.A., McCauley, E. and Wrona, F.J. (2001) Multiple dynamics in a single predator-prey system: Experimental effects of food quality. *P. Roy. Soc. Lond. B Bio.* **268**, 1223–1230.

Nisbet, R.M., Cunningham, A. and Gurney, W.S.C. (1983) Endogenous metabolism and the stability of microbial prey-predator systems. *Biotech. Bioeng.* **25**, 301–306.

Pickett-Heaps, J.D. (1975) *Green Algae: Structure, Reproduction and Evolution in Selected Genera.* Sinauer Associates, Sunderland, MA.

Scheuerell, M.D., Schindler, D.E., Litt, A.H. and Edmondson, W.T. (2002) Environmental and algal forcing of *Daphnia* production dynamics. *Limnol. Oceanogr.* **47**, 1477–1485.

Schliekelman, P. and Ellner, S.P. (2001) Egg size evolution and energetic constraints on population dynamics. *Theor. Pop. Biol.* **60**, 73–92.

Shertzer, K.W. and Ellner, S.P. (2002) Energy storage and the evolution of population dynamics. *J. Theor. Biol.* **215**, 183–200.

Shertzer, K.W., Ellner, S.P., Fussmann, G.F. and Hairston, N.G., Jr. (2002) Predator-prey cycles in an aquatic microcosm: Testing hypotheses of mechanism. *J. Anim. Ecol.* **71**, 802–815.

Simonoff, J.S. (1996) *Smoothing Methods in Statistics.* Springer-Verlag, New York, NY.

Tilman, D. (1982) *Resource Competition and Community Structure.* Princeton University Press, Princeton, NJ.

Tilman, D. (1996) Biodiversity: Population versus ecosystem stability. *Ecology* **77**, 350–363.

Turchin, P. (2003) Evolution in population dynamics. *Nature* **424**, 257–258.

Walz, N. (Ed.) (1993) *Plankton Regulation Dynamics.* Springer-Verlag, Berlin, Germany.

Yoshida, T., Jones, L.E., Ellner, S.P., Fussmann, G.F. and Hairston, N.G., Jr. (2003) Rapid evolution drives ecological dynamics in a predator-prey system. *Nature* **424**, 303–306.

Yoshida, T., Hairston, N.G., Jr. and Ellner, S.P. (2004) Evolutionary trade-off between defence against grazing and competitive ability in a simple unicellular alga, *Chlorella vulgaris. Proc. R. Soc. Lond. B* **271**, 1947–1953.

The Contribution of Laboratory Experiments on Protists to Understanding Population and Metapopulation Dynamics

MARCEL HOLYOAK AND SHARON P. LAWLER

I. SUMMARY

Several breakthroughs in ecological understanding of population dynamics have arisen through the use of protists as model systems. The combination of rapid generation times, being large enough to count, and small enough to manipulate make heterotrophic protozoa unparalleled in their utility for studies of population dynamics. Gains in understanding have come from three main areas: using laboratory experiments to test theory, developing new models to explain experimental results, and iterative rounds of model development and experimental testing. We review the contribution of protist experiments to population regulation, population responses to extrinsic perturbations, the role of space and habitat patchiness in creating predator and prey persistence, single species source-sink dynamics, and the stability of food chains with omnivores and intraguild predation, across productivity

ADVANCES IN ECOLOGICAL RESEARCH VOL. 37

0065-2504/05 $35.00
DOI: 10.1016/S0065-2504(04)37008-X

gradients. The results show that laboratory experiments with protists have played a key role in development and popular acceptance of a wide range of central theories in population ecology. We review these roles and also highlight new areas where protist model systems could provide similar advances.

II. INTRODUCTION

Laboratory experiments using heterotrophic protozoa have been central in the development of theory in population dynamics. There is a long tradition of ecologists using heterotrophic protists in microcosms to test ecological theory, beginning with classic research on competition models (e.g., Gause, 1934; Vandermeer, 1969; reviewed by Lawler, 1998). Recently, an increasing number of researchers have been constructing simple food webs of protists and bacteria that are helping us understand the dynamic consequences of common food web modules, like linear food chains (e.g., Kaunzinger and Morin, 1998), apparent competition (Lawler, 1993), and intraguild predation (Morin, 1999; Diehl and Feissel, 2000). More complex protist assemblages have also been used in important tests of the relationship between biodiversity and ecosystem functioning (e.g., McGrady-Steed *et al.*, 1997; Naeem and Li, 1998). Here we concentrate on the utility of heterotrophic protists for studying population dynamics and do not extend beyond studies of three-species food-web modules. Our main aim is to illustrate the kinds of gains in our understanding of population dynamics that have been made as a result of laboratory experiments using protists, largely from the Kingdom Protozoa. We do not attempt to cover the large body of excellent work using autotrophic protists (algae); these studies often address different branches of ecological theory and because of space constraints, we elected to cover fewer topics in more detail.

A major problem that has arisen in ecology, starting in the 1920s and still present today, is the increasing gap between the development of ecological theory and empirical knowledge. The root of this problem was captured by Fontenelle (1686; quoted in Kingsland, 1995): “Grant a mathematician one little principle, he immediately draws a consequence from it, to which you must necessarily assent, and from this consequence another, till he leads you so far (whether you will or no) that you have much ado to believe him.” As ecologists we might question whether the initial principle is correct and whether the elaboration of it is done in a biologically reasonable way. The outstripping of experimental tests by ecological theory is echoed by Kareiva (1989), who titled his paper “Renewing the Dialogue Between Theory and Experiments in Population Ecology,” and proceeded to present an agenda for doing just this.

Laboratory experiments are done for many reasons and they represent just one part of a larger ecological literature (e.g., Resetarits and Bernardo, 1998). In using laboratory experiments, ecologists may be endeavoring to understand particular field situations (e.g., Schindler, 1998), to test and develop general ecological theory, or to test the effects of novel treatments. Novel treatments often involve cases where there is no existing theory that is precise in its predictions, and the experimental results are intrinsically interesting because of interest in the treatment (e.g., effects of global warming on biodiversity and ecosystem functioning; Petchey *et al.*, 1999). For tests of mechanisms to explain particular phenomena observed in the field, recent papers suggest that dimensional analyses and scaling rules can be used to improve the design of small scale experiments (e.g., Petersen and Hastings, 2001; Englund and Cooper, 2003). We do not address this topic further here, rather, we focus on the use of microcosm experiments to test general theory. The challenge posed for microcosm tests of theory is whether results "scale" between experiments at different spatial and temporal scales or between the laboratory and field, thus providing good tests of general ecological principles (e.g., Gardner *et al.*, 2001; Naeem, 2001; Englund and Cooper, 2003). Many ecologists recognize that there is a real need for model systems in ecology that are realistic for the organisms being studied, so that they will reflect ecological principles that also act in more natural systems and at larger scales.

In this chapter we begin by describing why protists make appropriate study organisms for laboratory population experiments. Next, we describe some of the early work of Gause (1934), which provides a rich array of examples of testing and development of theories that are now regarded as the core of population ecology. We then discuss the influence of density-dependent feedback and extrinsic perturbations on populations, integrating early results with ongoing research. In addition, we expand to two species, predators and their prey, and review the effects of space and habitat patchiness on persistence and the insights gained from a search for persistence mechanisms. We briefly review laboratory experiments on source and sink dynamics, a topic that has proven difficult to explore in the field and offers much potential for future research. We conclude by discussing studies of intraguild predation and review a variety of experimental work that develops close links with mathematical models and leads us in the direction of community ecology. We attempt to capture the state of knowledge in these areas and indicate directions that might prove fruitful for future study, especially in those areas where protist model systems indicate that field tests of theory might be productive. The theme that emerges is that researchers start with a theory, test it with the protist system, and these tests lead either to field studies, improved models, or both.

III. PROTISTS AS MODEL SYSTEMS

The diversity of protists means that there is a rich variety of species and traits to "mine" for suitable model systems. A recent estimate puts the number of protist species at 213,000, and rising as their taxonomy progresses (Corliss, 2002). Taxonomically, protists are a heterogeneous group of single-celled (but sometimes colonial) eukaryotic organisms. According to Corliss (2002), these sophisticated cells are distributed among no less than five Kingdoms, two of which are dominated by non-colonial taxa. These are the Chromista (including zoosporic protists, opalinids, diatoms, some phytoflagellates) and the Kingdom Protozoa (which includes some of the best known, most diverse groups such as the ciliates, amoeboids, euglenoids, dinoflagellates, foramaniferans, and radiolarians). Other single-celled or colonial eukaryotes lie within Plantae (e.g., *Volvox*), Fungi (microsporidians), and Animalia (myxosporidians). The diversity of protists corresponds to a wealth of opportunity for study. Protists are also extremely important to the biosphere, because they are responsible for 40% of the earth's photosynthesis and almost all marine productivity (reviewed by Corliss, 2002). They have vital roles in terrestrial and freshwater ecosystems, especially in maintaining soil fertility and lake clarity through grazing bacteria.

The convenience of studying protists is difficult to overstate. Normal protist behaviors and even some intracellular processes can be easily observed under the microscope. Hundreds of generations of population dynamics may be observed within a few weeks. Thousands can be sacrificed in the name of science with nary an animal care protocol nor animal rights protestor in sight. Problems do occur, including contamination of cultures, cryptic species, difficulties in identifying field-collected materials, and laboratory artifacts extraneous to the hypothesis being tested. With extra care, effort and thought, most of these can be overcome. However, investigators would be wise to learn the biology of their species to guard against artifacts. Some aspects of biology differ from many other eukaryotes. For example, because they reproduce by fission, maternal effects are strong and can last more than one generation. Their sexual systems differ from many other organisms because they often involve several mating types, and gender is optional in more protists than in other eukaryotes (Bell, 1988).

We believe that ecologists have only scratched the surface of the ecological questions surrounding the use of protists. Protists make convenient analogs of multicellular organisms that use similar ecological strategies, and have some adaptations that are worth studying in their own right. Protists have solved an enormous number of ecological and evolutionary challenges, all using single cells and occasionally cooperative behavior. Many perform DNA repair to such a level that a cell is potentially immortal (Bell, 1988), while others senesce and die. Many form resting cysts, allowing the study of "storage effects" (Chesson and Huntly, 1997; e.g., McGrady-Steed *et al.*, 1997). Protists include producers, consumers, omnivores, internal and external parasites, and mutualists, and they possess a staggering number of adaptations to fill these trophic roles. Predators can select prey (Rapport *et al.*, 1972; review: Laybourn-Parry, 1984); prey can launch defenses that are specific to particular predation risks (Kusch, 1993). There are sit-and-wait predators, active hunters, bottom feeders, filterers, engulfers, and grazers. In short, convenient protistan models can be found for a host of important ecological traits, but this will only occur if ecologists learn more about them.

Of course not all of the ecological questions can be addressed with protists, and not all traits have protist analogs. Some questions can only really be answered within a particular system that one wants to understand. Nevertheless, there is a long tradition of successful protist models in ecology (Laybourn-Parry, 1984; Kingsland, 1995) which can be greatly expanded upon. The diversity of biologies possessed by protists sometimes causes them to have dynamics that are different from those predicted by simple models. However, these departures from theoretical expectation may reveal the importance of widespread aspects of biology that are absent from current models and which could usefully be added to them.

IV. EARLY EXPERIMENTS WITH PROTOZOA AND THE BIRTH OF POPULATION ECOLOGY

Experiments with protozoa were central to the early development of population dynamic theory and for providing field ecologists with more confidence in the validity of mathematically formulated theories, which they then tested in the field. The work of a Russian, G.F. Gause, was central to this early progress. Gause's work provides several examples of experimental tests of theories that are central to modern-day population ecology as well as at least two examples where experiments led to what is now recognized as increased realism in mathematical models.

During the 1920s the recently described logistic equation was controversial because the curve had not been tested even for short-term predictions and the main audience, demographers, were interested in long-term predictions. The curve was regarded as overly simplistic and detractors were suspicious because Pearl and Reed (1920) presented it as a generally applicable law with predictive ability. Gause set out to validate the logistic equation using bottle populations of a ciliate *Paramecium caudatum* and yeast, both of which he regarded as simplistic compared to field populations (Gause, 1934). Through repeatedly censusing 63 replicate growing populations and plotting numbers against time he produced a very smooth s-shaped curve (Gause, 1934). Hutchinson (1978) commented that an s-shaped curve was weak proof for the logistic theory and also likened populations of protists to an analogue computer: "What we have indeed done is to construct a rather inaccurate analogue computer for giving numerical solutions of our equation, using organisms for its moving parts. When we find that we have confirmed the logistic, what we have mainly confirmed is that the reduction in the rate of population growth is linearly dependent on the relative density of organisms. Actually, the beautiful s-shaped integral curve may be too insensitive a result to tell us how well we have established this conclusion." It is not clear what Hutchinson would have regarded as an adequate test of the theory.

Gause devised a creative and stronger test for the logistic theory. He did this by deriving equations with the same form as the logistic equation [$dN/dt = rN((K - N)/K)$], but which also included competition in a similar way to the recently described Lotka-Volterra competition model (Volterra, 1926; Lotka, 1932):

$$\frac{dN_1}{dt} = rN_1 \frac{K_1 - (N_1 + \alpha_{1,2})}{K_1},$$
$$\frac{dN_2}{dt} = rN_2 \frac{K_2 - (N_2 + \alpha_{2,1})}{K_2}$$

where N_i are numbers of species i, r_i are intrinsic growth rates, and K_i are carrying capacities. The terms $\alpha_{1,2}$ and $\alpha_{2,1}$ are competition coefficients that modify the logistic equation to scale the effects on the growth rate of the target species of individuals of the other species. Gause then used two ciliates, *Paramecium aurelia* and *Stylonychia pustulata*, to experimentally test the principle of competitive exclusion for species utilizing a single resource. By testing whether extinction of one species resulted when two species were grown on a single resource, Gause succeeded in simultaneously demonstrating the utility of the logistic model and the competitive exclusion principle. Kingsland (1995) suggested that this demonstration was important because it encouraged Lack and Hutchinson (two of the most productive ecologists in the 1950s) to attempt to test these ideas in the field.

Gause's work was also important because it showed the practical utility of the niche concept and linked the use of the term niche as an abstract space (Grinnell, 1917) with the idea that it was based on feeding relationships within a community (Elton, 1927). Gause experimentally demonstrated that *P. aurelia* and *P. bursaria* used different depths of bottles and suggested that this resulted in niche differentiation, which might explain their co-existence; there is no more precise work that we are aware of that tests the purported co-existence mechanism with these species. The idea that niches were measurable entities dramatically changed thinking about niches (Kingsland, 1995) and, together with the competitive exclusion principle, this led to community ecology being dominated by niche and competition theory up until the early 1970s (e.g., Dobzhansky, 1951; MacArthur, 1958; Hutchinson, 1961; Pianka, 1966).

Gause's other major contribution was to demonstrate the utility of Lotka and Volterra's predator-prey model (Lotka, 1925; Volterra, 1926), thus contributing to the growth of the predator and prey theory. Gause (1934) demonstrated short-term changes in density of a predatory ciliate, *Didinium nasutum*, and its prey, *P. aurelia*, that were consistent with the oscillatory dynamics of the Lotka-Volterra predator and prey equations. Interestingly, the test was rather weak by today's standards: short-term observed rates of population change were observed to be consistent with long-term population trajectories; this was assumed to represent sufficient support for the model (Kareiva, 1989). However, only after Gause (1934) showed the merits of the Lotka-Volterra equations did other ecologists explore their broader implications in field systems (Kareiva, 1989). Perhaps not satisfied with the link between the Lotka-Volterra equations and his own experiments, Gause performed further experiments where he varied predator and prey densities. In the Lotka-Volterra predator-prey equations, predator and prey densities have equal effects on the amplitude of oscillations, however Gause observed that the amplitudes of density oscillations depended more on the initial densities of predators than prey. This led Gause, Smaragdova, and Witt (1936) to introduce a nonlinear functional response into the Lotka-Volterra predator-prey model, thereby making it accord with the experimental results. This was also the first time a nonlinear functional response was used, whereas today we assume (with much evidence) that most functional responses are nonlinear.

V. POPULATION CONTROL AND ENVIRONMENTAL VARIATION

Few problems in population ecology have been as controversial as the search for empirical evidence for the causes of population regulation. In the 1950s two camps emerged, with Andrewartha and Birch (1954) stressing the importance of density-independent extrinsic factors, and Nicholson (1958), Lack (1954), Hutchinson (1957), and MacArthur (1958) stressing intrinsic factors (competition) and density dependence (Kingsland, 1995). This debate carried on through into the 1990s (Murdoch, 1994), fueled by some miscommunication and misunderstandings (reviewed by Sinclair, 1989) misleading statistical methods (e.g., Holyoak and Lawton, 1992) as well as a lack of long-term data from field populations (e.g., Woiwod and Hanski, 1992). While a consensus on the presence of density dependence in field populations was not reached until the 1990s (Sinclair, 1989; Woiwod and Hanski, 1992; Turchin, 1995), laboratory studies by Luckinbill and Fenton (1978) provided the first evidence from manipulative experiments for regulation in real populations (Harrison and Cappuccino, 1995). The experiments of Luckinbill and Fenton (1978) also provide useful information about the effects of extrinsic variation on populations.

Luckinbill and Fenton (1978) experimentally tested the effects of external perturbations on population fluctuations, which illustrated the effects of regulation and, hence, density-dependent feedback (Murdoch, 1970). Three species were used. Replicate populations of each were started either at their average density under experimental conditions, or either above or below this density. Each species returned to the average density at a speed inversely related to body size; this time period will be termed the characteristic return time. When replicate populations of two of the species were grown on resources that were sequentially perturbed downward and then upward with increasing frequency, the response varied between species. Densities were most constant for the largest-bodied species, *Paramecium primaurelia*, for this species they became less variable as perturbations became more frequent. The smaller-bodied ciliate *Colpidium campylum* showed larger amplitude oscillations than *P. primaurelia* and increased in amplitude with an increasing frequency of perturbation until the species became extinct. Luckinbill and Fenton (1978) related their results to the reduced time it took to starve when food levels were low for smaller-bodied *C. campylum* and the more rapid population growth when food was plentiful. The rapid changes in density of the small-bodied species is consistent with the occurrence of over-compensating density dependence. This also concurs in the theoretical results by May (1974), where the degree of overcompensation in a discrete time logistic model is dependent on the intrinsic growth rate, r, which eventually produces deterministic chaos when r is sufficiently large. As a tool, density perturbations have been conducted in a variety of other kinds of systems, but it is interesting that there was a long time period between Murdoch advocating them in 1970 (Murdoch, 1970) and their wider use (Harrison and Cappuccino, 1995). This indicates that the study by Luckinbill and Fenton (1978) may have been instrumental in the gaining popularity of density perturbation experiments in other kinds of study systems.

Inspired by the studies by Luckinbill and Fenton (1978), Orland and Lawler (2004) conducted an experiment of similar design, but used a different species, *Colpidium cf. striatum*, and included a wider range of frequencies of perturbation. When Orland and Lawler (2004) increased the frequency with which resources were perturbed to intervals less than the characteristic return time they found that populations became less temporally variable. This result is analogous to the findings by Luckinbill and Fenton (1978) on *P. primaurelia*; rather little effect was produced by resource perturbations that were too fast for the population density to match. Like Luckinbill and Fenton (1978), Orland and Lawler (in press) also found that resource perturbations at intervals greater than the characteristic return time increased the temporal variability of population densities. Under such circumstances the population would have had time to grow or decline in response to the altered conditions, which is a prerequisite for being able to overcompensate in response to changed conditions. This is consistent with *C. campylum* in the experiment by Luckinbill and Fenton (1978). The demonstration of increases and decreases in variability with different frequencies of perturbation in a single species shows that the results are not caused by interspecific differences.

So far we have only discussed resource perturbations. Extra insights are gained by considering the effects of other kinds of perturbations. The results of density perturbations can be seen in a study by Orland (2003), which was conducted under comparable conditions to Orland and Lawler (2004). Perturbing densities alternately upward and downward did not increase temporal variability in densities above that in unperturbed control treatments if perturbations were on the same frequency as the characteristic return time. However, perturbing densities more or less frequently produced greater temporal variability than controls. Density perturbations of a greater frequency than the return time are likely to have increased variability because the population could not respond sufficiently and rapidly to fully counteract the direct perturbation. By contrast, less frequent density perturbations could have produced density-dependent feedback that caused populations to overcompensate, thereby increasing temporal variability above the control. This explanation is supported by a further treatment where resources were renewed at an interval that was approximately one-half of that of the return time; all density perturbations then increased temporal variability in population densities. Renewing nutrients would have short-circuited density-dependent feedback that was mediated by resource availability. To date, these kinds of experiments investigating perturbations and long-term population dynamics are unique to protists in microcosms.

In the previous studies positive and negative perturbations occurred sequentially and were of similar magnitudes. In nature we are perhaps more likely to see random or structured sequences of perturbations that do not follow a strict balanced positive and negative sequence. A reddened spectrum is one where there is positive temporal (auto)correlation in environmental variation, such that events that have a negative effect on populations are likely to be followed by further negative events and vice-versa (positive events follow positive events). Reddening the spectra of environmental variability, therefore, is likely to increase the temporal variability of population densities and make extinction more likely. Reddened spectra appear to be frequent in ecology and are indicated by longer time series of abundances showing increased temporal population variability compared to shorter series (e.g., Pimm and Redfearn, 1988; Inchausti and Halley, 2001; Akcakaya *et al.*, 2003). Interest among ecologists in the effects of the reddening of spectra started with two laboratory experiments (Cohen *et al.*, 1998; Petchey, 2000). Cohen *et al.* (1998) programmed computer-controlled incubators to generate either random (white spectra) or autocorrelated (red spectra) series of temperature fluctuations while holding the long-term average cumulative temperature constant (Fig. 1A). Petchey (2000) then used this apparatus to examine the effects of different degrees of reddening on single species population dynamics using either *C. striatum* or *Paramecium tetraurelia*. They showed that both species tracked reddened temperature fluctuations more closely than random temperature fluctuations (Fig. 1B). Petchey (2000) also showed that, like in population models, the change in temporal variability in densities to the reddening of temperature variation depended on an interaction between the intrinsic growth rate and the degree of reddening (Fig. 1C); again this result corresponds to the frequency effects seen in the work of Luckinbill and Fenton (1978) and Orland and Lawler (in press). The spectra of weather patterns are usually reddened and time series of population data also appear to be reddened regardless of the source of environmental variation (e.g., Petchey, 2000; Akcakaya *et al.*, 2003). Therefore, it is likely that the color of environmental variation is of broad importance in ecology, and much work remains to be done exploring the phenomenon for other kinds of populations and species interactions. The only experimental evidence for the importance of reddening of spectra for population dynamics comes from protists in laboratory microcosms. These phenomena clearly merit study in other kinds of systems.

VI. STABILIZING PREDATOR-PREY INTERACTIONS IN POPULATIONS AND METAPOPULATIONS

Similar to single species, it has been difficult to experimentally study the long-term population dynamics of predators and prey in the field; the need for studies in simplified circumstances is augmented by the problem of studying species interactions and complex phenomena such as population cycles (e.g., Taylor, 1990, 1991; Turchin, 2003). Early laboratory experiments suggested that it was impossible to obtain persistence of predator and

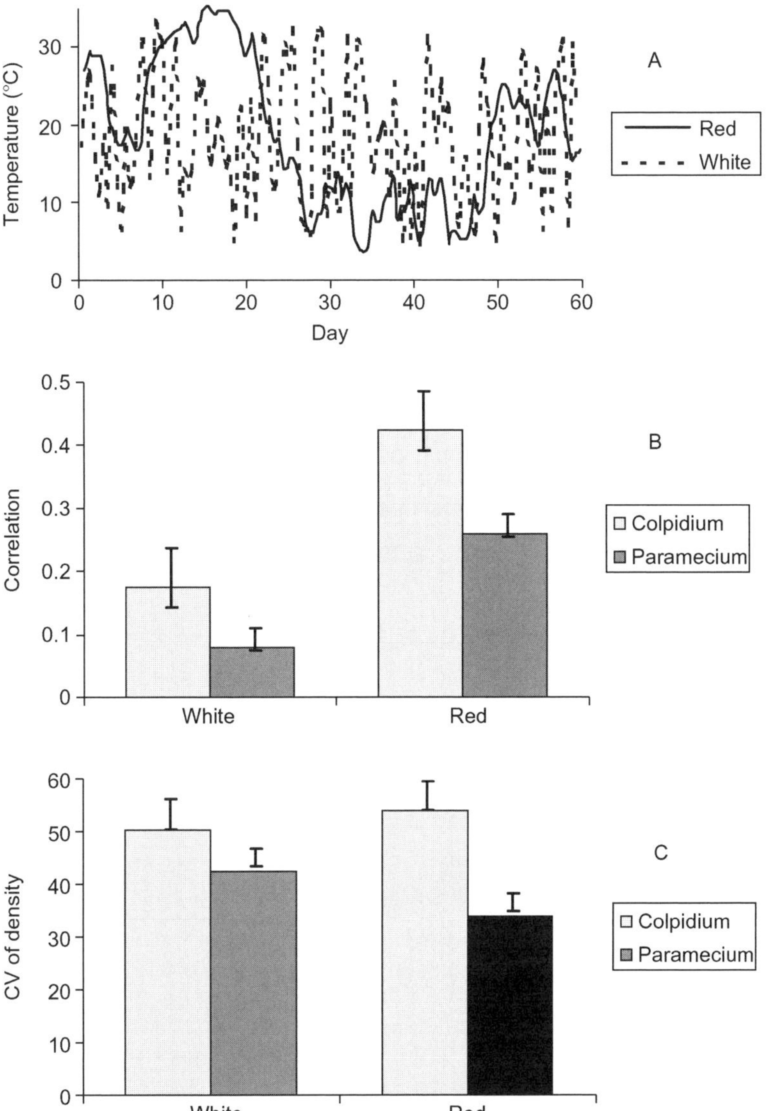

Figure 1 (A) The influence of reddened (autocorrelated) stochastic variation and white (uncorrelated) stochastic variation in temperature in computer-controlled incubators. (B) The correlation between population density and temperature in *Colpidium cf. striatum* and *Paramecium aurelia*. (C) The coefficient of variation (CN) of population density for populations grown under white and red noise. Bars represent standard errors in B and C. Further details in Petchey (2000). Redrawn from Petchey (2000).

prey invertebrates for more than one or two density oscillations, or more than about 10 generations (Gause, 1934; Luckinbill, 1973). This was likened to the neutral stability of the Lotka-Volterra model and the instability of the discrete-time Nicholson-Bailey model (Nicholson and Bailey, 1935). In both experiments and models longer-term persistence, however, could be obtained if something modified the basic predator-prey interaction. For example, Gause (1934) demonstrated that if a refuge was present for prey, or if a prey immigrant was added daily, then longer-term persistence was possible for predatory *Didinium nasutum* and *Paramecium caudatum*. This was the first demonstration of a predator-prey co-existence mechanism and came from protist microcosms. Huffaker (1958) also showed that predator and prey mites could persist for longer in a spatially complex (subdivided) habitat. In this section we briefly review the contributions of laboratory experiments with protists exploring whether predator and prey persistence could be obtained in a homogeneous environment without repeated intervention, and then describe investigations of metapopulation dynamics in subdivided environments.

Luckinbill (1973) intuited that laboratory predator and prey populations were short lived because the small size of experimental arena prevented prey from becoming sufficiently scarce that predators could not find them, which would have forced predators to decline in abundance through starvation. Prey could then become more abundant again allowing both species to persist. Luckinbill (1973) used an inert thickening agent, methyl cellulose, to simulate a larger environment by slowing the movement of the protists, which reduced the frequency of contact between predators and prey. Different concentrations of methyl cellulose reduced encounters to different degrees. Luckinbill's (1973) main result was that the addition of methyl cellulose reduced the swimming rates of both predators and prey, prolonged persistence of both species, and also reduced the maximum density that predators achieved. Luckinbill's (1973) paper has had a broad influence on theory about predator and prey encounters, subsequent dynamics and alternative methods of stabilizing predator and prey interactions (see also Luckinbill, 1974). Following Luckinbill and Fenton's lead, both Veilleux (1979) and Harrison (1995) investigated the form of functional responses caused by thickening the experimental medium with methyl cellulose and related changes in functional responses to gains in persistence in homogeneous environments using isoclines from simple models.

Early experiments in a variety of artificial systems showed the potential for habitat patchiness to prolong the persistence of specialist predators and their prey (e.g., Huffaker, 1958; Maly, 1978; Nachman, 1981, 1987). However, these experiments did not achieve long-term persistence nor tell us the mechanisms by which these species persisted. For example, was it through a refuge for prey, or through metapopulation dynamics, or through other changes in the predator and prey interaction that were caused by the different environment? By contrast a rich variety of predator and prey metapopulation models were described, often inspired by Huffaker's work (reviewed by: Kareiva, 1990; Taylor, 1990, 1991). Taylor (1990, 1991) failed to find any good examples of systems in the laboratory or field that could be said to persist via the mechanisms described in predator and prey metapopulation models; studies typically lacked evidence that movement was important or that local populations were prone to extinction. These deficiencies fueled us to try to test whether we could obtain long-term co-existence through predator and prey metapopulation dynamics.

Earlier experimental designs were improved on by using three replicated treatments: (1) Isolated bottles (patches) to test whether the basic predator-prey interaction was unstable; (2) Interconnected groups of 9 or 25 bottles (subdivided multipatch environments) to look for spatial dynamics (Holyoak and Lawler, 1996a,b); (3) The same total volume of habitat in undivided large bottles to test for effects of habitat size (e.g., Luckinbill, 1974). By using bottles that were connected with tubes we could use clamps to isolate bottles, allowing us to obtain samples without unduly disrupting any spatial structure. We elected to use ciliate predator and prey species (*Didinium nasutum* and *Colpidium cf. striatum*) that do not especially feed along surfaces since we could not vary habitat patchiness without changing the surface area of our microcosms. The distance between patches in our subdivided microcosms was based on the little field data that was available, where Taylor and Berger (1980) found distances of about 15 cm between aggregations of protist species in a Canadian pond. We found that spatial subdivision did prolong persistence beyond that seen in either the single patches or large undivided volumes (Fig. 2A); furthermore, the predator and prey did not show signs of becoming extinct from subdivided microcosms when the experiment was ended after 130 days (or about 602 prey and 437 predator generations). Further analysis showed that within individual bottles (patches) of subdivided microcosms, prey extinctions and recolonizations were frequent (Fig. 2B). For predators, local extinctions and recolonizations were recorded infrequently. However, there was evidence that immigration raised local population densities and thereby rescued predator populations from extinction; densities of predators were increased in bottles of subdivided microcosms with more connecting tubes and, therefore, would have permitted more immigration (Fig. 2C). An independent experiment showed that migration rates of prey between adjacent bottles in subdivided microcosms were low, as expected for repeated local extinction and colonization, and that predator movement rates were somewhat greater, which was consistent with rescue as opposed to local extinction. Density fluctuations of each species in different patches of subdivided microcosms were also only partly synchronous, which was consistent with metapopulation structure

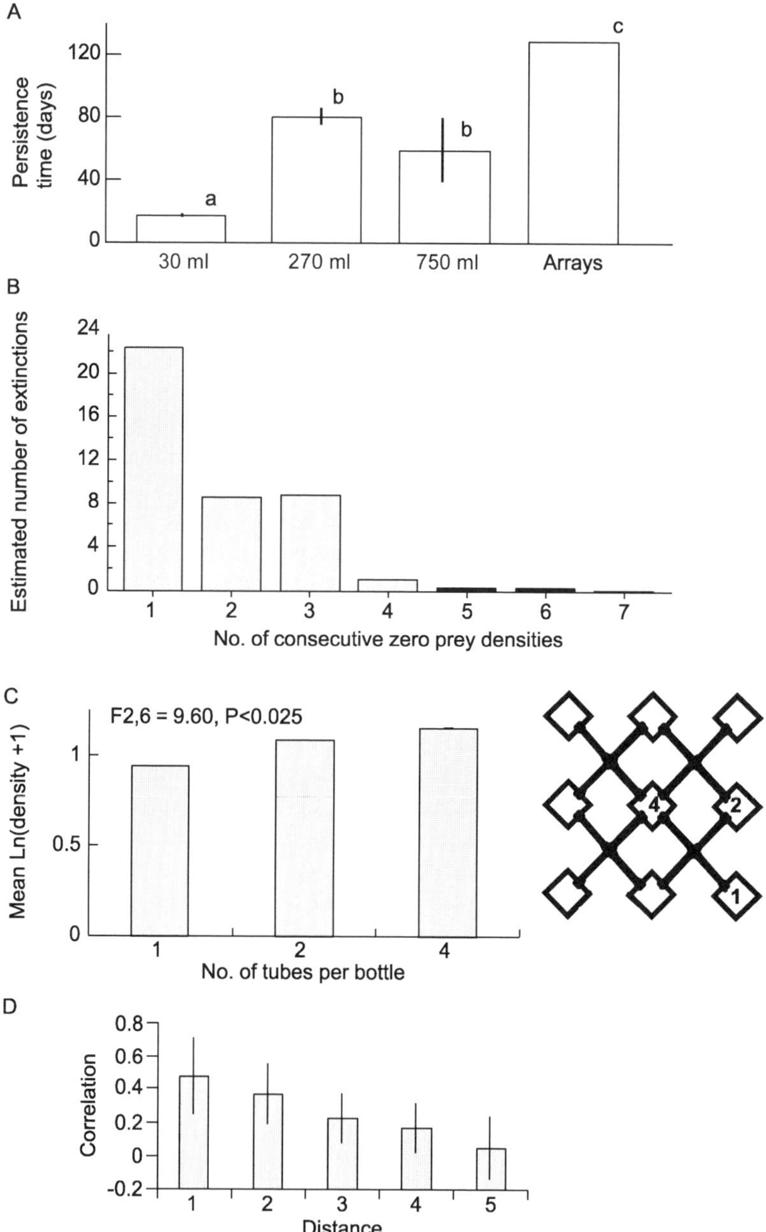

Figure 2 (A) Persistence times of predatory *Didinium nasutum* and its prey *Colpidium cf. striatum* in isolated 30 ml patches, large undivided 270 ml patches, large undivided 750 ml patches, and subdivided 9- or 25-patch microcosms ("arrays," e.g., see panel D). Both 9- and 25-patch subdivided microcosms are shown as a single

(Fig. 2D). Altogether the results provided the first clear example of how spatial dynamics generated predator and prey persistence in multipatch subdivided habitats. This was via the extinction-colonization dynamics of the prey and rescue effects in predators. A comparison with the predictions from mathematical models for predator and prey metapopulations (e.g., Crowley, 1981; Reeve, 1988) showed that the effects of dispersal on dynamics were entirely consistent with these models. This initial study prompted others to use similar experimental designs, opened the way for Holyoak and others to conduct some of the first tests of untested theory, and led to tighter model-empirical linkages. For example, concurrent development of a model and experiments demonstrated the utility of an individual-based model for predicting temporal dynamics and persistence of predator and prey populations, thereby increasing the testability of persistence mechanisms (Holyoak *et al.*, 2000).

A further experiment (Holyoak, 2000) evaluated the role of nutrient enrichment on predator and prey metapopulations. This experiment was based on a theoretical paper by Jansen (1995). Raising the carrying capacity within single patches (and patches that were connected by migration) was expected to produce predator and prey oscillations of larger amplitude, as predicted by Rosenzweig (1971), and which was termed the 'paradox of enrichment'. Jansen (1995) predicted that metapopulations could persist despite enrichment if local dynamics remained asynchronous in different patches. The experiment (Holyoak, 2000) provided support for Jansen's model with prolonged persistence in subdivided habitats and a large amount of asynchrony despite enrichment. One difference from Jansen's model was that in the experiment predator populations became more synchronous across patches with enrichment, whereas this did not happen in Jansen's simulations. The explanation was that predators were more likely to experience very low prey densities with enrichment. Independent experiments showed that predators had a delayed response and increased their interpatch movement rate when they were starving, which

---

bar because both species persisted for the duration of the experiment in all microcosms of these types. (B) The estimated number of prey extinctions occurring in individual 30 ml bottles within subdivided microcosms of 25 patches during a 40-day period. (C) Mean predator densities in bottles of 9-patch microcosms in relation to numbers of tubes connecting bottles; elevated densities with greater numbers of tubes indicate rescue effects as did a reduced number of local extinctions (data not shown) (D) The average correlation between prey densities in bottles of 25-patch subdivided microcosms that were separated by different numbers of tubes (1, 2 = 2 tubes and one bottle, 3 = 3 tubes and 2 bottles, and soon). Correlations for predators were broadly similar indicating substantial independence of dynamics in different patches. Error bars in A and D indicate standard errors. Redrawn from Holyoak and Lawler (1996a).

might cause more dispersal with enrichment and thereby increase the degree of synchrony across patches. This represents the kind of behavior that is often missing from simple models, and given that it is possessed by rather simple organisms, we might question whether it should be added to the models.

Altogether this series of experiments have helped to bring predator and prey metapopulation theory and empirical evidence somewhat closer together. There are a growing number of studies that have been exploring predator-prey (or other exploiter-victim) interactions in the field, e.g., Van Nouhuys and Hanski (1999, 2002), Weisser (2000), Berendonk and Bonsall (2002), Schops (2002). The experiments reviewed in this chapter provide greater congruence between experiments and theory, the first demonstrations of several widely accepted areas of population dynamic theory, and unexpected results that have suggested that theory might need to be modified.

VII. SOURCE AND SINK DYNAMICS

Source and sink dynamics represent another current area of ecological theory that has proven difficult to test fully in the field and where the existing tests examine only at short-term dynamics. In a literature review, Diffendorfer (1998) could not find a single example of a study that adequately distinguished between source-sink dynamics (Pulliam, 1988) and movement that resulted from optimal foraging ("balanced dispersal"; McPeek and Holt, 1992). A survey of the literature by Donahue *et al.* (2003) found that only about 28% of empirical studies about source and sink dynamics contained any data on movement, and in all of these cases the information was inadequate for distinguishing whether movement was from sources or sinks or was equal between different kinds of patches (balanced dispersal). A further complexity is that immigration may produce density-dependent reductions in population growth rates, making otherwise viable populations appear as sinks (termed pseudosinks by Watkinson and Sutherland, 1995). The difference between sinks and pseudosinks is that if interpatch movement were to cease, sink populations would decline to extinction whereas pseudosinks would remain viable. The complexities of source and sink dynamics are such that it is a topic where substantial progress has been made by coupling mathematical models with protists experiments.

Fueled by the absence of data, Donahue *et al.* (2003) investigated the patterns of movement and the influence of density changes mimicking movement on spatial dynamics. Experiments used *Colpidium cf. striatum*, a mixed flora of bacteria and millet seeds as resources (to alter habitat quality). Side experiments measuring dispersal rates predicted that throughout the main experiment there should be a net flux of protists from high to low resource patches. Simple source and sink models (Pulliam, 1988) predict that high dispersal will produce the greatest population abundance in sink patches (assuming that the source is not depleted by emigration). By contrast, in experimental treatments that permitted different levels of dispersal (with no tubes, long or short tubes), intermediate dispersal rates produced the greatest abundances in low resource patches. Donahue *et al.* (2003) parameterized a two-patch model and used this to test whether density-dependent feedback caused by immigration (pseudosink effects) could account for densities not being greatest in low resources patches at the highest dispersal rates. The results showed that pseudosink effects were feasible. There have been rather fewer attempts to examine pseudosink effects in field populations, although one field study did find some evidence for them (Boughton, 1999). A second novel aspect of this study was that over hundreds of generations sources and sinks converged in their population densities. It is likely that differences in the body size of protists leaving sources (large) versus sinks (small) created a large flow of nutrients from sources to sinks, which the altered sink conditions. This kind of feedback between movement and local conditions has been observed with concentrations of grazing herbivores in the Serengeti depositing nitrogenous wastes that fertilize already nutrient-rich areas (McNaughton *et al.*, 1997). There are few examples of changes in source and sink status (although see Boughton, 1999) and none that we are aware of that have examined long-term dynamics. Another set of investigations using protists in laboratory experiments and models showed that reddened environmental variability can inflate the abundances observed in sink populations, which could transform some sinks into sources (Gonzalez and Holt, 2002; Holt *et al.*, 2003). This phenomenon has no tests in other kinds of empirical systems. Clearly more field investigations of source and sink dynamics and their consequences are merited.

VIII. OMNIVORY AND STABILITY

Omnivory was once thought to be rare in food webs (e.g., Pimm, 1980), however, it is now recognized that many consumers feed at more than one trophic level (reviews: Polis, 1994; Morin and Lawler, 1995; Coll and Guershon, 2002; see also Link, 2002). Feeding at more than one trophic level is our operational definition of omnivory for this paper (Pimm and Lawton, 1977). In early papers, the apparent rarity of omnivory was attributed to either dynamic constraints (Pimm and Lawton, 1977), or to difficulties inherent in feeding on prey that differ in structure, size, and chemical composition (e.g., Yodzis, 1984; Stenseth, 1985; Lawton, 1989). Recognition of the prevalence of omnivory raised the challenge of establishing why these potential constraints on omnivores did not prevent them from succeeding.

Protists have been a valuable model system for understanding the dynamic consequences of omnivory. First, we outline a portion of the many theoretical treatments of omnivory, and then review empirical tests that used protist food chains. Early analysis of Lotka-Volterra style food web-models found that models that included omnivores tended to be unstable (prone to extinction of one or more species), or such models had a long return time to equilibrium behavior (Pimm and Lawton, 1977). They suggested that the "intraguild predation" (hereafter IGP) form of omnivory, where a top predator consumes both a prey and the prey's food, might be particularly unstable because it places the prey in a sort of double jeopardy (Pimm and Lawton, 1977; Polis and Holt, 1992). This form is called IGP because the top and intermediate consumers share prey. However, a review by Polis *et al.* (1989) found that IGP seems common and important to food-web dynamics in many systems. They modeled this subweb configuration, and found that omnivores could either stabilize food-chain dynamics if predation on the lowest trophic level was weak, or destabilize it if omnivores ate efficiently enough to out-compete intermediate prey for the resource. Additional work by McCann and Hastings (1997) and McCann *et al.* (1998) reinforced this theory using nonlinear models.

An initial study by Lawler and Morin (1993) tested Pimm and Lawton's model result that simple food chains containing omnivores would be less stable. Our empirical definition of stability was the combination of low population fluctuations and high population size (see Morin and Lawler, 1996, and Holyoak and Sachdev, 1998, for discussion of empirical stability). Lawler and Morin (1993) constructed three-link food chains of bacteria, bacterivorous *Colpidium cf. striatum*, and a predator. The top predator was either the omnivorous protist *Blepharisma americanum*, or either of the protist predators *Amoeba proteus* or *Actinosphaerium eichhornii*. The predators were not strict predators, in that they consume some bacteria. However, they were unable to divide if only given bacteria to consume, while *Blepharisma* could reproduce on a diet of either protists or bacteria. Results showed that bacterivores had no consistent population responses to the type of top predator, sometimes persisting at relatively steady abundance and sometimes varying widely or going extinct. However, the omnivore always had more abundant, less variable populations than the predators, when it fed as an omnivore. This result was intriguing because it showed that the purported dynamic limitations to omnivory in food webs might not be common in real systems (Yodzis and Innis, 1992). An extension of this study to include a second omnivore, *Tetrahymena vorax*, and four other predators, found similar results (Morin and Lawler, 1996). Effects of omnivores and predators on intermediate prey were idiosyncratic; however, omnivore populations were often more abundant and fluctuated less than populations of predators.

As previously mentioned, the predators in the studies by Morin and Lawler were actually omnivorous to some unknown extent, in that they probably consumed bacteria as well as smaller protists. Could "degree of omnivory" explain some of the idiosyncratic results? Holyoak and Sachdev (1998) did a more sophisticated study using many of the same species, but with the addition of a "ranking" of consumers by their degree of omnivory. If a species was known to eat bacteria, or took longer to starve when more bacteria were available, it was assigned a higher omnivory ranking. Cannibals also received a higher ranking, because they fed at their own trophic level as well as at least one below. In empirical terms it was expected that if omnivory was stabilizing, populations of more omnivorous species would persist better and fluctuate less in abundance, and that omnivory in predators would not necessarily cause prey to fluctuate more or become extinct sooner. As predicted, species that were more omnivorous fluctuated less and persisted longer than less omnivorous species. Intriguingly, effects of predators on prey persistence or fluctuations had no detectable relationship to omnivory.

To our knowledge, few if any field studies have attempted to relate "degree of omnivory" to population dynamics, and we hope that highlighting this study will encourage other ecologists to explore their systems for this effect. For example, it might be fruitful to focus on terrestrial food chains involving parasitoids and hyperparasitoids, because omnivory is known to affect the dynamics of these systems (e.g., Brodeur and Rosenheim, 2000; Montoya *et al.*, 2003).

Recent theory holds that although weak omnivory can be stabilizing, strong omnivorous links could still destabilize food webs (e.g., McCann and Hastings, 1997; McCann *et al.*, 1998; Mylius *et al.*, 2001; Kuijper *et al.*, 2003; and references therein). The original theory by Polis *et al.* (1989) also points out that the top omnivore cannot be a superior competitor for basal resources. Therefore, the positive empirical relationship between degree of omnivory and "stability" needs further exploration—a more detailed quantification of interaction strengths within the protist system would help us understand whether the strength of omnivorous links falls into the range that is thought to be stabilizing. For example, the omnivores we used might not consume bacteria efficiently enough to have much effect on bacterivores. Here, the main limiting factor is that it is empirically difficult to measure interaction strengths (Berlow *et al.*, 1999). Work by Morin (1999) and Diehl and Feissel (2001) do support this suggestion for one omnivore.

Several protistan omnivores switch between morphs that feed either mostly on bacteria or mostly on bacterivores, allowing a change in feeding efficiency on the different trophic levels that responds to the availability of prey at each level. This aspect of the system inspired an intriguing extension of IGP models to examine the dynamic consequences of including optimal foraging (Krivan, 2000). Krivan (2000) found that optimal foraging by intraguild predators could help stabilize food webs (see also Matsuda *et al.*, 1986). Not only can protists serve as models for adaptive changes in predators, they are excellent models for studying the population-dynamic effects of inducible defenses in prey (e.g., Kusch, 1998). Inducible defenses are being found in an increasing variety of organisms (review: Tollrian and Harvell, 1999; see also Johansson and Wahlstrom, 2002; Trussell and Nicklin, 2002; Relyea, 2002), yet their effects on population dynamics are rarely studied because of the long generation times of many organisms.

IX. THE INTERACTION BETWEEN PRODUCTIVITY AND IGP

The ability of the IGP system to persist is predicted to vary with the productivity of the system and the relative impacts of top and intermediate predators on basal prey (Polis *et al.*, 1989; Holt and Polis, 1997; Diehl and Feissel, 2000, 2001). For the system to persist under any condition, the intermediate species must be the better competitor for basal prey; otherwise this population cannot withstand the dual impacts of a superior competitor that eats them. At low levels of productivity, the intermediate species can outcompete the top omnivore, because it reduces basal prey to a low level, yet it is not itself abundant enough to sustain a population of the omnivore. At moderate productivity, all three can persist because populations of the intermediate can support the top omnivore, yet basal prey are sparse enough so that the top omnivore must rely on intermediate prey. At higher productivities, basal prey can sustain the top omnivore, and the top omnivore increasingly begins to suppress intermediate prey. With further enrichment under these circumstances, basal prey and the top omnivore increase until predation by the omnivore outstrips the intermediate prey's reproductive rate, and the system collapses to a predator-prey system.

Morin (1999) and Diehl and Feissel (2000, 2001) provided elegant tests of this theory. Morin (1999) grew the omnivore-consumer-basal prey IGP system *Blepharisma americanum*-*Colpidium cf. striatum*-bacteria under low and high nutrient conditions. At low-nutrient levels, the intermediate consumer *Colpidium* excluded the top omnivore, even though *Blepharisma* could eat *Colpidium*. Exclusion was proven by the fact that *Blepharisma* could persist in low-nutrient microcosms without *Colpidium*. A side experiment showed that *Colpidium* drove bacteria to lower levels than the top omnivore. The idea that a prey could exclude its predator was counter-intuitive at the time! At the higher nutrient level, both co-existed. This study established that enrichment caused most of the predicted changes in population levels, persistence, and dynamics. However, nutrient levels were not high enough in the Morin study to test the prediction that further enrichment would cause extinction of the intermediate consumer.

Diehl and Feissel (2000) used a very similar system (*Blepharisma americanum-Tetrahymena pyriformis*-bacteria) and a wider range of nutrient levels to see if all three conditions could be attained: competitive exclusion of the top omnivore, co-existence, and predatory exclusion of the intermediate consumer. The three conditions were indeed attained at the appropriate nutrient levels. A follow-up paper (Diehl and Feissel, 2001) demonstrated the mechanisms underlying these phenomena, that is, that competition was responsible for exclusion of the omnivore at low nutrients, predation caused lower abundances of *Tetrahymena* as the system was enriched, and the intermediate consumer benefited the top omnivore at intermediate levels of enrichment (see also Lawler and Morin, 1993). Two field studies now also show similar patterns of co-existence and exclusion across productivity gradients in parasitoid-host-plant systems (Amarasekare, 2000, Borer *et al.*, 2003), confirming that the theory and the protist model system have real-world relevance.

These results show that it may be possible to predict the outcome of IGP in advance, by measuring the relative competitive abilities of the top omnivore and intermediate consumer on shared prey. This could have implications for predicting the consequences of species invasions, when invasives are likely to cause IGP food-web configurations. Nutrient enrichment of the environment is one of the prevailing conservation challenges on the planet, and these experiments also show that excess nutrients could put some species of intermediate consumers at risk (Diehl and Feissel, 2001).

X. CONCLUSIONS

The combination of rapid generation times, being large enough to count and small enough to grow and manipulate in large numbers, make protozoa unparalleled in their utility for performing population experiments. The most potent use of these experiments is to use them to test general ecological theories that either cannot be tested or cannot be tested as thoroughly in other kinds of systems. Similarity of results between general models and laboratory studies adds generality to the laboratory findings and often allows additional tests of mechanisms that help to validate the models, thereby also strengthening inference. Parallel or iterative development of models and experiments, such as in the work on IGP and the cases of Holyoak *et al.* (2000) and Donahue *et al.* (2003), illustrated cases where models were necessary to interpret the original experiment and further experiments were necessary to parameterize the models. Some cases show that protist experiments provided the broader community of ecologists with confidence that theories were worth exploring in field systems. Experiments with protists in the laboratory clearly play a valuable role in a modern integrative population ecology that brings together theory, observations, and experiments in a wide variety of types of study system.

ACKNOWLEDGMENTS

We thank Robert Desharnais for inviting us to write this piece. Marcel Holyoak was supported by NSF DEB-0213026 to Marcel Holyoak and Alan Hastings.

REFERENCES

Akcakaya, H.R., Halley, J.M. and Inchausti, P. (2003) Population-level mechanisms for reddened spectra in ecological time series. *J. Animal Ecol.* **72**, 698–702.

Amarasekare, P. (2000) Coexistence of competing parasitoids on a patchily distributed host: Local vs. spatial mechanisms. *Ecology* **81**, 1286–1296.

Andrewartha, H.G. and Birch, L.C. (1954) *The Distribution and Abundance of Animals*. University of Chicago Press, Chicago, IL.

Bell, G. (1988) *Sex and Death in Protozoa: The History of an Obsession*. Cambridge University Press, Cambridge, England.

Berendonk, T.U. and Bonsall, M.B. (2002) The phantom midge and a comparison of metapopulation structures. *Ecology* **83**, 116–128.

Berlow, E.L., Navarrete, S.A., Briggs, C.J., Power, M.E. and Menge, B.A. (1999) Quantifying variation in the strengths of species interactions. *Ecology* **80**, 2206–2224.

Borer, E.T., Briggs, C.J., Murdoch, W.W. and Swarbrick, S.L. (2003) Testing intraguild predation theory in a field system: Does numerical dominance shift along a gradient of productivity? *Ecol. Lett.* **6**, 929–935.

Boughton, D.A. (1999) Empirical evidence for complex source-sink dynamics with alternative states in a butterfly metapopulation. *Ecology* **80**, 2727–2739.

Brodeur, J. and Rosenheim, J.A. (2000) Intraguild interactions in aphid parasitoids. *Entomologia Experimentalis et Applicata* **97**, 93–108.

Chesson, P. and Huntly, N. (1997) The roles of harsh and fluctuating conditions in the dynamics of ecological communities. *Am. Nat.* **150**, 519–553.

Cohen, A.E., Gonzalez, A., Lawton, J.H., Petchey, O.L., Wildman, D. and Cohen, J.E. (1998) A novel experimental apparatus to study the impact of white noise and 1/f noise on animal populations. *Proc. R. Soc. Lond. B Biol. Sci.* **265**, 11–15.

Coll, M. and Guershon, M. (2002) Omnivory in terrestrial arthropods. *Ann. Rev. Entomol.* **4**, 267–297.

Corliss, J.O. (2002) Biodiversity and biocomplexity of the protists and an overview of their significant roles in maintenance of our biosphere. *Acta Protozool.* **41**, 199–219.

Crowley, P.H. (1981) Dispersal and the stability of predator-prey interactions. *Am. Nat.* **118**, 673–701.

Diehl, S. and Feissel, M. (2000) Effects of enrichment on 3-level food chains with omnivory. *Am. Nat.* **155**, 200–218.

Diehl, S. and Feissel, M. (2001) Intraguild prey suffer from enrichment of their resources: A microcosm experiment with ciliates. *Ecology* **82**, 2977–2983.

Diffendorfer, J.E. (1998) Testing models of source-sink dynamics and balanced dispersal. *Oikos* **81**, 417–433.

Dobzhansky, T.G. (1951) *Genetics and the Origin of Species: Columbia Biological Series No. 11.* Columbia University Press, New York, NY.

Donahue, M.J., Holyoak, M. and Feng, C. (2003) Patterns of dispersal and dynamics among habitat patches varying in quality. *Am. Nat.* **162**, 302–317.

Elton, C. (1927) *Animal Ecology.* Macmillan, New York, NY.

Englund, G. and Cooper, S.D. (2003) Scale effects and extrapolation in ecological experiments. *Adv. Ecol. Res.* **33**, 161–213.

Gardner, R.H., Kemp, W.M., Kennedy, V.S. and Petersen, J.E. (2001) *Scaling Relations in Experimental Ecology.* Columbia University Press, New York, NY.

Gause, G.F. (1934) *The Struggle for Existence.* Williams and Wilkins, reprinted 1964 by Hafner, New York, Baltimore.

Gause, G.F., Smaragdova, N.P. and Witt, A.A. (1936) Further studies of the interaction between predators and prey. *Journal of Animal Ecology* **5**, 1–18.

Gonzalez, A. and Holt, R.D. (2002) The inflationary effects of environmental fluctuations in source-sink systems. *Proc. Natl. Acad. Sci. USA* **99**, 14872–14877.

Grinnell, J. (1917) The niche-relationships of the California Thrasher. *The Auk* **34**, 427–433.

Harrison, G.W. (1995) Comparing predator-prey models to Luckinbill's experiment with *Didinium* and *Paramecium. Ecology* **76**, 357–374.

Harrison, S. and Cappuccino, N. (1995) Using density-manipulation experiments to study population regulation. In: *Population Dynamics: New Approaches and Synthesis* (Ed. by N. Cappuccino and P.W. Price), pp. 131–147. Academic Press, New York, NY.

Holt, R.D., Barfield, M. and Gonzalez, A. (2003) Impacts of environmental variability in open populations and communities: "Inflation" in sink environments. *Theor. Pop. Biol.* **64**, 315–330.

Holt, R.D. and Polis, G.A. (1997) A theoretical framework for intraguild predation. *Am. Nat.* **149**, 745–764.

Holyoak, M. (2000) The effects of nutrient enrichment on predator-prey metapopulation dynamics. *J. Animal Ecol.* **69**, 985–997.

Holyoak, M. and Lawler, S.P. (1996a) Persistence of an extinction-prone predator-prey interaction through metapopulation dynamics. *Ecology* **77**, 1867–1879.

Holyoak, M. and Lawler, S.P. (1996b) The role of dispersal in predator-prey metapopulation dynamics. *J. Animal Ecol.* **65**, 640–652.

Holyoak, M., Lawler, S.P. and Crowley, P.H. (2000) Predicting extinction: Progress with an individual-based model of protozoan predators and prey. *Ecology* **81**, 3312–3329.

Holyoak, M. and Lawton, J.H. (1992) Detection of density dependence from annual censuses of bracken-feeding insects. *Oecologia* **91**, 425–430.

Holyoak, M. and Sachdev, S. (1998) Omnivory and the stability of simple food webs. *Oecologia* **117**, 413–419.

Huffaker, C.B. (1958) Experimental studies on predation: Dispersal factors and predator-prey oscillations. *Hilgardia* **27**, 343–383.

Hutchinson, G.E. (1957) Concluding remarks. In: *Population studies: Animal movement and demography*, pp. 415–427. Cold Spring Harbor Symposia on Quantitative Biology.
Hutchinson, G.E. (1961) The paradox of the plankton. *Am. Nat.* **95**, 137–145.
Hutchinson, G.E. (1978) *An Introduction to Population Ecology*. Yale University Press, New Haven, CT.
Inchausti, P. and Halley, J. (2001) Investigating long-term ecological variability using the Global Population Dynamics Database. *Science* **293**, 655–657.
Jansen, V.A.A. (1995) Regulation of predator-prey systems through spatial interactions: A possible solution to the paradox of enrichment. *Oikos* **74**, 384–390.
Johansson, F. and Wahlstrom, E. (2002) Induced morphological defence: Evidence from whole-lake manipulation experiments. *Can. J. Zool.* **80**, 199–206.
Kareiva, P. (1989) Renewing the dialogue between ecological theory and experiments in ecology. In: *Perspectives in Ecological Theory* (Ed. by J. Roughgarden, R.M. May and S.A. Levin), pp. 68–88. Princeton University Press, Princeton, NJ.
Kareiva, P. (1990) Population dynamics in spatially complex environments: Theory and data. *Philos. Trans. R. Soc. Lond., B* **330**, 175–190.
Kaunzinger, C.M.K. and Morin, P.J. (1998) Productivity controls food chain properties in microbial communities. *Nature* **395**, 495–497.
Kingsland, S.E. (1995) *Modeling Nature: Episodes in the History of Population Ecology*. The University of Chicago Press, Chicago, IL.
Krivan, V. (2000) Optimal intraguild foraging and population stability. *Theor. Popul. Biol.* **58**, 79–94.
Kuijper, L.D.J., Kooi, B.W., Zonneveld, C. and Kooijman, S.A.L.M. (2003) Omnivory and food web dynamics. *Ecol. Model* **163**, 19–32.
Kusch, J. (1993) Behavioural and morphological changes in ciliates induced by the predator. *Amoeba proteus. Oecologia* **96**, 354–359.
Kusch, J. (1998) Long-term effects of inducible defense. *Ecoscience* **5**, 1–7.
Lack, D. (1954) *The Natural Regulation of Animal Numbers*. Oxford University Press, Oxford, United Kingdom.
Lawler, S.P. (1993) Direct and indirect effects in microcosm communities of protists. *Oecologia* **93**, 184–190.
Lawler, S.P. (1998) Ecology in a bottle: Using microcosms to test theory. In: *Experimental Ecology: Issues and Perspectives* (Ed. by W.J. Resetarits, Jr. and J. Bernardo), pp. 236–253. Oxford University Press, New York, NY.
Lawler, S.P. and Morin, P.J. (1993) Food web architecture and population dynamics in laboratory microcosms of protists. *Am. Nat.* **141**, 675–686.
Lawton, J.H. (1989) Food webs. In: *Ecological Concepts: The Contribution of Ecology to an Understanding of the Natural World* (Ed. by J. Cherrett). Blackwell Scientific, Oxford, United Kingdom.
Laybourn-Parry, J. (1984) *A Functional Ecology of Free-Living Protozoa*. University of California Press, Berkely, CA.
Link, J. (2002) Does food web theory work for marine ecosystems? *Marine Ecol. Prog. Series* **230**, 1–9.
Lotka, A.J. (1925) Elements of Physical Biology. Williams and Wilkens, Baltimore, MD.
Lotka, A.J. (1932) The growth of mixed populations: Two species competing for a common food supply. *J. Washington Academy of Sci.* **22**, 461–469.
Luckinbill, L.S. (1973) Coexistence in laboratory populations of *Paramecium aurelia* and its predator *Didinium nasutum*. *Ecology* **54**, 1320–1327.

Luckinbill, L.S. (1974) The effects of space and enrichment on a predator-prey system. *Ecology* **55**, 1142–1147.

Luckinbill, L.S. and Fenton, M.M. (1978) Regulation and environmental variability in experimental populations of protozoa. *Ecology* **59**, 1271–1276.

MacArthur, R.H. (1958) Population ecology of some warblers of northeastern coniferous forests. *Ecology* **39**, 599–619.

Maly, E. (1978) Stability of the interaction between *Didinium* and *Paramecium*: Effects of dispersal and predator time lag. *Ecology* **59**, 733–741.

Matsuda, H., Kawasaki, K., Shigesada, N., Teramoto, E. and Riccairdi, L.M. (1986) Switching effect on the stability of the prey-predator system with three trophic levels. *J. Theor. Biol.* **122**, 251–262.

May, R.M. (1974) Biological populations with non-overlapping generations: Stable points, stable cycles and chaos. *Science* **186**, 645–647.

McCann, K. and Hastings, A. (1997) Re-evaluating the omnivory-stability relationship in food webs. *Proc. R. Soc. Lond. Ser. B* **264**, 1249–1254.

McCann, K., Hastings, A. and Huxel, G.R. (1998) Weak trophic interactions and the balance of nature. *Nature* **395**, 794–798.

McGrady-Steed, J., Harris, P.M. and Morin, P.J. (1997) Biodiversity regulates ecosystem predictability. *Nature* **390**, 162.

McNaughton, S.J., Banyikwa, F.F. and McNaughton, M.M. (1997) Promotion of the cycling of diet-enhancing nutrients by African grazers. *Science* **278**, 1798–1800.

McPeek, M.A. and Holt, R.D. (1992) The evolution of dispersal in spatially and temporally varying environments. *Am. Nat.* **140**, 1010–1027.

Montoya, J.M., Rodriguez, M.A. and Hawkins, B.A. (2003) Food web complexity and higher-level ecosystem services. *Ecol. Lett.* **6**, 587–593.

Morin, P.J. (1999) Productivity, intraguild predation, and population dynamics in experimental microcosms. *Ecology* **80**, 752–760.

Morin, P.J. and Lawler, S.P. (1995) Food web architecture and population dynamics: Theory and empirical evidence. *Ann. Rev. Ecol. Syst.* **26**, 505–529.

Morin, P.J. and Lawler, S.P. (1996) Effects of food chain length and omnivory on population dynamics in experimental food webs. In: *Food Webs: Integration of Patterns and Dynamics* (Ed. by G.A. Polis and K.O. Winemiller), pp. 218–230. Chapman and Hall, New York, NY.

Murdoch, W.W. and Lawler, S.P. (1970) Population regulation and population inertia. *Ecology* **51**, 497–502.

Murdoch, W.W. (1994) Population regulation in theory and practice. The Robert H. MacArthur Award Lecture. *Ecology* **75**, 271–287.

Mylius, S.D., Klumpers, K., de Roos, A.M. and Persson, L. (2001) Impact of intraguild predation and stage structure on simple communities along a productivity gradient. *Am. Nat.* **158**, 259–276.

Nachman, G. (1981) Temporal and spatial dynamics of an acarine predator-prey system. *J. Animal Ecol.* **50**, 435–451.

Nachman, G. (1987) Systems analysis of predator-prey interactions. II. The role of spatial processes in system stability. *J. Animal Ecol.* **56**, 267–281.

Naeem, S. (2001) Experimental validity and ecological scale as criteria for evaluating research programs. In: *Scaling Relations in Experimental Ecology* (Ed. by R.H. Gardner, W.M. Kemp, V.S. Kennedy and J.E. Petersen). Columbia, New York, NY.

Naeem, S. and Li, S. (1998) Consumer species richness and autotrophic biomass. *Ecology* **79**, 2603–2615.

Nicholson, A.J. (1958) Dynamics of insect populations. *Annual Review of Entomology* **3**, 107–136.
Nicholson, A.J. and Bailey, V.A. (1935) The balance of animal populations. Part 1. *Proc. Zool. Soc. London* **3**, 551–598.
Orland, M.C. (2003) Scale-dependent interactions between intrinsic and extrinsic processes reduce variability in protist populations. *Ecol. Lett.* **6**, 716–720.
Orland, M.C. and Lawler, S.P. (2004) Resonance inflates carrying capacity in protist populations with periodic resource pulses. *Ecology* **85**, 150–157.
Pearl, R. and Reed, L.J. (1920) On the rate of growth of the population of the United States since 1790 and its mathematical representation. *Proc. Natl. Acad. Sci. USA* **6**, 275–288.
Petchey, O.L. (2000) Environmental colour affects aspects of single-species population dynamics. *Proc. R. Soc. Biol. Sci. B* **267**, 747–754.
Petchy, O.L., McPhearson, P.T., Casey, T.M. and Morin, P.J. (1999) Environmental warming alters food-web structure and ecosystem function. *Nature* **402**, 69–72.
Petersen, J.E. and Hastings, A. (2001) Dimensional approaches to scaling experimental ecosystems: Designing mousetraps to catch elephants. *Am. Nat.* **157**, 324–333.
Pianka, E.R. (1966) Latitudinal gradients in species diversity: A review of concepts. *Am. Nat.* **100**, 33–46.
Pimm, S.L. (1980) Properties of food webs. *Ecology* **61**, 219–225.
Pimm, S.L. and Lawton, J.H. (1977) Number of trophic levels in ecological communities. *Nature* **268**, 329–331.
Pimm, S.L. and Redfearn, A. (1988) The variability of population densities. *Nature* **334**, 613–614.
Polis, G.A. (1994) Food webs, trophic cascades and community structure. *Austr. J. Ecol.* **19**, 121–136.
Polis, G.A. and Holt, R.D. (1992) Intraguild predation: The dynamics of complex trophic interactions. *Trends Ecol. Evol.* **7**, 151–154.
Polis, G.A., Myers, C.A. and Holt, R.D. (1989) The ecology and evolution of intraguild predation: Potential competitors that eat each other. *Ann. Rev. Ecol. Syst.* **20**, 297–330.
Pulliam, H.R. (1988) Sources, sinks, and population regulation. *Am. Nat.* **132**, 652–661.
Rapport, D.J., Berger, J. and Reid, D.B.W. (1972) Determination of food preference of *Stentor coeruleus*. *Biol. Bull.* **142**, 103–109.
Reeve, J.D. (1988) Environmental variability, migration, and persistence in host-parasitoid systems. *Am. Nat.* **132**, 810–836.
Relyea, R.A. (2002) Competitor-induced plasticity in tadpoles: Consequences, cues, and connections to predator-induced plasticity. *Ecol. Monogr.* **72**, 523–540.
Resetarits, W.J., Jr. and Bernardo, J. (1998) *Experimental ecology: Issues and perspectives.* Oxford University Press, Oxford, United Kingdom.
Rosenzweig, M.L. (1971) The paradox of enrichment: Destabilization of exploitation ecosystems in ecological time. *Science* **171**, 385–387.
Schops, K. (2002) Local and regional dynamics of a specialist herbivore: Overexploitation of a patchily distributed host plant. *Oecologia* **132**, 256–263.
Schindler, D.W. (1998) Replication vs. realism: The need for ecosystem-scale experiments. *Ecosystems* **1**, 323–334.
Sinclair, A.R.E. (1989) Population regulation in animals. In: *Ecological Concepts: The Contribution of Ecology to an Understanding of the Natural World* (Ed. by J.M. Cherrett, A.D. Bradshaw, F.B. Goldsmith, P.J. Grubb and J.R. Krebs), pp. 197–241. Blackwell, Oxford, United Kingdom.

Stenseth, N.C. (1985) The structure of food webs predicted from optimal food selection models: An alternative to Pimm's stability hypothesis. *Oikos* **44**, 361–364.

Taylor, A.D. (1990) Metapopulations, dispersal, and predator-prey dynamics: An overview. *Ecology* **71**, 429–433.

Taylor, A.D. (1991) Studying metapopulation effects in predator-prey systems. *Biol. J. Linnaean Society* **42**, 305–323.

Taylor, W.D. and Berger, J. (1980) Microspatial heterogeneity in the distribution of ciliates in a small pond. *Microbial Ecol.* **6**, 27–34.

Tollrian, R. and Harvell, C.D. (1999) *The Ecology and Evolution of Inducible Defenses*. Princeton University Press, Princeton, NJ.

Trussell, G.C. and Nicklin, M.O. (2002) Cue sensitivity, inducible defense, and trade-offs in a marine snail. *Ecology* **83**, 1635–1647.

Turchin, P. (1995) Population regulation: Old arguments and a new synthesis. In: *Population Dynamics: New Approaches and Synthesis* (Ed. by N. Cappucino and P. Price), pp. 19–40. Academic Press, New York, NY.

Turchin, P. (2003) *Complex Population Dynamics: A Theoretical Empirical Synthesis*. Princeton University Press, Princeton, NJ.

Vandermeer, J.H. (1969) The competitive structure of communities: An experimental approach with protozoa. *Ecology* **50**, 362–371.

Van Nouhuys, S. and Hanski, I. (1999) Host diet affects extinctions and colonizations in a parasitoid metapopulation. *J. Animal Ecol.* **68**, 1248–1258.

Van Nouhuys, S. and Hanski, I. (2002) Colonization rates and distances of a host butterfly and two specific parasitoids in a fragmented landscape. *Journal of Animal Ecology* **71**, 639–650.

Veilleux, B.G. (1979) An analysis of the predatory interaction between *Paramecium* and *Didinium*. *J. Animal Ecol.* **48**, 787–803.

Volterra, V. (1926) Variations and fluctuations of the numbers of animal species living together. In: *Animal Ecology* (Ed. by R.N. Chapman). McGraw-Hill, New York, NY.

Watkinson, A.R. and Sutherland, W.J. (1995) Sources, sinks and pseudo-sinks. *J. Animal Ecol.* **64**, 126–130.

Weisser, W.W. (2000) Metapopulation dynamics in an aphid-parasitoid system. *Entomologia Experimentalis et Applicata* **97**, 83–92.

Woiwod, I.P. and Hanski, I. (1992) Patterns of density dependence in moths and aphids. *J. Animal Ecol.* **61**, 619–630.

Yodzis, P. (1984) Energy flow and the vertical structure of real ecosystems. *Oecologia* **65**, 86–88.

Yodzis, P. and Innis, S. (1992) Body size and consumer-resource dynamics. *Am. Nat.* **139**, 1151–1175.

Microbial Experimental Systems in Ecology

CHRISTINE M. JESSUP, SAMANTHA E. FORDE AND BRENDAN J.M. BOHANNAN

I. SUMMARY

The use of microbial experimental systems has traditionally been limited in ecology. However, there has recently been a dramatic increase in the number of studies that use microbial experimental systems to address ecological questions. In this chapter, we discuss the strengths, limitations, and criticisms of microbial experimental systems in ecology. We highlight a number of recent studies in which microbial experimental systems have been used to address one of the primary goals in ecology—identifying and understanding the causes and consequences of life's diversity. Finally, we outline what we consider to be the characteristics of an ideal model system for advancing

ADVANCES IN ECOLOGICAL RESEARCH VOL. 37

0065-2504/05 $35.00
DOI: 10.1016/S0065-2504(04)37009-1

ecological understanding, and discuss several microbial systems we feel fit this description.

II. INTRODUCTION

Ecology has been described as the "most intractable of the biological sciences" (Slobdokin, 1988). Ecologists are faced with the challenge of understanding the distribution and abundance of organisms that are embedded in complex, dynamic systems of interactions. To make sense of such complexity, ecologists have relied on a number of approaches, including mathematical modeling, observations of natural systems, and field manipulations. Increasingly, ecologists are turning to another tool—laboratory experimental systems. Microorganisms, in particular, have become increasingly popular as experimental systems with which to explore ecological processes (Fig. 1). These experimental systems include viruses, bacteria, and single-celled eukaryotes (e.g., protists, algae and fungi). Detailed reviews of many of these experimental systems have been published previously on issues such as the generation and maintenance of diversity (Rainey *et al.*, 2000) and the interplay between genetic change, ecological interactions, and community evolution (Bohannan and Lenski, 2000a), among others (e.g., Lawler,

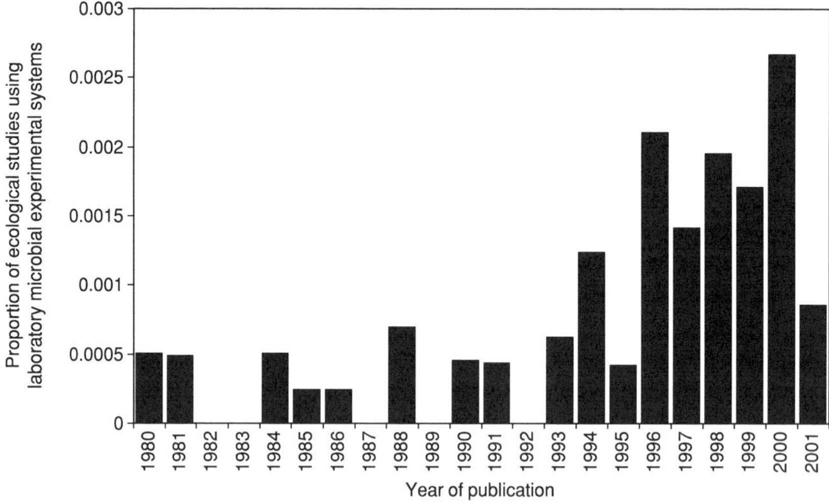

Figure 1 Increase in the proportion of published ecological studies using laboratory-based microbial experimental systems from 1980 through 2001 in the journals Nature, Science, Proceedings of the Royal Society, Proceedings of the National Academy of Sciences, Ecology, American Naturalist, Ecology Letters, Oikos, Oecologia and Evolution.

1998). In this chapter, we take a broader perspective and discuss the history of microbial experimental systems in ecology, the strengths and limitations of such systems, and controversies surrounding their use in ecology. We highlight the use of microbial experimental systems to explore the causes and consequences of ecological diversity. We conclude this chapter with a discussion of microbial systems that may be ideal model systems in ecology, spanning theory, laboratory manipulation, and field experimentation.

III. THE HISTORY OF MICROBIAL EXPERIMENTAL SYSTEMS IN ECOLOGY

A. G.F. Gause and His Predecessors

Microbial experimental systems have played a central, if sometimes underappreciated, role in ecological history. The earliest published record of the use of a microbial experimental system in population biology is that of Rev. W. D. Dallinger who, in his address as the president of the Royal Microscopy Society in 1887, described his attempt to discover "whether it was possible by change of environment, in minute life-forms, whose life-cycle was relatively soon completed, to superinduce changes of an adaptive character, if the observations extended over a sufficiently long period" (Dallinger, 1887). Dallinger addressed this question using populations of protists as an experimental system, altering their environment by varying the temperature of the cultures. He demonstrated with these experiments that ecological specialization can incur a cost of adaptation (a decline in competitive fitness in environments other than the one to which the organisms have specialized) and that it was possible to study such phenomena with laboratory experimental systems.

Over 20 years later, L.L. Woodruff (1911, 1912, 1914) used protists in hay infusions as an experimental system to study density-dependent growth and succession. He concluded that interactions between organisms were an important driving force in the successional sequence of the protozoan community that he observed. Similar studies by Skadovsky (1915) and Eddy (1928) were published in the decade following Woodruff's work.

In what were to become arguably the most famous microbial experiments in ecology, G.F. Gause built upon this earlier work by using experimental systems containing bacteria, yeast, and protists to ask "why has one species been victorious over another in the great battle of life?" (Gause, 1934). He coupled his laboratory experiments with the mathematical models of competitive and predator-prey interactions first proposed by Alfred Lotka and Vito Volterra, and later popularized by Raymond Pearl (Kingsland, 1995). This combination of experiment and theory was especially powerful, allowing him to make quantitative predictions of the outcome of interactions between species, and then to test the plausibility of these predictions through carefully designed experiments with microorganisms. Several key principles in ecology are attributed to these studies. For example, Gause demonstrated that he was able to predict which of two species of *Paramecium* would be competitively dominant by estimating the growth parameters for each of the species grown alone. Subsequent interpretation of this work by Garrett Hardin led to the development of the principle of competitive exclusion, or "Gause's Axiom" (Hardin, 1960). In another experiment using *Didinium nasutum* and *Paramecium caudatum*, Gause demonstrated the importance of both spatial refugia and immigration for the co-existence of predators and prey.

B. Studies Since Gause

Since Gause's pioneering experiments, microbial experimental systems have been used to study a number of topics in ecology. For example, laboratory systems containing eukaryotic microorganisms (protists and/or microalgae) have been used to study succession (Gorden *et al.*, 1979), the diversity-stability relationship (Hairston *et al.*, 1968; Van Voris *et al.*, 1980), predator-prey interactions (Luckinbill, 1973; van den Ende, 1973; Luckinbill, 1974, 1979), and the co-existence of competitors (Vandermeer, 1969; Tilman, 1977; Sommer, 1984; Tilman and Sterner, 1984). The microbial experimental systems used by Tilman and his colleagues have particularly influenced the development of ecological theory. Although not the first to demonstrate the predictable nature of resource competition (this distinction would probably go to Levin, 1972), Tilman's experiments with diatoms elegantly demonstrated the value of the resource-explicit approach to competition theory first proposed by MacArthur (1972) and later developed by Tilman (1982).

Experimental populations and communities of bacteria have also been used to explore ecological concepts such as resource competition (Hansen and Hubbell, 1980), the relationship between growth kinetics and population dynamics (Contois, 1959), specialist versus generalist strategies (Dykhuizen and Davies, 1980), optimality theory (Wang *et al.*, 1996; Bull *et al.*, 2004) and ecological diversification (Rosenzweig *et al.*, 1994). The interaction between bacteria and bacteriophage (viruses that infect bacteria) has been utilized in many ecological and evolutionary studies of predator-prey and parasite-host interactions (Horne, 1970, 1971; Chao *et al.*, 1977; Levin *et al.*, 1977; Levin and Lenski, 1985; Lenski, 1988a; Schrag and Mittler, 1996; Bohannan and Lenski, 1997, 1999). These include studies of predator-mediated co-existence of competitors, trophic cascades, heterogeneity in prey vulnerability, the role of spatial refuges, and optimal foraging, among others. Studies of the interaction between bacteria and bacteriophage have made particularly important contributions to the study of antagonistic co-evolution.

Recently, microbial systems containing eukaryotic microorganisms and/or bacteria have been used to study the causes and consequences of ecological diversity; these studies have significantly influenced ecological thought. Before highlighting these recent contributions of microbial experimental systems to the study of diversity, we first consider the strengths and limitations of these experimental systems, as well as the controversies surrounding their use.

IV. STRENGTHS AND LIMITATIONS OF MICROBIAL EXPERIMENTAL SYSTEMS

A. Strengths of Microbial Experimental Systems

The growing popularity of microbial laboratory experiments in ecology is due, in part, to the control afforded to the ecologist over all aspects of the experimental system. The abundance of genetic and physiological information available for the most commonly used microorganisms allows the researcher to choose particular organisms to test a given hypothesis. The small size and short generation time of microorganisms facilitate replicated experiments across a wide range of spatial and temporal scales. Furthermore, microorganisms are amenable to genetic manipulation and to long-term storage in a state of "suspended animation." These advantages allow ecologists to dissect the complexity of nature into its component parts, analyzing each part's role in isolation and then in combination. Microbial experimental systems thus provide an important link between theory and the complexity of nature.

1. Experimental Control

Microbial experimental systems offer the researcher an unusually high degree of experimental control. Experiments are usually initiated with defined communities of organisms and defined resources under controlled laboratory conditions. Populations can be unambiguously identified and tracked over time, resources can be monitored accurately, and environmental conditions can be precisely controlled. This allows the exploration of ecological and evolutionary phenomena at a fine scale. For example, Bennet and Lenski (1993) used laboratory populations of *Escherichia coli* as an experimental system to study the ecology and evolution of the thermal niche. They observed that populations of *E. coli* adapted to 37 °C could increase in abundance at a temperature of 42.1 °C, but that raising the temperature only 0.2° further prevented growth. Thus, they were able to define the thermal niche boundary for this organism at a very fine scale. Bennet and Lenski (1993) could then rigorously study evolution at this upper thermal niche boundary. They observed fixation of mutants with higher competitive fitness at this temperature, but evolution did not result in the extension of the niche boundary. These observations were in conflict with theories widely held at the time that predicted that extension of the thermal niche boundary was a necessary correlate of organismal evolution at high temperature (Huey and Kingsolver, 1989).

2. *Ease of Replication*

Laboratory-based microbial experimental systems can be maintained in relatively small growth vessels facilitating replication. The ability to highly replicate ecological systems allows for the study of variation in ecological processes, and increases the statistical power of an ecological experiment, which can allow a researcher to test models with subtle differences in their predictions. For example, Bohannan and Lenski (1997) tested the predictions of two competing predator-prey models using microcosm experiments with predators (bacteriophage) and prey (bacteria) at different levels of productivity. There has been debate among food-web ecologists as to whether predator-prey interactions are better represented using models that assume that the attack rate of predators depends only on the instantaneous density of the prey ("prey-dependent") or using models that assume that the attack rate is a function of the ratio of prey to predator density ("ratio-dependent"). Although there have been several attempts to address this issue in field systems, the results have been inconclusive. Bohannan and Lenski (1997) maintained six communities of bacteriophage and bacteria in continuous culture devices (chemostats), with three communities receiving a lower nutrient medium, and three receiving a higher nutrient medium. Populations of bacteriophage and bacteria were monitored twice daily and the resulting time series were compared to the predictions of simulations using prey- and ratio-dependent models. By comparing equilibrium densities in replicate chemostats as well as population stability, they observed that a prey-dependent model was better than a ratio-dependent model in predicting population abundances and dynamics in this system. The prey-dependent model was also better supported by observations such as the rate of invasion by predator-resistant mutants.

3. *Short Generation Times*

Organisms in microbial experimental communities have relatively short generation times, making it possible to study hundreds of generations in experiments as short as one week. The traditional distinction between ecology and evolution, which is arguably arbitrary (Antonovics, 1976), is often blurred in microbial laboratory experiments (Lenski, 2001). Mutations readily arise in microbial populations because of the short generation times and large population sizes, and these unavoidably lead to evolution. Such long-term dynamics allow researchers to explore both "the ecological theater and the evolutionary play" (Hutchinson, 1965), which is difficult in most field studies. For example, whether a bacteriophage-resistant mutation in a bacterial population persists in glucose-limited communities is a function of the cost of resistance, which depends on the particular mutation and the environment (Bohannan and Lenski, 2000a). If a resistant mutant invades, the concentration of limiting resource decreases, which has important consequences for both the equilibrium densities, rate of turnover and dynamics of the bacteriophage-resistant, bacteriophage-sensitive, and bacteriophage populations. The ecological changes prompted by the invasion of a bacteriophage-resistant population in turn, can affect subsequent evolution within the community. Invasion by a bacteriophage-resistant population affects the densities of bacteriophage and bacteriophage-resistant populations and the rate of turnover of the sensitive population. These rates then govern the emergence of new mutations and the rate of adaptation in each population (Gerrish and Lenski, 1998; de Visser *et al.*, 1999; Bohannan and Lenski, 2000a). Thus, community properties affect the tempo of evolution, which subsequently changes community structure and the responsiveness of the community to environmental change over ecological and evolutionary time scales (Bohannan and Lenski, 2000a).

4. The Ability to Archive Organisms

Populations from many microbial experimental systems can be readily archived by freeze-drying, storage in cryoprotectant at very low temperature (e.g., −80°C), or storage of resistant life stages (such as spores or cysts). Archiving can allow a researcher to dissect an experiment *a posteriori*, providing a detailed understanding of the mechanisms underlying the ecological and evolutionary processes being studied. Rainey and Travisano (1998) showed that populations of the bacterium *P. fluorescens* rapidly diversify genetically and ecologically in physically heterogeneous (static) environments but not in environments lacking physical heterogeneity (those that are shaken; see section VI.B.2). The genotypes that emerge in the heterogeneous environments have different colony morphologies on agar plates that make them readily distinguishable by the eye. Throughout the experiment, the different morphs of *P. fluorescens*, as well as the ancestors, were archived for future use. This allowed the researchers to study directly the role of competition in adaptive radiation and to determine how the relative competitive abilities of the morphs changed through time to produce the patterns observed in the original experiment.

The ability to archive organisms from microbial experiments also enables a researcher to explore unexpected or surprising results in more detail. In a study of predator-prey communities, Schrag and Mittler (1996) documented the stable co-existence of bacteriophage and bacteria over approximately 50 generations under continuous culture conditions, a situation where theory predicted that co-existence was not possible. Because both bacteria and bacteriophage cultures had been archived throughout the experiment, it was possible to explore the mechanisms underlying this unexpected co-existence. The investigators found that a small fraction (~5% or less) of the bacterial population was phage-sensitive, allowing the bacteriophage to persist. Furthermore, the bacteriophage-sensitive population was sustained because the culture environment was much less homogeneous than was first thought—populations of bacteria grew on the walls of the culture vessel as well as in the liquid media contained therein. Growth on the walls of the culture vessel acted as a spatial refuge from bacteriophage predation, stabilizing the predator-prey interaction.

5. *Well-Characterized Constituents*

Another benefit of microbial experimental systems lies in their often well-characterized constituents. Some organisms used in laboratory experimental communities (e.g., *E. coli, Saccharomyces cervisiae*) have been used as model organisms in physiologic and genetic studies for decades. The wealth of information available on such organisms is advantageous in several respects. First, it permits manipulation of organismal traits on a genetic level. For example, populations can be tagged with neutral physiological markers (such as the ability to utilize a particular sugar or resistance to a virus) facilitating the tracking of population dynamics in laboratory systems. Second, the genetic and physiological basis of ecological phenomena can be identified, and the feedback between ecological and evolutionary processes can thus be understood. These attributes of microbial experimental systems have been exploited by a number of scientists to increase our understanding of ecological processes at multiple scales of biological organization, from genes to ecosystems. For example, Lenski and colleagues studied the trade-off in *E. coli* between competitive ability and resistance to predation by bacteriophage (Lenski and Levin, 1985; Lenski, 1988b, reviewed in Bohannan and Lenski, 2000a). They demonstrated that mutations in different genes that conferred resistance to predation incurred different competitive costs, and that these costs affected the dynamics of the bacterial and bacteriophage populations in predictable ways. These different mutations also altered ecosystem level properties such as sensitivity to invasion and response to environmental change.

B. Limitations of Microbial Experimental Systems

There are experimental "costs" associated with many of the advantages previously outlined. The small scale of microorganisms can make it difficult to explicitly impose and maintain environmental heterogeneity at small scales. Rapid evolution can lead to changes in populations or population dynamics before such changes have been characterized by the researcher. Studies of small population sizes are challenging. The unicellular biology of most microorganisms can make it hard to study learned behavior. Finally, there are limits to extrapolating from microbial experimental systems. We discuss each of these, in turn. Some of these costs can be reduced, although not eliminated, through careful choice of experimental organisms and experimental protocols. Few of these limitations are unique to microbial experimental systems. Not all questions are appropriate for all experimental systems; the major challenge all experimenters face is matching research questions with appropriate experimental systems, and this is as true for microbial experimental systems as for any other.

1. Scale of the Environment

Although heterogeneity in resources or conditions does emerge in microbial experimental systems (e.g., gradients in resource availability and waste products), and this heterogeneity is important for subsequent ecological and evolutionary dynamics, researchers are usually forced to describe this heterogeneity after the fact. The small size of microorganisms makes it experimentally difficult to explicitly manipulate and maintain environmental heterogeneity at the spatial scale of an individual microorganism, or even a small population; such manipulation would require resolution on the order of millimeters or even microns. Although a number of studies have used microbial experimental systems to understand the role of environmental heterogeneity on population and community dynamics (e.g. Bell and Reboud, 1997), this heterogeneity is usually at a much larger scale relative to body size than the analogous scale in studies of macroorganisms.

2. Rapid Evolution

The fact that organisms in microbial experimental systems grow rapidly to such large population sizes can also yield practical problems. Evolution of organisms in some microbial microcosms can occur on the order of days, often changing interactions and population dynamics before the researcher has completed their characterization. This can be reduced by using microorganisms with longer generation times and smaller population sizes, such as protists rather than bacteria or viruses, or by maintaining the experimental populations using conditions that slow evolution. For example, the evolution of resistance to some bacteriophages can be slowed by adding the antibiotic novobiocin to the growth medium (Lenski, 1988b). Resistance to these bacteriophages is accompanied by increased sensitivity to this antibiotic, which will select against these mutants.

3. Large Population Sizes

Microbial systems allow for a high degree of replication and the study of the integrated effects on individuals by looking at large, often clonal, populations. However, some questions in ecology are fundamentally based on the small populations. These questions become difficult to address in microbial experimental systems where, for example, a standard 10 ml culture of *E. coli* cells can contain 10 billion cells after <1 day of growth, and a viral plaque on a lawn of bacteria can contain 10 million virions. Questions about genetic drift, population bottlenecks, and extinction risk may not be readily addressed with the large population sizes of most microorganisms. However, Chao (1990) showed that Muller's ratchet (a theory that explains that, without recombination caused by sex, the most fit class of a population is gradually lost due to the accumulation of deleterious mutations) operated in RNA viruses that were forced through a bottleneck of one individual. Thus, some questions particular to small population sizes, in fact, may be answered by choosing the appropriate microorganism and suitable protocols.

4. Unique Aspects of Microbial Biology

Some of the unique aspects of microorganisms, such as clonal reproduction, similar morphologies, and unicellular biology make microbial experimental systems inappropriate for addressing some ecological questions. For example, studies of life-span and age-based phenomena are not easily addressed using microorganisms (Andrews, 1991; but also see Ackermann *et al.*, 2003). In addition, many questions based on learned behavior, such as sexual selection or territoriality, cannot be addressed with microbial systems.

5. Limits to Extrapolation

There are also limits to our ability to extrapolate from microbial experimental systems to larger and often more complex systems. Such limitations are not unique to microbial systems however, but are shared by ecological studies in general. Identifying the appropriate experimental scale and the limits of extrapolation from this scale are critical aspects of conducting research in all areas of biology, not just microbial systems (Carpenter, 1996; Hastings, 2000).

V. THE ROLE OF MICROBIAL EXPERIMENTAL SYSTEMS IN ECOLOGY: CONSENSUS AND CONTROVERSY

A. The Role of Microbial Experimental Systems

There is considerable debate about the role laboratory experimental systems should play in ecology, and the relationship of laboratory studies to ecological theory and field experiments. Carpenter (1996) and Diamond (1986) have argued that laboratory studies should be primarily supportive of field studies. In other words, the purpose of laboratory studies is to provide supporting information for field studies that is impossible or impractical to gather in the field. Others have argued that a major role of laboratory studies is to provide "clean tests" of ecological theory (Daehler and Strong, 1996; Drake *et al.*, 1996). Still others have argued that laboratory ecology should act as a bridge between theory and the field rather than being in the service of theoretical or field ecology (Lawton, 1995). In the words of Slobodkin (1961):

> In one sense, the distinction between theoretician, laboratory worker and field worker is that the theoretician deals with all conceivable worlds while the laboratory worker deals with all possible worlds and the field worker is confined to the real world. The laboratory ecologist must ask the theoretician if his possible world is an interesting one and must ask the field worker if it is at all related to the real one.

Another way to state this is that the role of theory is to define what is logically *possible* (given a set of assumptions), the role of the laboratory experiment is to determine what is biologically *plausible* and the role of the field study is to delineate what is ecologically *relevant*. Microbial experimental systems have played all three of these roles: in support of field studies, in support of theory, and as a bridge between theory and the field.

Just as plant and animal ecologists use controlled laboratory experiments to better understand the organisms that are the focus of their field research programs (e.g., glasshouse experiments designed to explore the effects of elevated CO_2 or nutrient enrichment on plant physiology, or food preference experiments to determine diet breadth), microbiologists also use laboratory experiments to augment and inform field studies. For example, field observations of biogeochemical transformations are supported by the laboratory isolation of the microorganisms involved in these transformations. In the laboratory, the responses of these microorganisms to resources (e.g., growth kinetics) and abiotic conditions (e.g., temperature optima) can be determined in order to better understand their ecology in the field. This approach dates to the early microbial ecologists Martinus Beijerinck and Sergei Winogradsky (Beijerinck, 1889; Winogradsky, 1889; as cited in Brock, 1999). Although this approach was central to the development of the subdiscipline of microbiology known as "microbial ecology," it has had little influence on the development of general ecology (i.e., plant, animal and theoretical ecology) with the possible exception of some aspects of ecosystem ecology.

In contrast, the traditional use of laboratory study of microorganisms in general ecology has been in support of ecological theory (e.g., Gause, 1934). Theoretical predictions are often tested using microbial experimental systems by comparing the predictions of a verbal or mathematical model with the observations of a microbial population or community in the laboratory. The microbial system is usually perturbed in some way and the behavior of the system is compared to behaviors predicted by different models using different theoretical assumptions. Although a mathematical model can never be "wrong" (unless a mathematical error is made; Levins, 1966), a model's theoretical assumptions can certainly be incorrect. By challenging alternative models with experimental observations one can assign degrees of plausibility to the models (Oreskes *et al.*, 1994; Hilborn and Mangel, 1997).

Microbial experimental systems are particularly well suited to exploring theory in this way. Because such systems can be relatively simple and controlled, researchers can not only compare observations rigorously to theoretical predictions, but also determine *why* a given model may fail to accurately predict the behavior of a population or community. The ultimate test of our ecological understanding (whether in the form of a mathematical model or not) is whether we can predict the behavior of an ecological system. For example, in the study mentioned in Section IV.A.2, Bohannan and Lenski (1997) were able to determine why the ratio-dependent model failed: it was unable to predict the destabilization of the populations due to increased productivity (the "paradox of enrichment" predicted by Rosenzweig, 1971).

As previously described, this use of microbial experimental systems to test theory dates back at least to G.F. Gause, and such studies have had a significant influence on the development of general ecology, although not without controversy (see section V.B).

More recently, these two approaches—using microbial experimental systems in support of field research and in support of theory—have merged. The use of microbial experimental systems to create a bridge between theory and field studies have become more common, particularly through the use of experimental systems that bring relevant microbial communities (as opposed to particular individuals) into the laboratory for further exploration. For example, Zhou *et al.* (2002) observed that the structure of bacterial communities varied with soil depth; specifically, they observed that communities from deeper soil samples tended to have a few numerically dominant taxa and many rare taxa, while communities from shallower samples had more even distributions of taxa. They hypothesized that this difference in taxon distribution was due to different intensities of competition at different depths; the higher water content of the deeper soils reduced the spatial isolation of populations and intensified competition. They created a mathematical model based on this hypothesis and tested its plausibility using a laboratory experimental system. The experimental system consisted of pairs of competing bacterial populations in sand environments that differed in water content. They observed that water content altered the outcome of competition and the relative abundance of the populations, as predicted by their mathematical model. Other examples of the use of microbial experimental systems to bridge theory and field studies include studies of the dynamics of bacteriophage and bacteria in marine systems (Waterbury and Valois, 1993), the population dynamics of antibiotic resistant bacteria (Levin *et al.*, 2000; Lipsitch, 2001), and the maintenance of genetic diversity in yeast (Greig *et al.*, 2002) and *E. coli* (Kerr *et al.*, 2002; Riley and Wertz, 2002). The increase in the popularity of this approach in recent years is due to an increased understanding of the field ecology of microorganisms, and an increased appreciation for the utility of ecological theory in the study of microbial ecology.

B. Controversies Surrounding Experimental Systems and Microbial Experimental Systems

Throughout the long history of research with microbial experimental systems in ecology, their use has remained controversial (Kingsland, 1995). Mertz and McCauley (1982) have argued that the controversy surrounding laboratory studies in general ecology increased steadily from the 1960s to the 1980s. This claim is supported by the observation that the number of laboratory studies published in major ecological journals declined (as a percentage of the total papers published in these journals) during this period (Ives *et al.*, 1996). Mertz and McCauley (1982) attributed the decline in the popularity of laboratory studies to the influence of Robert MacArthur on the development of general ecology during this period. MacArthur was highly critical of laboratory studies as being generally uninteresting and unimportant, referring to them as "bottle experiments" (MacArthur, 1972). MacArthur was not the only eminent ecologist to weigh in against laboratory studies during this period. G. Evelyn Hutchinson criticized laboratory experimental systems as being highly artificial and essentially "a rather inaccurate analogue computer ... using organisms as its moving parts" (Hutchinson, 1978).

More recent critics of laboratory studies have expanded on these earlier critiques, raising concerns about the validity of laboratory microcosm research in general and microbial experimental systems in particular (Peters, 1991; Carpenter, 1996; Lawton, 1996). Laboratory experimental systems have been criticized for being too simple, too artificial, and too small in spatial and temporal scale to be useful for most ecological questions. In addition, microbial experimental systems have been criticized for using model organisms that are not representative of most other organisms. These critiques suffer from confusion about the purpose of an experimental system, misconceptions about laboratory experiments, and a general ignorance of microbial biology.

1. Are Laboratory Systems Too Simple?

Laboratory experimental systems have been criticized for lacking in realism and/or generality because of their simplicity (Diamond, 1986; Hairston, 1989; Carpenter, 1996; Lawton, 1996). They have been referred to as "analogies" (i.e., overly simplistic representations of natural systems; Peters, 1991) because they do not necessarily incorporate complexities such as seasonality, disturbance, or immigration (Lawton, 1996). But these critiques suffer from confusion about the purpose of laboratory experimental systems. Laboratory systems are not intended to be miniature versions of field systems. Laboratory ecologists do not intend to reproduce nature in a laboratory experimental system any more than theoreticians intend to reproduce nature with a mathematical model. Rather, the purpose is to simplify nature so that aspects of it can be better understood (Lawton, 1995, 1996; Drake *et al.*, 1996). Like mathematical models, laboratory model systems are necessary because we do not have full access, in time or space, to phenomena in nature (Oreskes *et al.*, 1994). Thus, simplicity is a strength of laboratory experimental systems, not a weakness (Drake *et al.*, 1996).

There is no reason why complexities such as immigration, disturbance, or seasonality cannot be included in a laboratory experimental system if the research question being addressed warrants the inclusion of these complexities. One of the advantages of laboratory systems is that the decision of whether to include particular complexities is up to the experimenter, and not imposed upon the experiment by the vagaries of nature (Lawton, 1995; Drake *et al.*, 1996). With a laboratory system, it is possible for complexity to be first reduced and then increased in a controlled fashion. Many examples of microbial studies in which ecological complexity was directly manipulated are described in section VI, including studies that manipulated dispersal (Donahue *et al.*, 2003), trophic structure (Kaunzinger and Morin, 1998), and disturbance (Buckling and Rainey, 2002b).

2. *Do Laboratory Systems Lack Realism and Generality?*

Two misconceptions are that laboratory experiments lack realism and that they behave like computers; neither the experimental organisms nor their interactions are creations of the experimenter, nor are they under the direct control of the experimenter (Mertz and McCauley, 1982). Few laboratory studies use truly artificial communities. Most studies use species that co-occur in a particular habitat; in that sense, they are arguably no more artificial than exclosure experiments in the field (Lawton, 1996). There is no reason to believe that nature operates any differently within the walls of a laboratory than outside.

It is also a misconception that laboratory systems lack generality. Laboratory experiments usually address fundamental ecological questions using simple systems. Because of their simplicity, laboratory systems potentially offer more generality than studies of more complex and thus more idiosyncratic field systems (Drake *et al.*, 1996; Lawton, 1999).

3. *Are Laboratory Systems Too Small in Scale?*

Laboratory studies have been criticized for being too small in spatial and temporal scale (Lawton, 1996). It is a misconception that laboratory experiments usually occur on small temporal scales. One of the advantages of using microorganisms in laboratory studies is that large relative temporal and spatial scales are possible. Based on a literature search, Ives *et al.* (1996) concluded that microcosm studies may actually have longer average duration, in terms of generations of the organisms involved, than field studies. For example, as discussed in section IV.A, it is possible to maintain microbial experimental systems for hundreds and even thousands of generations and at spatial scales that are orders of magnitude larger than the individual organisms.

4. Are Microorganisms Inappropriate as Experimental Organisms?

The criticism that the use of microorganisms in laboratory studies is suspect because of their unique biology (Diamond, 1986; Hairston, 1989; Carpenter, 1996) reflects an ignorance of microbial biology. While there are unique aspects to being of small size (e.g., living in a world primarily governed by intermolecular forces rather than by gravity) and being prokaryotic (e.g., parasexuality), prokaryotic and eukaryotic microorganisms share the fundamental properties of larger organisms. They must overcome the same basic challenges that larger organisms face, including competing for resources, avoiding consumption by predators, and maximizing reproductive success, while facing many of the same constraints on resource allocation that larger organisms face (Andrews, 1991). Microorganisms are valid model organisms for questions that are concerned with these fundamental properties, such as the role of trade-offs in ecological processes, the ecology and evolution of interspecies interactions, and the evolutionary ecology of life-history strategies.

The use of microorganisms as laboratory experimental systems in ecology has also been criticized because the organisms used in such studies must be easily cultured and thus tend to be "adaptable weeds" that bear little resemblance to most populations in nature (Hairston, 1989). Furthermore, it has been argued that because so little is known about the ecology of such species in the field that "it is not to be expected that they would reveal much of interest that could be applied to field systems" (Hairston, 1989). It is certainly true that relative to our understanding of the ecology of many plants and animals we know very little about the ecology of microorganisms in the field—it is estimated that <1% of microorganisms have been isolated and described (Amann *et al.*, 1995). However, it is not true that microbial laboratory systems do not "reveal much of interest that could be applied to field systems." Precisely because so little is known about microbial ecology in the field, laboratory systems have been very successful in revealing ecological phenomena with relevance to field systems. For example, our understanding of the interactions between bacteriophage and bacteria in field systems has been greatly influenced by laboratory studies of bacteriophage-bacteria communities. The importance of bacteriophage-resistant bacteria in the interactions between bacteriophage and bacteria was first demonstrated in the laboratory (Lenski and Levin, 1985; Lenski, 1988b; Bohannan and Lenski, 1997, 1999, 2000b), and stimulated the study of such effects in the field (Waterbury and Valois, 1993). Our understanding of the ecology of drug resistance in bacteria was also preceded by laboratory studies of microbial populations. Laboratory studies with model organisms demonstrated that a trade-off could exist between competitive ability and drug resistance (a "cost of resistance") and that the magnitude of this trade-off could be lowered by evolution (Andersson and Levin, 1999; Levin *et al.*, 2000). Subsequent research with major pathogenic bacteria and clinically important drugs suggests that these phenomena are also important in the field (Lipsitch, 2001; Nagaev *et al.*, 2001; Maisnier-Patin *et al.*, 2002).

5. *Is There Any Consensus?*

There are at least two points that both proponents and critics of laboratory experimental systems appear to agree on (Diamond, 1986; Lawton, 1995, 1996; Carpenter, 1996; Drake *et al.*, 1996). First, all agree that laboratory experimental systems have the definite advantages of replicability, reproducibility, mastery of environmental variables, and ease of manipulation. While there is debate over how important these advantages are, the existence of these advantages is not in question. Second, all parties agree that laboratory systems are just one of the many tools available to ecologists. According to Lawton (1995), such systems are "one part of a rich, interrelated web of approaches to understanding and predicting the behavior of populations and systems." Just as there are trade-offs between different approaches to ecological modeling (Levins, 1966), there are trade-offs between different approaches to ecological experimentation (Diamond, 1986). Many ecologists have suggested that the ideal strategy is to address ecological questions by utilizing multiple approaches, because conclusions tested by different methodologies become more robust (Diamond, 1986; Lawton, 1995; Drake *et al.*, 1996). This is particularly crucial when the stakes are high, for example when addressing questions of great societal importance or with important public policy implications. In the following section, we focus on the use of microbial experimental systems to study the generation, maintenance, and consequences of ecological diversity—subjects with important implications for conservation and environmental management.

VI. RECENT STUDIES OF MICROBIAL EXPERIMENTAL SYSTEMS

What determines the extraordinary diversity of life on earth? What are the consequences of this diversity? Identifying and understanding the causes and consequences of life's diversity is a primary goal of ecology. There is a rich theoretical literature on these topics, but experimental studies of the plausibility of these various hypotheses have lagged behind the development of theory, primarily because of the inherent difficulty of addressing these questions in field systems and with long-lived organisms. Recently, microbial experimental systems have been shown to be particularly powerful tools for studying these questions.

A. The Ecological Causes of Diversity

1. The Role of Trade-Offs

Ecological trade-offs are central to most theories on the maintenance of biological diversity (but not all, see Bell, 2001; Hubbell, 2001). Yet trade-offs have been notoriously difficult to observe and measure in the field because of the large sample sizes required and the fact that the magnitude of trade-offs can vary in different environments (Bergelson and Purrington, 1996; Bohannan *et al.*, 2002). Microbial experimental systems have provided some of the most compelling evidence for trade-offs mediating the co-existence of populations (Dykhuizen and Davies, 1980; Travisano and Lenski, 1996; Rainey *et al.*, 2000; Bohannan *et al.*, 2002; Kassen, 2002; MacLean and Bell, 2002). In these systems, trade-offs can be measured directly through competition experiments between marked strains, the effects of genotype and environment on the magnitude of the trade-off quantified, and the importance of these parameters for the structure and function of communities explored (Bohannan *et al.*, 2002).

Experiments with bacteria and bacteriophage, for example, have demonstrated that the trade-off between the abilities of organisms to compete for resources and to resist predators allows for the predator-mediated co-existence of populations competing for a single limiting resource (Fig. 2), as predicted by theory (Leibold, 1996) (see figure legend for details). Furthermore, the magnitude of the trade-off has important direct and indirect implications for community structure and function (Bohannan and Lenski, 2000a; Bohannan *et al.*, 2002). Field observations of bacteria and bacteriophage are consistent with the results of these laboratory studies. For example, studies of natural populations of the marine cyanobacterium *Synechococcus* suggest that resistance to bacteriophage is common and that bacteriophage persist in these environments by feeding on minority populations of sensitive cells. These sensitive cells likely co-exist with resistant cells due to a trade-off between resistance and competitive ability (Waterbury and Valois, 1993).

Microbial experimental systems have also demonstrated that costs of adaptation may not always explain trade-offs in ecological systems. In a recent review of experiments with microbial experimental systems, Kassen

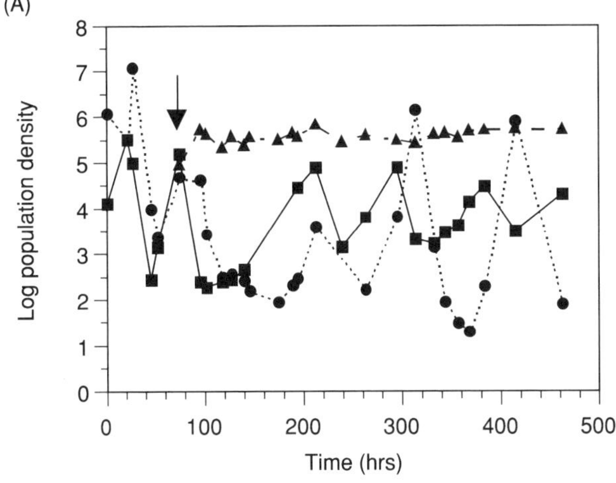

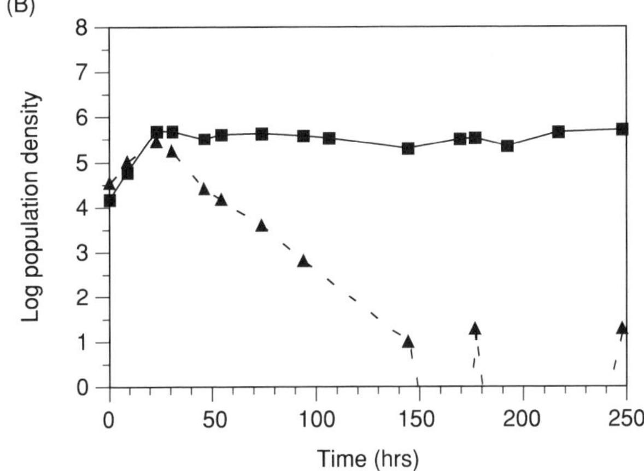

Figure 2 (A) Predator-mediated co-existence of phage-sensitive (squares) and phage-resistant (triangles) bacteria in a chemostat community consisting of phage T4 (circles) and *Escherichia coli*. The phage-resistant mutant was introduced at the time indicated by the arrow. (From Bohannan and Lenski, 1999, Fig. 2B, with permission of the American Naturalist, © 1999 by The University of Chicago Press.) (B) Population dynamics of T4-sensitive (squares) and T4-resistant (triangles) *E. coli* in a representative chemostat lacking phage. The decline of the resistant population is due to the trade-off between resistance and competitive ability. (From Bohannan and Lenski, 2000, Fig. 8F, with permission of The American Naturalist, © 2000 by The University of Chicago Press.)

(2002) reported the existence of trade-offs following adaptation to different environments, but only a fraction of these were underlain by a true cost of adaptation; the majority were due to the specificity of local adaptation, where direct responses to selection are often greater than correlated responses. Thus, while costs of adaptation may go undetected in many systems, these highly controlled studies demonstrate that costs of adaptation can indeed be absent.

When costs of adaptation *are* observed, determining the underlying causes is often difficult in the field. Observed costs of adaptation may be the result of antagonistic pleiotropy or mutation accumulation (Kawecki *et al.*, 1997). Microbial experimental systems have been used to explore which of these underlying causes is most likely. Cooper and Lenski (2000) studied the decay of unused catabolic functions in multiple populations of *E. coli* maintained in glucose-limited culture for 20,000 generations of growth. They observed that most of the catabolic losses of function occurred early in the experiment when beneficial mutations were rapidly being fixed, an observation consistent with antagonistic pleiotropy but not mutation accumulation. Furthermore, while several populations evolved higher mutation rates during this experiment, these populations did not lose catabolic function at a faster rate than those with lower mutation rates, also inconsistent with the mutation accumulation hypothesis. Separating these underlying causes of trade-offs would be extraordinarily difficult in other experimental systems.

2. *The Importance of Space*

Space has long been suspected of playing a major role in the maintenance of diversity, but it has proven to be difficult to untangle the effects of space from other factors. Spatial effects can take two forms. First, spatial effects can be generated from within a community. Constraints on dispersal and interaction can result in spatial aggregation of populations even if the underlying abiotic environment is completely homogeneous. Such aggregation can allow the co-existence of competitors that would otherwise not co-exist. Second, spatial effects can be imposed on a community externally. The external environment can be discontinuous in space (i.e., "patchy") or it can vary continuously such that gradients in resources or conditions can form. Such heterogeneity can allow co-existence via niche partitioning. Microbial experimental systems have recently proven to be powerful tools for studying both of these spatial processes.

The importance of spatial structure was demonstrated in a recent study by Kerr *et al.* (2002) of toxin-producing, -sensitive and -resistant *E. coli.* All three of these populations can be isolated from natural environments, but it is unclear how they co-exist. Kerr *et al.* (2002) showed that these three populations can exhibit non-transitive competitive relationships, similar to the game rock-paper-scissors. Furthermore, they showed that the co-existence of all three types is favored when competition and dispersal occur locally (e.g., when the community is grown on the surface of solid media and propagated in a way that preserves patterns of spatial aggregation), while diversity is rapidly lost when these processes occur globally (e.g., in a well-mixed flask where spatial structure does not develop). Research on mice infected with similar strains has demonstrated the importance of spatial structure in the maintenance of microbial diversity *in vivo* as well (Kirkup and Riley, 2004).

Microbial systems have also been used to study the role of environmental heterogeneity in the maintenance of diversity. For example, Rainey and Travisano (1998) observed that when initially identical populations of *P. fluorescens* were propagated in two environments that differed solely in the amount of physical heterogeneity present, a diversity of morphologically distinct types evolved in the heterogeneous environments but not in the physically homogeneous environments. Physical heterogeneity was maintained in these experiments by incubating vessels containing liquid media without shaking, whereas homogeneous environments were generated by incubating identical vessels with continuous shaking. The diversity of morphological types was maintained because the different types were niche-specialists, adapted to specific combinations of resources and conditions present in different regions of the unshaken vessels. The importance of physical heterogeneity was further demonstrated by placing the specialist populations in physically homogeneous environments. Diversity was rapidly lost when heterogeneity was reduced (Rainey and Travisano, 1998).

3. The Co-Existence of Predators and Prey

Many models of predator-prey interactions, including the classic Lotka-Volterra model, fail to predict the stable co-existence of predator and prey populations that is observed in many natural populations. Laboratory studies with *E. coli* and virulent bacteriophage have revealed the importance of spatial refuges from predators in maintaining the co-existence of bacteria and bacteriophage. When mixed populations of bacteria and bacteriophage were propagated by daily transfer into fresh media, bacteriophage-resistant bacteria evolved and bacteriophage were lost from the system (Schrag and Mittler, 1996). However, when propagated in a continuous culture system, surprisingly, bacteria and bacteriophage co-existed, despite a high abundance of resistant bacteria. Upon sampling growth on the walls of the chemostats, Schrag and Mittler (1996) found that this wall-associated population was predominantly bacteriophage-sensitive. To demonstrate the importance of wall growth as spatial refuge from predator and prey, bacteria and bacteriophage were propagated by continuous culture, with daily replacement of the culture vessels to prevent wall growth, resulting in the rapid loss of bacteriophage from the system.

Our understanding of food chains and webs and their responses to various manipulations have been hampered by the challenges of delineating the constituents of a web, identifying interactions between guilds, and dealing with long response times. A complementary way to approach questions about food webs is to construct such networks in the laboratory. Kauzinger and Morin (1998) explored the effect of productivity on food chain length and stability using a three-level microbial food chain. By manipulating the resource concentration available to the primary producer, the effects of productivity were explored. Only high productivity environments supported the longest food chains (three-level). Furthermore, food-chain length affected the response of within-trophic-level densities to productivity (Fig. 3). Thus, Morin and colleagues have been able to directly investigate the population dynamic consequences of different food-web architectures. Their results have provided some of the first experimental evidence related to a number of important issues in food-web theory such as the importance of primary productivity in controlling the length of food chains (Morin, 1999),

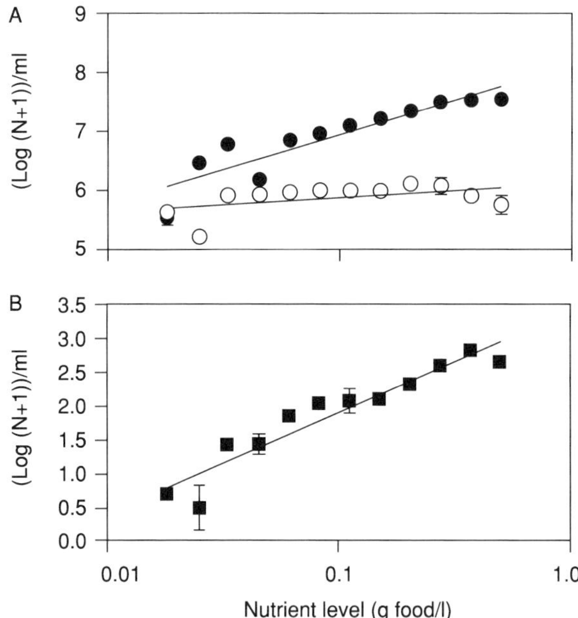

Figure 3 The effects of productivity on population abundances in trophic levels 1 and 3 of a microbial food chain. (A) Response of the bacterium *Serratia marcescens* abundance to a productivity gradient when cultured alone (solid circles) and with the bacterivorous ciliate predator *Colpidium striatum* (open circles). (B) Response of *C. striatum* abundance to the same productivity gradient. (From Kaunzinger and Morrin, 1998, Fig. 2 with permission of Nature, © 1998 by Nature Publishing Group.)

the instability of longer food chains, and the stabilizing effects of omnivory on omnivore population dynamics (Lawler and Morin, 1993).

Experiments with bacteria and bacteriophage have also explored the effects of resource enrichment and prey diversity on population dynamics. Studies with *E. coli* B and bacteriophage T4 in glucose-limited chemostats have shown that resource enrichment yields large increases in the density of the prey and destabilization of both predator and prey populations, as predicted by theoretical models (Bohannan and Lenski, 1997). Additional studies revealed that when the prey population consisted of two types—one bacteriophage-sensitive ("edible") and the other bacteriophage-resistant ("inedible")—resource enrichment resulted only in an increase in the density of the inedible prey, but not the edible prey. While predator density increased with enrichment in the presence of only sensitive prey, it did not significantly increase when the prey population was heterogeneous (Bohannan and Lenski, 1999).

4. The Effect of Positive Interactions

Traditionally, ecologists have focused the majority of their attention on competitive and exploitative interactions, paying less attention to positive interactions between organisms. Recently, however, there has been an upsurge in interest in the role of positive interactions in ecology (Bruno, 2003). Microorganisms, like macroorganisms, exhibit a range of positive interactions in natural communities. Bacteria (Rainey and Rainey, 2003; Velicer and Yu, 2003), microbial eukaryotes (Strassman *et al.*, 2000; Greig *et al.*, 2002), and even viruses (Turner and Chao, 1999) have been shown to act altruistically with one another. The role of positive interactions in the maintenance of diversity has been explored with a number of microbial experimental systems.

Grieg *et al.* (2002) recently used experimental populations of the yeast *Saccharomyces cerevisiae* to study the role of positive interactions in the maintenance of genetic diversity. They studied a population that produced and exported an enzyme (invertase) that cleaves sucrose into monomers that can be easily imported and metabolized by other cells. The export of this enzyme resulted in the release of nutrients into the surrounding environment, where any microbial cell in the vicinity could use these nutrients. The production of this enzyme also came at a physiological cost to the producing cell. The production and export of this molecule was considered a form of cooperation. The authors demonstrated that the cooperative yeast were competitively superior to "cheaters" that do not produce invertase when maintained at low density. However, when maintained at high density, cheaters could competitively displace the cooperators. They suggested that the diversity (in both number and activity) of invertase genes observed in natural populations of *S. cerevisiase* could be explained by the instability of cooperation across environments.

B. The Ecological Consequences of Diversity

Equally important to understanding the mechanisms underlying the diversity of life is identifying the consequences of this diversity. However, the controlled manipulations of diversity required to test these relationships are nearly impossible in many field systems (Morin, 2000), and it can be difficult to separate transient from sustained impacts of diversity manipulations with long-lived organisms. Recently, ecologists have turned to microbial experimental systems to address these questions more rigorously.

1. Diversity and Ecosystem Function

Since Odum (1953), Elton (1958), and MacArthur (1955) first suggested that more complex communities are more stable, researchers (e.g., Tilman and Downing, 1994) have sought empirical tests of these predictions and a more mechanistic understanding of the diversity-stability relationship (reviewed in McCann, 2000). While these large-scale field experiments are undoubtedly central to understanding the relationship between diversity and many important ecosystem properties, their experimental design and execution presents enormous logistical challenges. The control offered by experimental microbial systems allows researchers to assemble communities differing in diversity and to then test the effects of perturbations on indices of community stability. For example, McGrady-Steed *et al.* (1997) used algae, bacteria, protists, and small metazoans to assemble laboratory communities that differed in their initial diversity, both within and across trophic levels. The different communities rapidly stabilized within 40 to 80 generations of the dominant organisms at levels of diversity that were lower than initial levels. Measuring the functional attributes of the stable communities after six weeks revealed that the more diverse communities showed less variation in CO_2 flux and more resistance to invasion by an exotic species than less diverse communities.

In another experiment, Naeem and Li (1997) used algae, bacteria, and protists to create replicated experimental systems with differing degrees of diversity within functional groups to test the hypothesis that more diverse communities should exhibit more predictable ecosystem properties. The authors found that measures of biomass and density were more predictable as the number of species per functional group increased (i.e., the standard deviation of density measures decreased as the number of species per functional group increased, irrespective of environmental conditions).

Both of these studies observed a positive relationship between diversity and stability and provided support for the "biological insurance hypothesis," which posits that the redundancy within functional groups is important to overall ecosystem performance (McCann, 2000).

2. *Patterns of Diversity Across Scales of Space and Time*

A central aim in ecology is to understand the processes governing the spatial and temporal distribution of diversity. However, the mechanisms underlying these patterns have proven difficult to identify. Studies of microbial experimental systems have provided important insights into these questions.

Productivity (the rate of energy flow through an ecological system) is thought by many ecologists to be a key determinant of diversity (Rosenzweig, 1995; Gaston, 2000). Several studies have demonstrated a hump-shaped relationship between diversity and productivity in macroorganisms and some microorganisms (Rosenzweig, 1995; Leibold, 1999; Horner-Devine *et al.*, 2003). However, the mechanisms underlying such trends are not well understood. Recent work on the relationship between diversity and productivity in a microbial experimental system (Kassen *et al.*, 2000) revealed that a unimodal relationship between productivity and diversity was explained by selection for niche specialization in a phy] sically heterogeneous environment. When populations of the bacterium *Pseudomonas fluorescens* were allowed to compete in a heterogeneous environment (static environment, as previously described), local competition resulted in the unimodal productivity–diversity pattern. This pattern was not observed in homogeneous (shaken) environments, demonstrating the importance of physical heterogeneity affecting local interactions and generating this larger-scale diversity pattern.

Physical disturbance is also believed by many ecologists to be an important determinant of diversity. Disturbances of intermediate frequency and intensity should support the highest diversity, an assertion that has come to be known as the "intermediate disturbance hypothesis" (Connell, 1978; Sousa, 1984). This is exactly the relationship observed in experimental systems containing *Pseudomonas fluorescens* (Buckling *et al.*, 2000). Diversity peaks at intermediate frequencies of physical disturbance (i.e., shaking). As with the productivity–diversity relationship previously discussed, the mechanism underlying this relationship appears to be local competition among niche specialists in a heterogeneous environment.

C. The Response of Diversity to Environmental Change

Human activity is profoundly altering ecological systems. These alterations include habitat fragmentation and land use change, with subsequent increases in atmospheric CO_2 and air temperature (Vitousek *et al.*, 1997; Shaw *et al.*, 2002). Understanding the consequences of these global changes is an important goal of ecologists. Because of their rapid generation times, microbial experimental systems are capable of addressing ecological questions over long temporal scales, an important requirement for studying community-level responses to environmental change (Lawton, 1995; Bohannan and Lenski, 2000b).

Microbial experimental systems are also effective tools for studying the effect of habitat fragmentation on the dynamics of populations. Metapopulation models predict a lower probability of extinction for many small patchily distributed populations connected by dispersal than for a single large population of the same size because dispersal enables locally extinct populations to be recolonized (Levins, 1969; Hanski, 1991; Harrison, 1991). However, demonstrating that populations meet the criteria of classical metapopulation models remains challenging (Harrison and Taylor, 1997). By comparing predator-prey communities of protists in small patches linked by dispersal (i.e., arrays of connected vessels) with communities in isolated large patches (single undivided vessels of the same total volume as the array of small vessels), Holyoak and Lawler (1996) demonstrated that both predator and prey persisted longer in spatially subdivided arrays of vessels than they did in undivided cultures. Furthermore, by quantifying densities of predators and prey in individual patches, the authors identified many of the features predicted by metapopulation theory, including extinction-prone patches and asynchronous population dynamics among patches (Hanski and Gilpin, 1991; Hanski, 1999).

Recently, the effects of patch quality on population dynamics have been explored using a microbial experimental system. Donahue *et al.* (2003) used communities of protists arranged in patches that differed in resource quantity to test the population dynamics and dispersal predictions of source-sink theory and optimal foraging theory. They propagated communities of the bactivorous ciliate *Colpidium striatum* and the bacterium *Serratia marcescens* in microcosms consisting of several bottles that differed in resource concentration. These bottles were connected by tubes in dispersal treatments (low and high dispersal treatments were generated by long and short distances between adjacent bottles, respectively) and isolated from one another in the no dispersal treatment. They observed a net flux of individuals from high- to low-resource patches, which broadly supports the predictions of the source-sink theory instead of optimal foraging and balanced dispersal theories, which predict the no-net flux. However, several observations did not fit the predictions of either model and call for further exploration of many of the underlying assumptions of these models. This work is an excellent example of the importance of testing the plausibility of theoretical assumptions with laboratory systems.

D. Directions for Future Research

The experiments we have reviewed here underscore the utility of microbial experimental systems for studying ecology. These systems offer the ability to answer ecological questions that would be difficult to address with other organisms. Moreover, the rapidly increasing pool of genetic information for many microorganisms provides the opportunity to understand ecological processes at all scales of biological organization, from genes to ecosystems. A few studies have begun to take advantage of this opportunity. For example, work on adaptive radiation in *P. fluorescens* has been extended to identify the genes responsible for niche specialization (Spiers *et al.*, 2002), with the aim of ultimately providing a comprehensive explanation for adaptive radiation in this system. Continuing this integration of scales is an important future direction for research with microbial experimental systems.

Although microbial populations and communities are excellent laboratory experimental systems for studying ecology, they have been historically poor "model" systems for ecology (*sen su* Ehrlich, 2003). An ideal model system in ecology is a system that is especially amenable to field and laboratory study, that can be studied in great depth and that can act as an example of other ecological systems. Because of our relative ignorance of microbial field biology, it has been difficult to directly link the results of microbial laboratory studies to field observations. However, this has begun to change, as discussed in section V.A. Continuing to develop microbial model systems is an important future direction for ecology. This will require not only creative approaches to studying microorganisms in the field, but also thoughtful approaches to bridging the historical divide between the studies of microbial and general ecology.

There are three systems that we feel are particularly promising as microbial model systems for ecology. The first is *E. coli*. Colicinogenic *E. coli* (those that produce the antimicrobial compounds called colicins) are especially well suited as a model group. There are interesting patterns in their distribution in the field, they are amenable to experimental manipulation both within experimental animals (Kirkup and Riley, 2004) and in artificial media (Kerr *et al.*, 2002) and there is an extensive body of theoretical literature on their ecology and evolution (Durrett and Levin, 1997; Czaran *et al.*, 2002). Furthermore, the wealth of genetic and physiological information available for *E. coli* is unparalleled.

The second potential model system is chemoautotrophic ammonia-oxidizing bacteria (reviewed in Kowalchuk and Stephen, 2001). This group of bacteria has been suggested as an ideal model group for studies in microbial ecology for a number of reasons: they are ubiquitous, they are of great environmental importance, they have been the subject of intense study over the past decade, and there is a growing body of knowledge regarding their genetics, physiology, and distribution. In addition, molecular tools have been developed for tracking their dynamics in the field and in laboratory microcosms.

The third group with great potential as a model system for ecological studies is the marine cyanobacterium *Prochlorococcus* (Rocap *et al.*, 2003). This group is important in global carbon cycling, its population dynamics can be tracked in the field and it exhibits interesting patterns in its distribution. In addition, predators of this group (i.e., bacteriophage) have recently been isolated and characterized (Sullivan *et al.*, 2003).

VII. CONCLUSIONS

The studies we have discussed represent only a small subset of the many recent applications of microbial experimental systems to ecological questions. For example, microbial experimental systems have also been successfully used to explore the effect of nonlinear interactions on population dynamics (Fussmann *et al.*, 2000), ecosystem-level selection (Swenson *et al.*, 2000), the effect of resource supply ratios on the outcome of competition (Grover, 2000), the ecology and evolution of mutualisms (Rosenzweig *et al.*, 1994), and co-evolution between predators and prey (Buckling and Rainey, 2002a,b). As ecologists select tools ranging from analytical and simulation models to field manipulations and statistical analyses in order to understand the natural world, microbial experimental systems offer an especially powerful approach.

ACKNOWLEDGMENTS

Portions of the review presented in Sections V and VI were first presented in a review article in *Trends in Ecology and Evolution*; we are grateful to our coauthors of that article, R. Kassen, B. Kerr, A. Buckling and P. Rainey. We thank two anonymous referees for their comments.

REFERENCES

Ackermann, M., Stearns, S.C. and Jenal, U. (2003) Senescence in a bacterium with asymmetric division. *Science* **300**, 1920.

Amann, R.I., Ludwig, W. and Schleifer, K.-H. (1995) Phylogenetic identification and *in situ* detection of individual microbial cells without cultivation. *Microbiological Reviews* **59**, 143–169.

Andersson, D.I. and Levin, B.R. (1999) The biological cost of antibiotic resistance. *Curr. Opin. Microbiol.* **2**, 489–493.

Andrews, J. (1991) *Comparative Ecology of Microorganisms and Macroorganisms.* Springer-Verlag, New York, NY.

Antonovics, J. (1976) The input from population genetics: 'The new ecological genetics.' *Systematic Botany* **1**, 233–245.

Beijerinck, M. (1889) L'auxanographie, ou la methode de l'hydrodiffusion dans la gélatine appliquée aux recherches microbiologiques. *Archives Néerlandaises des Sciences Exactes et Naturelles* **23**, 367–372.

Bell, G. and Reboud, X. (1997) Experimental evolution in *Chlamydomonas*. II. Genetic variation in strongly contrasted environments. *Heredity* **78**, 498–506.

Bell, G. (2001) Neutral macroecology. *Science* **293**, 2413–2418.

Bennet, A.F. and Lenski, R.E. (1993) Evolutionary adaptation to temperature. 2. Thermal niches of experimental lines of *Escherichia coli*. *Evolution* **47**, 1–12.

Bergelson, J. and Purrington, C.B. (1996) Surveying patterns in the cost of resistance in plants. *Am. Nat.* **148**, 536–558.

Bohannan, B.J.M. and Lenski, R.E. (1997) Effect of resource enrichment on a chemostat community of bacteria and bacteriophage. *Ecology* **78**, 2303–2315.

Bohannan, B.J.M. and Lenski, R.E. (1999) Effect of prey heterogeneity on the response of a model food chain to resource enrichment. *Am. Nat.* **153**, 73–82.

Bohannan, B.J.M. and Lenski, R.E. (2000a) Linking genetic change to community evolution: Insights from studies of bacteria and bacteriophage. *Ecol. Lett.* **3**, 362–377.

Bohannan, B.J.M. and Lenski, R.E. (2000b) The relative importance of competition and predation varies with productivity in a model community. *Am. Nat.* **156**, 329–340.

Bohannan, B.J.M., Kerr, B., Jessup, C., Hughes, J. and Sandvik, G. (2002) Trade-offs and coexistence in microbial microcosms. *Antonie Van Leeuwenhoek* **81**, 107–115.

Brock, T.D. (1999) *Milestones in Microbiology: 1546 to 1940.* ASM Press, Washington, DC.

Bruno, J.F. (2003) Inclusion of facilitation into ecological theory. *Trends Ecol. Evol.* **18**, 119–125.

Buckling, A., Kassen, R., Bell, G. and Rainey, P.B. (2000) Disturbance and diversity in experimental microcosms. *Nature* **408**, 961–964.

Buckling, A. and Rainey, P.B. (2002a) Antagonistic coevolution between a bacterium and a bacteriophage. *Proc. Roy. Soc. Lond. Ser. B* **269**, 931–936.

Buckling, A. and Rainey, P.B. (2002b) The role of parasites in sympatric and allopatric host diversification. *Nature* **420**, 496–499.

Bull, J.J., Pfennig, D.W. and Wang, I.N. (2004) Genetic details, optimization and phage life histories. *Trends Ecol. Evol.* **19**, 76–82.

Carpenter, S.R. (1996) Microcosm experiments have limited relevance for community and ecosystem ecology. *Ecology* **77**, 677–680.

Chao, L., Levin, B.R. and Stewart, F.M. (1977) A complex community in a simple habitat: Experimental study with bacteria and phage. *Ecology* **58**, 369–378.

Chao, L. (1990) Fitness of RNA virus decreased by Muller's ratchet. *Nature* **348**, 454–455.

Connell, J.H. (1978) Diversity in tropical rain forests and coral reefs: high diversity of trees and corals is maintained only in a non-equilibrium state. *Science* **199**, 1302–1310.

Contois, D.E. (1959) Kinetics of bacterial growth: Relationship between population density and specific growth rate of continuous cultures. *J. Gen. Microbiol.* **21**, 40–50.

Cooper, V.S. and Lenski, R.E. (2000) The population genetics of ecological specialization in evolving *Escherichia coli* populations. *Nature* **407**, 736–739.

Czaran, T.L., Hoekstra, R.F. and Pagie, L. (2002) Chemical warfare between microbes promotes biodiversity. *Proc. Natl. Acad. Sci. USA* **99**, 786–790.

Daehler, C.C. and Strong, D.R. (1996) Can you bottle nature? The roles of microcosms in ecological research. *Ecology* **77**, 663–664.

Dallinger, W.D. (1887) The president's address. *J. R. Microsc. Soc.* **7**, 184–199.

de Visser, J.A.G.M., Zeyl, C.W., Gerrish, P.J., Blanchard, J.L. and Lenski, R.E. (1999) Diminishing returns from mutation supply rate in asexual populations. *Science* **283**, 404–406.

Diamond, J. (1986) Overview: laboratory experiments, field experiments and natural experiments. In: *Community Ecology* (Ed. by J. Diamond and T. Case), pp. 3–22. Harper and Row, New York, NY.

Donahue, M.J., Holyoak, M. and Feng, C. (2003) Patterns of dispersal and dynamics among habitat patches varying in quality. *Am. Nat.* **162**, 302–317.

Drake, J.A., Huxel, G.R. and Hewitt, C.L. (1996) Microcosms as models for generating and testing community theory. *Ecology* **77**, 670–677.

Durrett, R. and Levin, S. (1997) Allelopathy in spatially distributed populations. *J. Theor. Biol* **185**, 165–171.

Dykhuizen, D. and Davies, M. (1980) An experimental model: Bacterial specialists and generalists competing in chemostats. *Ecology* **61**, 1213–1227.

Eddy, S. (1928) Succession of Protozoa in cultures under controlled conditions. *Transact. Am. Micr. Soc.* **47**, 283.

Ehrlich, P.R. (2003) Introduction: Butterflies, test systems, and biodiversity. In: *Butterflies: Ecology and Evolution Take Flight* (Ed. by C.L. Boggs, W.B. Watt and P.R. Ehrlich), pp. 1–6. University of Chicago Press, Chicago, IL.

Elton, C.S. (1958) *The Ecology of Invasions by Animals and Plants*. Methuen and Co., Ltd., London.

Fussmann, G.F., Ellner, S.P., Shertzer, K.W. and Hairston, N.G. (2000) Crossing the Hopf bifurcation in a live predator-prey system. *Science* **290**, 1358–1360.

Gaston, K.J. (2000) Global patterns in biodiversity. *Nature* **405**, 220–227.

Gause, G.F. (1934) *The Struggle for Existence*. Dover, New York, NY.

Gerrish, P. and Lenski, R.E. (1998) The fate of competing beneficial mutations in an asexual population. *Genetica* **102/103**, 127–144.

Gorden, R.W., Beyers, R.J., Odum, E.P. and Eagon, R.G. (1979) Studies of a simple laboratory microecosystem: Bacterial activities in a heterotrophic succession. *Ecology* **50**, 86–100.

Greig, D., Louis, E., Borts, R. and Travisano, M. (2002) Hybrid speciation in experimental populations of yeast. *Science* **298**, 1773–1775.

Grover, J. (2000) Resource competition and community structure in aquatic microorganisms: Experimental studies of algae and bacteria along a gradient of organic carbon to inorganic phosphorus supply. *J. Plank. Res.* **22**, 1591–1610.

Hairston, N.G., Allan, J.D., Colwell, R.K., Futuyma, D.J., Howell, J., Lubin, M.D., Mathias, J. and Vandermeer, J.H. (1968) The relationship between species diversity and stability: An experimental approach with protozoa and bacteria. *Ecology* **49**, 1091–1101.

Hairston, N.G. (1989) *Ecological Experiments: Purpose, Design and Execution*. Cambridge University Press, Cambridge, UK.

Hansen, S.R. and Hubbell, S.P. (1980) Single-nutrient microbial competition: Qualitative agreement between experimental and theoretically forecast outcomes. *Science* **207**, 1491–1493.

Hanski, I. (1991) Single-species metapopulation dynamics: Concepts, models and observations. *Biol. J. Linn. Soc.* **42**, 17–38.

Hanski, I. and Gilpin, M. (1991) Metapopulation dynamics: Brief history and conceptual domain. *Biol. J. Linn. Soc.* **42**, 3–16.

Hanski, I. (1999) *Metapopulation Ecology*. Oxford University Press, Oxford.

Hardin, G. (1960) The competitive exclusion principle. *Science* **131**, 1292–1297.

Harrison, S. (1991) Local extinction in a metapopulation context: An empirical evaluation. *Biol. J. Linn. Soc.* **42**, 73–88.

Harrison, S. and Taylor, A.D. (1997) Empirical evidence for metapopulation dynamics. In: *Metapopulation Biology: Ecology, Genetics and Evolution* (Ed. by I. Hanski and M.E. Gilpin), pp. 27–42. Academic Press, Inc., San Diego, CA.

Hastings, A. (2000) Ecology: The lion and the lamb find closure. *Science* **290**, 1712–1713.

Hilborn, R. and Mangel, M. (1997) *The Ecological Detective: Confronting Models with Data*. Princeton University Press, Princeton, NJ.

Holyoak, M. and Lawler, S.P. (1996) Persistence of an extinction-prone predator-prey interaction through metapopulation dynamics. *Ecology* **77**, 1867–1879.

Horne, M.T. (1970) Coevolution of *Escherichia coli* and bacteriophages in chemostat culture. *Science* **168**, 992–993.

Horne, M.T. (1971) Characterization of virulent bacteriophage infections of *Escherichia coli* in continuous culture. *Science* **172**, 405.

Horner-Devine, M., Leibold, M., Smith, V. and Bohannan, B. (2003) Bacterial diversity patterns along a gradient of primary productivity. *Ecol. Lett.* **6**, 613–622.

Hubbell, S.P. (2001) *The Unified Neutral Theory of Biodiversity and Biogeography*. Princeton University Press, Princeton, NJ.

Huey, R.B. and Kingsolver, J.G. (1989) Evolution of thermal sensitivity of ectotherm performance. *Trends Ecol. Evol.* **4**, 131–135.

Hutchinson, G.E. (1965) *The Ecological Theater and the Evolutionary Play*. Yale University Press, New Haven, CT.

Hutchinson, G.E. (1978) *An Introduction to Population Ecology*. Yale University Press, New Haven, CT.

Ives, A.R., Foufopoulos, J., Klopfer, E.D., Klug, J.L. and Palmer, T.M. (1996) Bottle or big-scale studies: How do we do ecology? *Ecology* **77**, 681–685.

Kassen, R., Buckling, A., Bell, G. and Rainey, P.B. (2000) Diversity peaks at intermediate productivity in a laboratory microcosm. *Nature* **406**, 508–512.

Kassen, R. (2002) The experimental evolution of specialists, generalists, and the maintenance of diversity. *Journal of Evolutionary Biology* **15**, 173–190.

Kaunzinger, C.M.K. and Morin, P.J. (1998) Productivity controls food-chain properties in microbial communities. *Nature* **395**, 495–497.

Kawecki, T., Barton, N. and Fry, J. (1997) Mutational collapse of fitness in marginal habitats and the evolution of ecological specialisation. *Journal of Evolutionary Biology* **10**, 407–429.

Kerr, B., Riley, M.A., Feldman, M.W. and Bohannan, B.J.M. (2002) Local dispersal promotes biodiversity in a real-life game of rock-paper-scissors. *Nature* **418**, 171–174.

Kingsland, S.E. (1995) *Modeling Nature: Episodes in the History of Population Ecology*, 2nd ed. University of Chicago Press, Chicago, Ill.

Kirkup, B.C. and Riley, M.A. (2004) Antibiotic-mediated antagonism leads to a bacterial game of rock-paper-scissors *in vivo*. *Nature* **428**, 412–414.

Kowalchuk, G. and Stephen, J. (2001) Ammonia oxidizing bacteria: A model for molecular microbial ecology. *Annu. Rev. Microbiol.* **55**, 485.

Lawler, S.P. and Morin, P.J. (1993) Food web architecture and population dynamics in laboratory microcosms of protists. *Am. Nat.* **141**, 675–686.

Lawler, S.P. (1998) Ecology in a bottle: Using microcosms to test Theory. In: *Experimental Ecology: Issues and Perspectives* (Ed. by W.J. Resetarits Jr. and J. Bernardo), pp. 236–253. Oxford University Press, Oxford, UK.

Lawton, J.H. (1995) Ecological experiments with model systems. *Science* **269**, 328–331.

Lawton, J.H. (1996) The Ecotron facility at Silwood Park: The value of "big bottle" experiments. *Ecology* **77**, 665–669.

Lawton, J.H. (1999) Are there general laws in ecology? *Oikos* **84**, 177–192.

Leibold, M. (1999) Biodiversity and nutrient enrichment in pond plankton communities. *Evol. Ecol. Res.* **1**, 73–95.

Leibold, M.A. (1996) A graphical model of keystone predators in food webs: Trophic regulation of abundance, incidence, and diversity patterns in communities. *Am. Nat.* **147**, 784–812.

Lenski, R.E. and Levin, B.R. (1985) Constraints on the coevolution of bacteria and virulent phage: A model, some experiments, and predictions for natural communities. *Am. Nat.* **125**, 585–602.

Lenski, R.E. (1988a) Dynamics of interactions between bacteria and virulent bacteriophage. *Adv. Microb. Ecol.* **10**, 1–44.

Lenski, R.E. (1988b) Experimental studies of pleiotropy and epistasis in *Escherichia coli*. I. Variation in competitive fitness among mutants resistant to virus T4. *Evolution* **42**, 425–432.

Lenski, R.E. (2001) Testing Antonovics' five tenets of ecological genetics: Experiments with bacteria at the interface of ecology and genetics. In: *Ecology: Achievement and Challenge* (Ed. by M.C. Press, N.J. Huntly and S. Levin), pp. 25–45. Blackwell Science, Oxford, UK.

Levin, B.R. (1972) Coexistence of two asexual strains on a single resource. *Science* **175**, 1272–1274.

Levin, B.R., Stewart, F.M. and Chao, L. (1977) Resource-limited growth, competition, and predation: A model and experimental studies with bacteria and bacteriophage. *Am. Nat.* **111**, 3–24.

Levin, B.R. and Lenski, R.E. (1985) Bacteria and phage: A model system for the study of the ecology and co-evolution of hosts and parasites. In: *Ecology and Genetics of Host-Parasite Interactions* (Ed. by D. Rollinson and R.M. Anderson), pp. 227–242. Academic Press, London, UK.

Levin, B.R., Perrot, V. and Walker, N. (2000) Compensatory mutations, antibiotic resistance and the population genetics of adaptive evolution in bacteria. *Genetics* **154**, 985–997.

Levins, R. (1966) The strategy of model building in population biology. *Am. Sci.* **54**, 421–431.

Levins, R. (1969) Some demographic and genetic consequences of environmental heterogeneity for biological control. *Bulletin of the Entomological Society of America* **15**, 237–240.

Lipsitch, M. (2001) The rise and fall of antimicrobial resistance. *Trends in Microbiology* **9**, 438–444.

Luckinbill, L.S. (1973) Coexistence in laboratory populations of *Paramecium aurelia* and its predator *Didinium nasutum*. *Ecology* **54**, 1320–1327.

Luckinbill, L.S. (1974) The effects of space and enrichment on a predator-prey system. *Ecology* **55**, 1142–1147.

Luckinbill, L.S. (1979) Regulation, stability, and diversity in an model experimental microcosm. *Ecology* **60**, 1098–1102.

MacArthur, R. (1955) Fluctuations of animal populations, and a measure of community stability. *Ecology* **36**, 533–536.

MacArthur, R. (1972) *Geographical Ecology*. Harper and Row, New York, NY.

MacLean, R. and Bell, G. (2002) Experimental adaptive radiation in *Pseudomonas*. *Am. Nat.* **160**, 569–581.

Maisnier-Patin, S., Berg, O.G., Liljas, L. and Andersson, D.I. (2002) Compensatory adaptation to the deleterious effect of antibiotic resistance in *Salmonella typhimurium*. *Molecular Microbiology* **46**, 355–366.

McCann, K. (2000) The diversity-stability debate. *Nature* **405**, 228–233.

McGrady-Steed, J., Harris, P.M. and Morin, P.J. (1997) Biodiversity regulates ecosystem predictability. *Nature* **390**, 162–165.

Mertz, D.B. and McCauley, D.E. (1982) The domain of laboratory ecology. In: *Conceptual Issues in Ecology* (Ed. by E. Saarinen), pp. 229–244. D. Reidel, Dordrecht.

Morin, P. (1999) Productivity, intraguild predation, and population dynamics in experimental food webs. *Ecology* **80**, 752–760.

Morin, P.J. (2000) Ecology: The complexity of co-dependency. *Nature* **403**, 718–719.

Naeem, S. and Li, S.B. (1997) Biodiversity enhances ecosystem reliability. *Nature* **390**, 507–509.

Nagaev, I., Bjorkman, J., Andersson, D.I. and Hughes, D. (2001) Biological cost and compensatory evolution in fusidic acid-resistant *Staphylococcus aureus*. *Molecular Microbiology* **40**, 433–439.

Odum, E.P. (1953) *Fundamentals of Ecology*. W.B. Saunders Co., Philadelphia, PA.

Oreskes, N., Shraderfrechette, K. and Belitz, K. (1994) Verification, validation, and confirmation of numerical models in the earth sciences. *Science* **263**, 641–646.

Peters, R.H. (1991) *A Critique for Ecology*. Cambridge University Press, Cambridge, UK.

Rainey, P.B. and Travisano, M. (1998) Adaptive radiation in a heterogeneous environment. *Nature* **394**, 69–72.

Rainey, P.B., Buckling, A., Kassen, R. and Travisano, M. (2000) The emergence and maintenance of diversity: Insights from experimental bacterial populations. *Trends Ecol. Evol.* **15**, 243–247.

Rainey, P.B. and Rainey, K. (2003) Evolution of cooperation and conflict in experimental bacterial populations. *Nature* **425**, 72–74.

Riley, M.A. and Wertz, J.E. (2002) Bacteriocins: Evolution, ecology, and application. *Annual Review of Microbiology* **56**, 117–137.

Rocap, G., Larimer, F., Lamerdin, J., Malfatti, S., Chain, P., Ahlgren, N., Arellano, A., Coleman, M., Hauser, L., Hess, W., Johnson, Z., Land, M., Lindell, D., Post, A., Regala, W., Shah, M., Shaw, S., Steglich, C., Sullivan, M., Ting, C., Tolonen, A., Webb, E., Zinser, E. and Chisholm, S. (2003) Genome divergence in two *Prochlorococcus* ecotypes reflects oceanic niche differentiation. *Nature* **424**, 1042–1047.

Rosenzweig, M.L. (1971) Paradox of enrichment: Destabilization of exploitation ecosystems in ecological time. *Science* **171**, 385–387.

Rosenzweig, M.L. (1995) *Species Diversity in Space and Time*. Cambridge University Press, Cambridge, UK.

Rosenzweig, R.F., Sharp, R.R., Treves, D.S. and Adams, J. (1994) Microbial evolution in a simple unstructured environment: Genetic differentiation in *Escherichia coli*. *Genetics* **137**, 903–917.

Schrag, S.J. and Mittler, J.E. (1996) Host-parasite coexistence: The role of spatial refuges in stabilizing bacteria-phage interactions. *Am. Nat.* **148**, 348–377.

Shaw, M., Zavaleta, E., Chiariello, N., Cleland, E., Mooney, H. and Field, C. (2002) Grassland responses to global environmental changes suppressed by elevated CO2. *Science* **298**, 1987–1990.

Skadovsky, S.N. (1915) The alteration of the reaction of environment in cultures of Protozoa. *Transact. Univ. Shaniavsky. Biol. Lab.* **1**, 157.

Slobdokin, L.B. (1988) Intellectual problems of applied ecology. *Bioscience* **38**, 337–342.

Slobodkin, L.B. (1961) Preliminary ideas for a predictive theory of ecology. *Am. Nat.* **95**, 147–153.

Sommer, U. (1984) The paradox of the plankton: Fluctuations of phosphorus availability maintain diversity of phytoplankton in flow-through cultures. *Limnology and Oceanography* **29**, 633–636.

Sousa, W.P. (1984) The role of disturbance in natural communities. *Annu. Rev. Ecol. Syst.* **15**, 353–391.

Spiers, A., Kahn, S., Bohannon, J., Travisano, M. and Rainey, P. (2002) Adaptive divergence in experimental populations of *Pseudomonas fluorescens*. I. Genetic and phenotypic bases of wrinkly spreader fitness. *Genetics* **161**, 33–46.

Strassman, J.E., Zhu, Y. and Queller, D.C. (2000) Altruism and social cheating in the social amoeba *Dictyostelium discoideum*. *Nature* **408**, 965–967.

Sullivan, M.B., Waterbury, J.B. and Chisholm, S.W. (2003) Cyanophages infecting the oceanic cyandoactenum. *Prochlorococcus. Nature* **424**, 1047–1051.

Swenson, W., Wilson, D.S. and Elias, R. (2000) Artificial ecosystem selection. *Proc. Natl. Acad. Sci. USA* **97**, 9110–9114.

Tilman, D. (1977) Resource competition between planktonic algae: Experimental and theoretical approach. *Ecology* **58**, 338–348.

Tilman, D. (1982) *Resource Competition and Community Structure*. Princeton University Press, Princeton, NJ.

Tilman, D. and Sterner, R.W. (1984) Invasions of equilibria: Tests of resource competition using two species of algae. *Oecologia* **61**, 197–200.

Tilman, D. and Downing, J. (1994) Biodiversity and stability in grasslands. *Nature* **367**, 363–365.

Travisano, M. and Lenski, R. (1996) Long-term experimental evolution in *Escherichia coli*. IV. Targets of selection and the specificity of adaptation. *Genetics* **143**, 15–26.

Turner, P.E. and Chao, L. (1999) Prisoner's dilemma in an RNA virus. *Nature* **398**, 441–443.

van den Ende, P. (1973) Predator-prey interactions in continuous culture. *Science* **181**, 562–564.

Van Voris, P., O'Neill, R.V., Emanuel, W.R. and Shugart, H.H. (1980) Functional complexity and ecosystem stability. *Ecology* **61**, 1352–1360.

Vandermeer, J.H. (1969) The competitive structure of communities: An experimental approach with protozoa. *Ecology* **50**, 362–371.

Velicer, G.J. and Yu, Y.T.N. (2003) Evolution of novel cooperative swarming in the bacterium *Myxococcus xanthus*. *Nature* **425**, 75–78.

Vitousek, P.M., Mooney, H.A., Lubchenco, J. and Melillo, J.M. (1997) Human domination of Earth's ecosystems. *Science* **277**, 494–499.

Wang, I.N., Dykhuizen, D.E. and Slobodkin, L.B. (1996) The evolution of phage lysis timing. *Evol. Ecol.* **10**, 545–558.
Waterbury, J.B. and Valois, F.W. (1993) Resistance to co-occurring phages enables marine *Synechococcus* communities to coexist with cyanophages abundant in seawater. *Applied and Environmental Microbiology* **59**, 3393–3399.
Winogradsky, S. (1889) Recherches physiologiques sur les sulfobactéries. *Annales le l'Institut Pasteur* **3**, 49–60.
Woodruff, L.L. (1911) The effect of excretion products of Paramecium on its rate of reproduction. *Journal of Experimental Zoology* **10**, 551.
Woodruff, L.L. (1912) Observations on the origin and sequence of the protozoan fauna of hay infusions. *Journal of Experimental Zoology* **12**, 205–264.
Woodruff, L.L. (1914) The effect of excretion products of Infusoria on the same and on different species, with special reference to the protozoan sequence in infuensions. *Journal of Experimental Zoology* **14**, 575.
Zhou, J., Xia, B., Treves, D.S., Wu, L.-Y., Marsh, T.L., O'Neill, R.V., Palumbo, A. V. and Tiedje, J. (2002) Spatial and resource factors influencing high microbial diversity in soil. *Applied and Environmental Microbiology* **68**, 326–334.

Parasitism Between Co-Infecting Bacteriophages

PAUL E. TURNER

I. SUMMARY

Co-infection of a single host by multiple virus genotypes or species is common in nature, facilitating studies of ecological interactions between viruses at the cellular level. When two or more viruses co-infect the same host cell, this can have profound consequences for the fitness (growth performance) of an individual virus. Co-infection may be advantageous to an individual virus due to increased pathogenesis, enhanced transmission, or the opportunity for genetic exchange (sex) which produces the raw material for natural selection. In contrast, co-infection may be disadvantageous to an individual virus because it increases the likelihood of competition for proteins and other resource products available within the cell. One intriguing cost of co-infection is the recent evidence that intra-cellular interactions between viruses can be antagonistic, where a virus genotype evolves to specialize in parasitizing other co-infecting viruses. Here I review laboratory experiments involving the RNA bacteriophage $\phi 6$, which demonstrate the evolution of parasitism when viruses are propagated in environments where co-infection is common. Frequency-dependent selection is shown to govern the fitness of these parasitic (cheater) genotypes, because their benefit of cheating depends on the relative abundance of ordinary and cheater genotypes encountered within the host cell. I relate why the evolution of parasitic

ADVANCES IN ECOLOGICAL RESEARCH VOL. 37 0065-2504/05 $35.00

DOI: 10.1016/S0065-2504(04)37010-8

viruses may be relevant for observed limits to the absolute number of phages that can simultaneously infect a single cell. The evolution of parasitic interactions in viruses infecting animal and plant hosts is briefly discussed. I suggest directions for future research on the evolutionary ecology of virus co-infection, especially the need to study phage interactions under natural (non laboratory) conditions.

So, naturalists observe, a flea
Has small fleas that on him prey;
And these have smaller still to bite 'em
And so proceed ad infinitum.
—Jonathan Swift

II. INTRODUCTION

Viruses usurp the biomolecular machinery of living cells in order to replicate their genomes. Depending on the virus, reproduction may occur within a unicellular host organism such as a bacterium or alga, or inside specific cells (tissues) that comprise a multicellular host. When multiple genotypes or species of virus co-infect the same host cell, ecological interactions can occur between viruses within the cellular milieu. Virus ecology (relationships between viruses, their hosts and the abiotic environment; Hurst, 2000) includes these intracellular virus encounters, which can span the entire interaction continuum ranging from mutualism to parasitism (Petney and Andrews, 1998; Hammond *et al.*, 1999; Lopez-Ferber *et al.*, 2003). Mutualistic interactions lead to enhanced fitness (reproductive success) of the co-infecting viruses, such as their improved reproduction due to an overall weakening of the host or the host's immune system. In contrast, viruses can experience antagonistic encounters within the cell, which decrease the fitness of one or more virus strains in the co-infection group. Lowered fitness can arise due to competition for limiting intra-host resources, interference competition that causes one virus to disrupt the reproduction of another, and apparent competition where one virus stimulates a host response (e.g., immunity) that acts against both strains. Thus, multiple infections can have either positive or negative consequences for the fitness of an individual virus genotype.

Theoretical models predict that within-host ecological interactions between viruses (or other parasites) should profoundly impact selection for traits relating to virulence (damage to the host). May and Nowak (1995) examined the evolution of virulence in light of co-infection, an ecological complexity that was often missing in previous modeling efforts (but see e.g.,

Levin and Pimentel, 1981). Their work shows that competing strains simultaneously infecting the same host are selected to rapidly exploit the host before resources are depleted, causing escalation to greater virulence; similar theoretical outcomes are shown by other authors (Nowak and May, 1994; van Baalen and Sabelis, 1995; Frank, 1996; Mosquera and Adler, 1998). Empirical support comes from human immunodeficiency virus-1, where within-host evolution of competing virus strains contributes to the increased virulence associated with the onset of AIDS (Antia *et al.*, 1996). Additional models have examined the evolution of virulence, under differing genetic relatedness of co-infecting parasites. The general prediction is that antagonism between genetically diverse parasites favors rapid exploitation of within-host resources and, hence, increased virulence (Frank, 1992). However, in controlled experiments with viruses (and other parasites) avirulent genotypes can outcompete virulent ones (see Read and Taylor, 2000, for many examples). We argued that this discrepancy between theory and data may be due to incorrect assumptions regarding the nature of competition, because faster rates of host exploitation need not be the only parasite adaptation favored by within-host competition (Chao *et al.*, 2000). Rather, traits that involve exploitation or inhibition of competing genotypes (such as production of anti-competitor toxins) can also be selected, potentially resulting in less effective exploitation of the host and reduced virulence. Data supporting this hypothesis include experiments on bacteria and their viruses (Turner and Chao, 1998, 1999), which are the primary focus of this chapter. Overall, the various models generally agree that multiple infections tend to select for antagonistic interactions between unrelated parasites, whether or not this result leads to increased or decreased evolution of virulence. A thorough account of the mathematical theory relating to multiple infections extends beyond the scope of this chapter, but interested readers should consult recent papers and books dealing with the subject (Nowak and May, 2000; Read and Taylor, 2001; Frank, 2002). Despite the large body of theoretical work, relatively few experiments have directly addressed the predictions of the various models.

One powerful method that can be used to test these theoretical predictions is laboratory experiments using model organisms, where evolutionary ecology can be studied under strictly controlled conditions. Natural selection is the differential survival and reproduction of genotypes in a population, due to the challenges imposed by the prevailing environment. Therefore, laboratory environments can be manipulated while still allowing evolution to proceed through natural selection, because the experimental habitats determine which genotypes contribute their genes to subsequent generations. This process should not be confused with artificial selection (such as animal breeding), where the experimenter explicitly chooses which individuals contribute genes to the descendants in order to obtain a population dominated by genotypes featuring a desired trait (such as increased milk production in dairy cattle). Drosophila fruit flies are well-known as study organisms in experimental evolution (Kohler, 1994), but microbes (bacteria, viruses, fungi) are increasingly used due to their large population sizes, short generation times, and ease with which their environments and genetic systems can be manipulated (Lenski, 2002). Most importantly, microbes can be stored in a freezer for indefinite periods of time, permitting direct comparisons between an ancestral genotype and its evolved descendants.

This chapter will summarize the data from recent laboratory experiments on bacteria and the RNA virus $\phi 6$, which show that antagonistic interactions can evolve when viruses are propagated under high levels of co-infection. These studies demonstrate how parasite genotypes can be selected to selfishly interfere with the reproduction of other strains during co-infection, at the expense of reduced performance in alternate habitats where multiple infections are less common. This chapter also describes how frequency-dependent selection underlies within-host competition for resources in these experiments, and relate why the data may be relevant for observed limits to the number of viruses that can simultaneously infect a single host cell. Then the evolution of parasitic interactions in viruses that infect animal and plant hosts is briefly discussed. Throughout, ideas for future research are suggested that deal with the consequences of co-infection in viruses, especially in terms of the largely unexplored natural ecology of viruses that infect bacteria.

III. PHAGE BIOLOGY AND INTRACELLULAR CONFLICTS

Natural ecosystems abound with viruses, especially the bacteriophages (literally "eaters of bacteria") that infect bacteria (Paul and Kellogg, 2000). For example, bacteriophage (or simply phage) particle counts ranging from 70,000 to 15 million per milliliter have been measured in the open ocean (Bergh *et al.*, 1989; Wommack and Colwell, 2000), and it is suggested that phages can influence global ecological processes by regulating the population sizes of their less numerous bacterial hosts (Fuhrman, 1999). Aside from the continual ecological influence exerted by their teeming presence in the biosphere, phages were highly influential in the development of basic laboratory research during the past century. In particular, they factored prominently in experiments that laid the groundwork for molecular biology as a separate discipline (Cairns *et al.*, 1992). Examples include bacteria/phage studies that demonstrated that mutations occur spontaneously (at random) and not in direct response to environmental change (Luria and Delbruck, 1943), and experiments that proved a triplet genetic code underlies protein synthesis (Crick *et al.*, 1961). Because phages possess a rich history as subjects of laboratory study, there is a wealth of information on their basic biology, genetics, and techniques for genome manipulation. For these reasons, phages provide powerful systems for the study of experimental ecology and evolution in laboratory settings (Turner, 2003; Lenski, 2002).

Many phages are lytic, featuring infection cycles that result in death (lysis) of the host cell. The basic reproductive cycle of a lytic virus is shown in Fig. 1. The lytic cycle features three discrete stages: (i) attachment of the virus to a host cell, and injection into the cell of the phage genetic material (whether RNA or DNA); (ii) replication of virus genetic material to create viral enzymes, structural components and other products within the cell; and

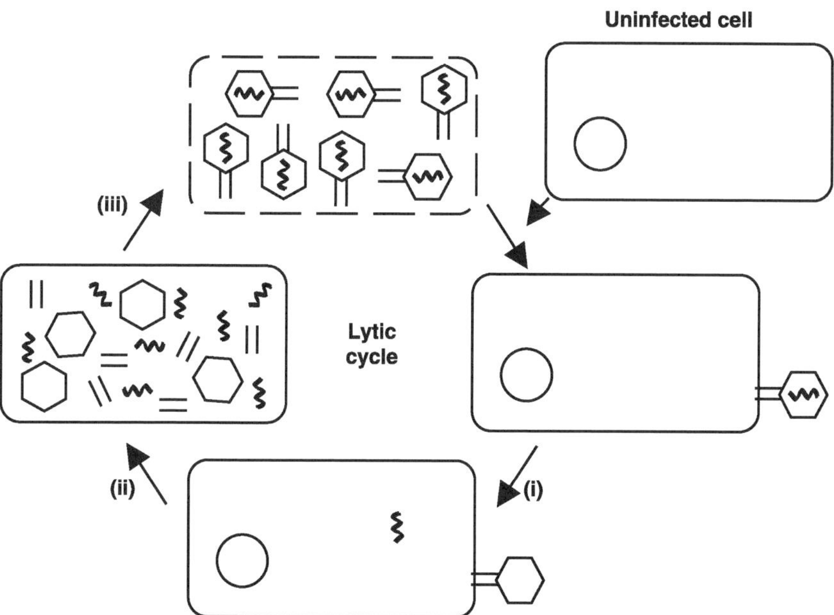

Figure 1 The reproductive cycle of lytic bacteriophages includes intracellular production of sharable resources. (i) The cycle begins with phage attachment to an uninfected host cell, and injection of the virus genetic material (RNA or DNA) into the cell. (ii) The genome is replicated and transcribed, to create virus products which diffuse within the cell to create a sharable resource pool (e.g., replicase enzymes, virus structural proteins). Other phages that simultaneously co-infect the cell may contribute equally to the resource pool, a process analogous to cooperative behavior. In contrast, the co-infecting phages might contribute fewer resources, but specialize in sequestering products from the pool; an antagonistic process analogous to selfish behavior. (iii) The cycle ends with assembly of progeny viruses, which are released into the extra-cellular environment through cell lysis (death).

(iii) assembly of progeny phages that are released into the environment through cell lysis, setting the stage for further rounds of infection.

One key to parasitic interactions between lytic viruses is the intracellular manufacturing of viral products (stage ii in Fig. 1). These products diffuse within the cell to create a common resource pool available to all co-infecting strains. Representative products include replication enzymes, and structural proteins that are used to construct the capsid which surrounds virus genetic material. We have argued that the resource pool tends to prevent an individual virus genotype from exclusive access to any of the gene products it contributes to the pool (Turner and Chao, 1999, 2003). In turn, the resource pool can create a 'conflict-of-interest' between viruses over available resources. Thus, the resource pool should promote selection for virus genotypes that parasitically interact with other co-infecting strains, through traits where one virus interferes with the ability of other viruses to use or acquire essential intracellular products (Nee and Maynard Smith, 1990; Nee, 2000; Brown, 2001).

Viruses do not exhibit behavior *per se*, but terminology borrowed from behavioral ecology is useful (though decidedly anthropomorphic) in describing how viruses utilize products in the resource pool. A virus that makes large (excess) amounts of useful products stands to benefit other co-infecting genotypes; hence, the virus can be defined as a cooperator (Szathmary, 1993; Turner and Chao, 1999). In contrast, a virus might synthesize fewer products but specialize in appropriating a larger share of the products available within the pool, a cheating strategy that exemplifies parasitism (Turner and Chao, 1999). The recent literature suggests that cheating (defection) is a universal phenomenon in biological populations; nature provides numerous examples of the temptation for individuals to cheat for personal reward, in lieu of cooperative behavior that promotes the common good (Axelrod, 1985; Dugatkin, 1997). Viruses also appear subject to the temptation, as shown in the following experiments where evolved genotypes of phage $\phi 6$ resort to cheating strategies to obtain products from the intracellular resource pool.

IV. PARASITISM IN CO-INFECTING RNA PHAGES

Phage $\phi 6$ is a member of the family *Cystoviridae*, bacterial viruses containing RNA genomes that are divided into three smaller segments (termed large, medium and small). Cystoviruses infect certain bacterial pathogens of wild and cultivated plants (Mindich *et al.*, 1999), explaining why $\phi 6$ was originally isolated from bacterial infections of bean straw (Vidaver *et al.*, 1973). Great care has been taken to identify the infected plant from which each of the nine viruses in the family *Cystoviridae* were isolated (Vidaver *et al.*, 1973; Mindich *et al.*, 1999), but the far more difficult task of determining the primary bacterial host for these viruses in the wild remains a goal for future research. Nevertheless, the typical host of $\phi 6$ in the laboratory is *Pseudomonas syringae* pathovar *phaseolicola*, a plant pathogen that causes bean halo-blight disease (Tsiamis *et al.*, 2000). The reproductive cycle of $\phi 6$ is typical of a lytic phage (Fig. 1). Phage $\phi 6$ has been the subject of extensive research on molecular virology (Mindich, 1999), and has successfully been used as a model for experimental ecology and evolution (Burch and Chao, 1999; Turner and Chao, 1999; Lythgoe and Chao, 2003), especially in testing theories for the evolution of genetic exchange between co-infecting viruses (Chao, 1990; Chao *et al.*, 1997; Turner and Chao, 1998; see review by Turner, 2003).

Parasitism and other competitive interactions can only occur when viruses co-infect cells. Rates of co-infection in $\phi 6$ are controlled by manipulating the multiplicity of infection (moi) or ratio of infecting phages to host cells in laboratory culture (Turner and Chao, 1998). Assuming Poisson sampling (Sokal and Rohlf, 1995), the probability that a cell will be infected with zero viruses is:

$$P(0) = \mathrm{e}^{-\mathrm{moi}}.$$

Similarly, the probability that a cell is infected by a single virus is:

$$P(1) = (\mathrm{e}^{-\mathrm{moi}} \times \mathrm{moi})/1.$$

The probability of co-infection (two or more viruses per cell) can be calculated as:

$$\mathrm{P}(\geq 2) = 1 - P(0) - P(1).$$

To examine evolutionary ecology of $\phi 6$ in the presence and absence of co-infection, we conducted an experiment that manipulated moi (Turner and Chao, 1998). A single clone of wild type $\phi 6$ was used as an ancestral virus to initiate three high moi (moi = 5) and three low moi (moi = 0.002) populations, which were then allowed to evolve on *P. phaseolicola*. Following the above logic, at moi = 5, co-infection by two or more viruses is common and 97% of cells should experience multiple infections. In contrast, only 0.1% of all infected cells contain two or more viruses at moi = 0.002. Viruses and host cells were mixed together in a test tube at high or low moi, and the resultant virus progeny were then diluted and plated onto agar containing a lawn (superabundance) of host cells, where each virus that hits the lawn grows to form a visible plaque overnight. The plaques were then harvested and filtered to obtain a bacteria-free lysate. Finally, a dilution of the lysate was mixed with a fresh stock of naïve host cells at high or low moi. This propagation cycle was repeated for 50 consecutive days, which is equivalent to 250 generations of viral evolution (1 generation in the tube plus 4 generations on the plate per day) (Turner and Chao, 1998). Thus, moi differences between treatments were imposed every fifth generation.

At the end of the experiment, population samples (stored in the freezer at $-20\,^{\circ}$C) were competed against a common competitor of the ancestral (wild type) genotype to measure changes in fitness (defined as growth performance on *P. phaseolicola*). To do so, an evolved population and a genetically marked ancestor were mixed together, and assayed for their ability to grow on *P. phaseolicola* cells in a 24-hour period. The ratio of phages in the starting mixture (R_0) and after 24 hours (R_1) was monitored by plating on selective agar, where genetically marked and unmarked viruses are easily distinguished by their plaque morphology (Chao, 1990). Fitness (W) was defined as:

$$W = R_1/R_0.$$

Thus, $W \neq 1.0$ indicates that the competitors differ in fitness because they produce dissimilar numbers of progeny during the assay.

Fitness trajectories of high-moi– and low-moi–evolved populations were obtained by plotting the grand mean fitness of each treatment group over time (in 50 generation increments). Figure 2A shows that all of the derived viruses increased in fitness relative to their common ancestor, indicating that beneficial mutations led to fitness improvements in the populations. The low-moi populations increased linearly in fitness over time to achieve the highest end point fitness values. In contrast, the fitness trajectory of high-moi populations was concave; these populations appeared to quickly reach a selective plateau that led to a fitness decline. At face value, these data suggested that co-infection hindered the rate of fitness enhancement. But the odd fitness trajectory observed in the high-moi lineages hinted that a more complex explanation existed.

We realized that the standard assay (Chao, 1990) used to measure fitness in Fig. 2A compares virus growth rates during strictly clonal infections, where viruses do not ecologically interact within the cell (Turner and Chao, 1998). During clonal infections, selection is primarily for a virus that best exploits the host cell. But intracellular interactions can occur during co-infection; thus, the viruses evolved at high moi experienced an ecological habitat radically different from that of the standard assay. Co-infecting viruses may be selected for host exploitation *and* for within-host competition of limited products available in the resource pool. Adaptation to intra-host competition may occur through novel virus traits that detract from the ability of the virus to exploit its host (Lewontin, 1970), and this within-host selection may then create a cost for co-infection.

To examine the importance of ecological differences between high- and low-moi treatments, we conducted additional fitness measurements for the evolved populations at moi = 5 (Turner and Chao, 1998). Figure 2B shows that the high-moi–evolved viruses perform much better relative to the ancestor during mixed infections, in comparison to their fitness during clonal

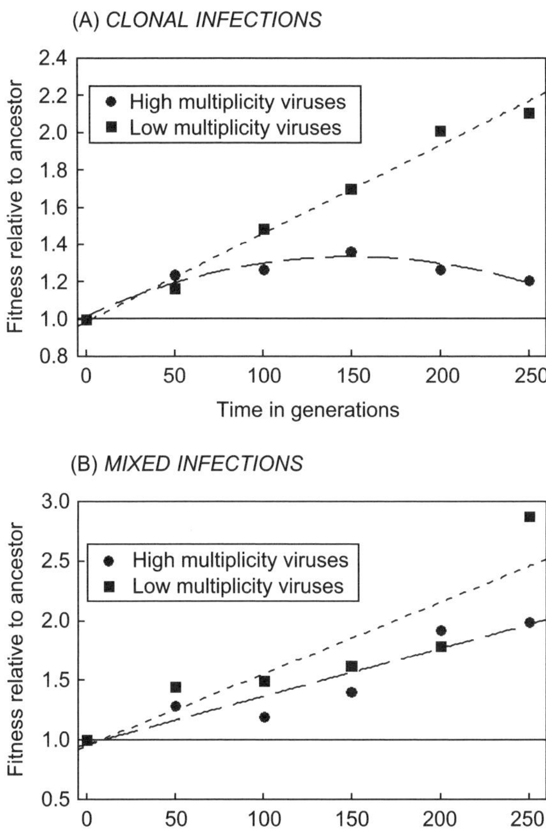

Figure 2 Results of a laboratory evolution experiment, where populations of the RNA phage ϕ6 were propagated for 250 generations in the presence and absence of co-infection. Viruses evolved under high multiplicity of infection (moi) experienced frequent co-infection of *Pseudomonas phaseolicola* cells, whereas those grown at low moi infected cells alone. Each point represents the grand mean fitness (relative to the ancestral wild-type virus) of three populations. (A) Traditional assays measure fitness by comparing the growth rates of evolved and ancestral viruses in strictly clonal infections. Here, the low-moi–evolved viruses (filled squares) show rapid fitness improvement through time, but their counterparts evolved at high moi perform very poorly. (B) When the fitness assays were modified to allow for co-infection, it was revealed that the high-moi–evolved viruses perform much better in assay habitats resembling their evolutionary environment. The combined results indicate that phages propagated at high moi evolve traits that are beneficial for co-infection, but which cause a performance tradeoff in environments where mixed infections are disallowed. These data suggest that the high-moi–evolved viruses adapted by acquiring traits to parasitically interact with co-infecting genotypes, at the cost of reduced fitness when infecting cells on their own. Adapted from Turner and Chao, 1998.

infections. These data clearly indicate a fitness trade-off; the viruses evolved traits specifically for co-infection, but detrimental for clonal infection (see also Sevilla *et al.*, 1998). The performance trade-off is quite obvious by late in the experiment when a downturn occurs in the fitness trajectory of high-moi–evolved viruses during clonal infections, compared with their continuing improvement in the ability to compete during mixed infections (compare Figs 2A and 2B).

The combined results can be explained by the evolution of parasitic traits in the phages propagated at high moi. The viruses obviously gain an added advantage in mixed infections, strongly suggesting that they evolved traits to selfishly sequester products that other co-infecting viruses contribute to the resource pool. The drawback to this evolved strategy is that genes for parasitism (or genes at other loci that are closely linked to parasitism genes) are inferior when the viruses compete in alternate environments where co-infection is uncommon and, hence, ecological interactions between viruses are less important. It is important to emphasize that these viruses are able to infect and replicate within cells completely on their own (Fig. 2A). Therefore, the cheating strategy is not simply due to a shortened genome, where the virus has eliminated key genes whose products are provided in trans by co-infecting helper strains (as seen in many plant and animal virus systems; Vogt and Jackson, 1999). Because the benefit of virus traits for parasitism depends on the likelihood of encountering parasitic versus cooperative genotypes during co-infection, frequency-dependent selection should determine the relative fitness of a selfish genotype.

V. FREQUENCY-DEPENDENT SELECTION AND VIRUS PARASITISM

Frequency-dependent selection occurs when the fitness of a genotype is not constant, but depends on the relative frequencies of other genotypes in the population. In inverse frequency-dependent selection, the rarer a genotype is in the population, the greater its fitness. Many biological phenomena give rise to this form of selection. In a classic example involving predation, Popham (1942) observed that aquarium fish prey selectively on one of three color morphs of the aquatic bug, *Sigara distincta*—namely, the morph that was most abundant at a given time. However, the complexity of natural environments often makes it difficult to identify the ecological mechanisms responsible for frequency-dependent selection. Laboratory experiments with microbes have been important in documenting frequency dependence (e.g., Rosenzweig *et al.*, 1994; Turner *et al.*, 1996; Elena *et al.*, 1997; Rozen and Lenski, 2000), and the tractability of these systems offers hope that researchers can decipher the underlying causes.

In environments where co-infection is common, one can predict that the fitness of parasitic viruses should be governed by inverse frequency-dependent selection. This outcome is expected because the selfish viruses should gain their greatest fitness advantage when rare; situations where the parasitic viruses are most likely to co-infect cells with ordinary genotypes and gain a large competitive advantage by sequestering products from the resource pool. In contrast, when cheater viruses are very common, they will most often co-infect cells with other cheaters and experience no benefit.

To examine this idea, we isolated individual cheater clones from the high-moi populations (Turner and Chao, 1999). We then measured the fitness of a cheater virus relative to its ancestor, when the two competitors were mixed at different initial frequencies in competition on *P. phaseolicola*. To ensure that co-infection was common, all of these fitness assays occurred at moi = 5. Results showed that the fitness of a cheater was a decreasing function of its initial frequency in competition (Fig. 3). That is, when the cheaters are rare they mostly co-infect cells with ancestral genotypes and gain a large fitness advantage, whereas when they are common they co-infect cells with other parasitic viruses and are less advantaged on average. Because all of the observed fitness values in this experiment exceed unity ($W > 1.0$), the data indicate that the cheating strategy (i.e., parasitic genotypes) should spread to fixation regardless of initial frequency. In turn, this result explains how the cheaters were able to sweep through the virus populations evolved at high moi, despite the fitness trade-off experienced by these viruses in alternate environments (Figs. 2A and 2B).

Although we demonstrated that the cheaters should be able to completely replace the ancestral genotype (Fig. 3), it is unknown whether the cheaters were the only genotypes present in the high-moi lineages at generation 250. That is, the possibility exists that the majority of cheaters co-existed with a minority population of cooperator viruses during all or a portion of the experiment. Exploring this hypothesis is extremely difficult because there is currently no genetic marker that distinguishes cheaters from other viruses that may be present in small numbers in the population. Future efforts could involve sequence analyses to look for genetic polymorphisms in the high-moi lineages, and to examine the possibility of co-existence between cheaters and helper viruses in these populations.

We conducted additional experiments at moi = 5, where the cheaters competed against viruses that evolved at low moi (i.e., in the virtual absence of co-infection according to the Poisson) (Turner and Chao, 2003). Once again, frequency dependence was observed (Fig. 3). But the decreasing fitness function for parasitic viruses in these assays was observed to cross a value of $W = 1.0$. Thus, the cheaters are shown to be advantaged when rare, but the magnitude of their fitness disadvantage when common is now so large that they are prevented from sweeping through the population. Rather,

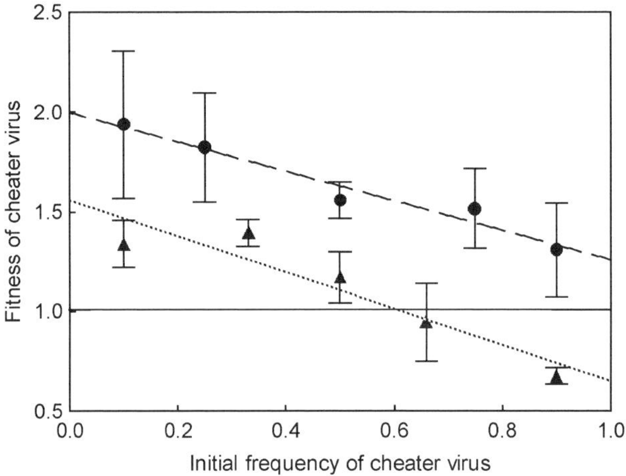

Figure 3 A parasitic virus specializes in sequestering intracellular resources provided by co-infecting ordinary genotypes. Because the benefit of parasitism depends on the relative abundance of parasitic versus ordinary viruses encountered inside the cell, the fitness of a cheater virus is expected to be frequency-dependent. Laboratory experiments prove that when co-infection is common, the fitness of an evolved parasitic phage is governed by inverse frequency-dependent selection; each point represents the mean and standard error of replicate fitness assays, initiated at different initial frequencies of competing viruses. The dashed and dotted lines indicate the best linear fit to each respective dataset. When the cheater phage is competed against its wild type ancestor (filled circles), it is most advantaged when rare because it primarily encounters the ordinary genotype inside the cell; however, the advantage diminishes at higher initial frequencies as the cheaters more often encounter one another during mixed infections. Because all of these data exceed 1.0, parasitic viruses (i.e., the cheating strategy) should always fix when invading a population of wild types. In contrast, when the parasitic phage is competed against an evolved cooperator (filled triangles) the frequency-dependent result crosses 1.0, indicating that each genotype is expected to invade when rare. For this reason, both viruses (strategies) should co-exist in a mixed polymorphism. Adapted from Turner and Chao, 1999, 2003.

if the two genotypes are together in the same population either can increase in frequency when rare, resulting in their co-existence in a mixed polymorphism. The data in Fig. 3 can also be used to predict the equilibrium frequencies of the two genotypes in the mixed population; the linear regression for these data intercept $W = 1.0$ when the initial frequency of cheaters equals 0.62. Therefore, it is predicted that the equilibrium percentage for cheaters in the population should be 62% (Turner and Chao, 2003).

Why are the parasitic viruses able to displace their ancestor but unable to sweep through a population of low-moi–evolved viruses? As previously described, viruses that specialize in overproducing intracellular products are effectively cooperators, whereas those that specialize in sequestering products are cheaters. The previous experiment allowed the viruses in the low-moi treatment to be grown under strictly clonal conditions (absence of competitive interactions), creating the opportunity for greater cooperation to evolve (Turner and Chao, 1998). One possibility for the polymorphism observed in Fig. 3 is that the low-moi viruses evolved to synthesize extremely large quantities of intracellular products that are made available in the intracellular resource pool. These over-abundant resources could decrease the sensitivity of the low-moi viruses to parasitism. Therefore, the ancestral population might succumb to invasion by cheaters, whereas the derived cooperators could tolerate a subpopulation of selfish individuals.

The mechanism for cheating in the high-moi–evolved genotypes is unknown. One possibility is that during intracellular replication the RNA of cheater genotypes is preferentially packaged into protein capsids of the virus progeny. The cheaters may be rather inefficient at producing capsids (or other key virus products) when they infect cells on their own, as suggested by their reduced competitiveness in clonal infections (Fig. 2A). However, the cheaters may gain a competitive edge during mixed infections if their genetic material experiences biased entry into capsids produced by other co-infecting strains. The encapsidation process in ϕ 6 is well-described (Qiao *et al.*, 2003), and a packaging preference might result from a duplication or other change in the *pac* region that governs binding and entry of RNA to capsids. Through this hypothesized mechanism, a selfish virus would reduce the probability that other co-infecting genotypes contribute the expected share of their genes to the progeny. This idea is supported by data from animal virus systems, because certain defective (non full-length) RNA viruses feature gene duplications involved in encapsidation, which provide these cheaters with a replication advantage over their co-infecting helper viruses (Holland, 1991).

VI. HOW MANY VIRUSES SHOULD ENTER A CELL?

Our studies and others suggest that co-infection can be costly to an individual virus, due to the presence of parasitic viruses that selfishly appropriate intracellular products (Turner and Chao, 1998; Hammond *et al.*, 1999; Vogt and Jackson, 1999). However, there are clear benefits associated with co-infection, such as enhanced pathogenicity or greater transmissibility of the co-infecting strains (Cohen, 1998; Lopez-Ferber *et al.*, 2003). Perhaps the most intriguing benefit of co-infection is that it allows viruses to generate novel variants through genetic exchange otherwise known as sex (Chao *et al.*, 1997; Nagy and Simon, 1997), where sex can be broadly defined as any exchange of genetic material between individuals (Michod and Levin,

1988). Under this definition, sex extends beyond the framework of obligately sexually reproducing populations composed of males and females, to include macro- and micro-organisms that asexually reproduce but also feature mechanisms for inter-individual genetic exchange (such as transduction, transformation, and conjugation in bacteria). In any such population, sex promotes linkage equilibrium, free association between alleles at two or more loci. The presumed advantage of sex is that it reduces the frequency of suboptimal allele combinations, and can potentially increase the frequency of superior allele combinations (Fisher, 1930; Muller, 1932; see also Barton and Charlesworth, 1998, and West *et al.*, 1999 for reviews of debates on the origin and maintenance of sex). Thus, along with spontaneous mutations, sex generates the population genetic diversity which provides the raw material for evolution by natural selection.

Viruses experience genetic exchange through two basic mechanisms. In DNA viruses, sex is promoted exclusively by recombination (generation of a new nucleotide strand from two or more parental strands). Some RNA viruses also undergo recombination, but others do not. Instead, genetic exchange occurs because the genome is divided into several smaller RNA molecules (segments), allowing formation of hybrid progeny containing a random re-assortment of segments descending from two or more co-infecting parent viruses. For example, ϕ6 and its relatives in the family *Cystoviridae* feature genomes divided into three segments, and co-infection of the same cell by multiple genotypes yields reassortants (Mindich *et al.*, 1999). Interestingly, recombination between segments is rare or non-existent in many segmented RNA viruses including ϕ6 and the majority of cystoviruses (Horiuchi, 1975; Mindich *et al.*, 1976, 1999), fueling arguments that segment reassortment might have evolved as an alternative to recombination for the purpose of promoting sex in some viruses (Pressing and Reanney, 1984; Chao, 1988).

If sex is advantageous in virus populations (Turner, 2003), one adaptive compromise would be virus evolution which would balance the positive and negative consequences of co-infection; viruses would limit the number of co-infecting particles to reduce intra-host competition, but allow entry by more than one virus to achieve genetic exchange. This hypothesis is intriguing, but has not yet been explicitly tested. Indirect support for an adaptive compromise could be the exclusion mechanism in phage ϕ6 where three viruses (on average) are allowed to enter the cell. Olkkonen and Bamford (1989) used radioactive labeling of viruses to show that a limit to co-infection exists in phage ϕ6. In particular, their experiments used the amount of incorporated carbon-14 label as a measure of the number of ϕ6 phages entering a *P. phaseolicola* cell.

We sought to confirm the estimated limit using a simpler method that is easily adapted to a variety of viruses and culture conditions (Turner *et al.*,

1999). To do so, we devised a method that compares the frequency of hybrids produced by two marked phage strains to that predicted by a mathematical model based on differing limits to co-infection. Wild type $\phi 6$ may be symbolized as +/+/+, corresponding to segments L (large), M (medium), and S (small). We studied two $\phi 6$ strains, +/β/+ and β/+/+, with an engineered marker (β) (Onodera *et al.*, 1993) on the M and L segments, respectively (Turner *et al.*, 1999). The β marker encodes the alpha-subunit of the beta-galactosidase (β-gal) gene from the bacterium *Escherichia coli* (Horwitz *et al.*, 1964). The marker is highly stable when inserted into the M or L segment, evidenced by its very low rate of mutational reversion (Turner *et al.*, 1999). Marked phages form blue plaques on selective plates containing the chemical X-gal and *P. phaseolicola* carrying a plasmid that encodes the beta-subunit of the β-gal gene (Onodera *et al.*, 1993).

To measure the limit to co-infection in $\phi 6$, we conducted experiments that crossed the β-*gal*–marked strains at moi = 0.02, 1, 2, 3, 4, 5, 10, and 25 on *P. phaseolicola* (Turner *et al.*, 1999). As moi increases, hybrid frequency also should increase to a maximum that is determined by the limit of co-infection. A cross between marked strains produces six possible hybrids, but only the +/+/+ reassortants were monitored because they are easily distinguished on selective plates. In theory, if phage fitnesses are equal, a maximum frequency of +/+/+ progeny is reached when cells are infected with an equal number of marked phage [frequency (+/β/+) = frequency (β/+/+) = 0.5]. Here, the probability that progeny will acquire both + segments is obtained by simply multiplying the frequencies of the two segments: $0.5 \times 0.5 = 0.25$. Only if the limit is infinitely large (i.e., no limit) will the ratio of marked viruses within a given cell approach 1:1. Thus, as the limit decreases, so will the maximum frequency of +/+/+ progeny. At a limit of one, no hybrids are formed regardless of moi. To determine the limit in $\phi 6$, we compared our observed hybrid frequencies to those expected based on a theoretical model (Turner *et al.*, 1999).

Our model (Turner *et al.*, 1999) predicts the frequency (H) of any one hybrid over a range of multiplicities in crosses between two marked viruses according to the Poisson (Sokal and Rohlf, 1995). Because genetic markers are often deleterious and may be prone to revert, the model incorporates several fitness parameters that adjust H if marked segments experience a replication disadvantage relative to wild type, and if marker reversion (even at low rates) occurs. Figure 4 shows a series of curves generated by the model for expected frequencies of +/+/+ hybrids over a range of multiplicities in crosses between beta-marked phages of $\phi 6$. Expected values of H asymptote above 0.25 because of the fitness disadvantage suffered by marked phage and segments (Turner *et al.*, 1999). Observed data reveal that hybrid frequency increases with moi but reaches a plateau of $H = 0.23$. The observed maximum matches the mathematical model for a limit to co-infection between

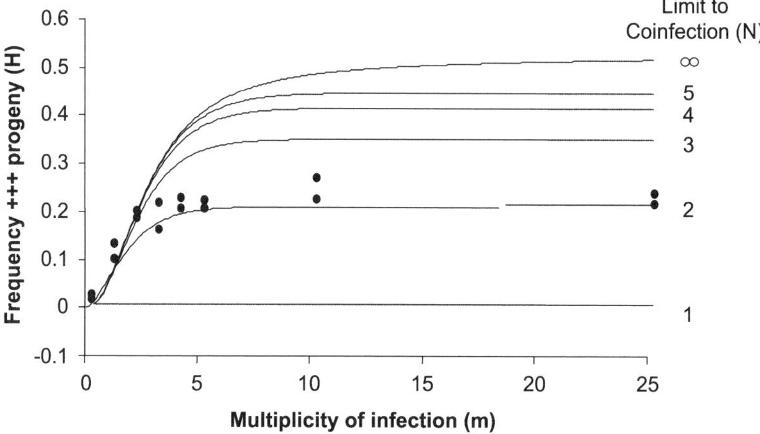

Figure 4 Laboratory experiments reveal that a limit to co-infection exists in phage $\phi 6$. A comparison was made between expected and observed values of H, the frequency of +/+/+ hybrid reassortants generated in crosses between two marked genotypes (β/+/+ and +/β/+). Theoretical curves were generated by a mathematical model that predicts H over a range of multiplicities of infection (m), assuming a defined limit to co-infection (N). Expected values for H are adjusted if marked segments experience a replication disadvantage relative to wild-type (see text for details). Observed values of H were generated in two replicate experimental crosses between the marked phages at 8 moi; each point represents an independent estimate. Observed H saturates at a value corresponding to a limit of between two and three phages per cell. Adapted from Turner *et al.*, 1999.

two and three phages per cell, and confirms the previous estimate of three phages per cell by Olkkonen and Bamford (1989).

Although the limit is firmly established in phage $\phi 6$, details of the phenomenon are fodder for future studies. If the limit is an adaptive compromise, mean fitness of a virus population should be maximal when an intermediate number of viruses co-infect a cell. By definition, this requires that variation in the co-infection number be at least partly due to genetic differences among viruses (i.e., due to a genetically determined viral mechanism). In contrast, the phenomenon could be due to ecological interactions between viruses and the host cell. For instance, it might be that too many viruses attaching to a cell breaks down the integrity of the cell wall, thus destroying the cell before virus replication can be completed. If the latter is true, benefits of co-infection may lead to directional selection to increase co-infection number, but the upper threshold for virus entry prevents balancing selection because large numbers of viruses never enter the cell. However, this scenario seems unlikely for $\phi 6$ because as many as 50 viruses can attach to a cell, but only 3 can enter (Olkkonen and Bamford, 1989). Limits are widely observed in viruses (e.g., Lu and Henning, 1994; Singh *et al.*, 1997), although it is sometimes very difficult to elucidate the underlying mechanism that prevents an overabundance of viruses from entering. When the mechanisms have been unraveled they are demonstrated to be under strict virus control, and this overwhelmingly suggests that the limit in ϕ6 is virus determined as well.

Future experiments on viral limits could involve a comparative approach using ϕ6 and its relatives. For example, one could hold the host bacterium (such as *P. phaseolicola*) constant, and measure the limit for all nine members of the family *Cystoviridae*. These assays would inform whether the limit on a single host is highly conserved within this evolutionarily related group; if not, the results would suggest that the particular ecology of the virus-host cell interaction is very important for defining the limit. For example, some cystoviruses (such as ϕ6) appear to use the type IV pilus of a host bacterium as a site of initial attachment, whereas others enter directly through the cell membrane (Mindich *et al.*, 1999). Perhaps the difference in surface area available for attachment also translates to a difference in the number (limit) of viral particles able to infect the cell. As previously mentioned, the natural host for each member of the *Cystoviridae* has yet to be determined. But measurements of the limit for each virus in its primary host would be most informative for elucidating how the natural ecology impacts virus population genetics. For example, in some viruses there may be restrictions to the genetic variation created through reassortment simply because the limit to co-infection lies close to a value of one (on average). The ultimate goal would be to study limits to co-infection when virus growth occurs under more natural conditions, such as in infections of plant pathogenic bacteria that are themselves parasitizing plants grown in a greenhouse.

Because segment reassortment occurs in ϕ6, possibilities for genetic exchange can impact the interpretation of our data in the study where virus lineages were propagated in the laboratory at high moi (Turner and Chao, 1998). In particular, it is a challenge to hypothesize how a selfish *genotype* of the virus can evolve in a high-moi environment where reassortment readily occurs. That is, parasitism might occur through some gene that promotes cheating, and the mutated gene would be tightly linked to other genes residing on the same segment because recombination appears absent in ϕ6. However, the mutated gene should not be tightly linked to genes located on the other two RNA segments, due to the phenomenon of reassortment. For this reason, we have suggested that it may be more appropriate to speak of cheater 'segments' in ϕ6 rather than cheater 'viruses' (Turner and Chao, 2003). The encapsidation mechanism previously proposed would still apply because parasitism might evolve if a cheater gene biased the entry of its resident segment into available capsids, a process analogous to meiotic drive in eukaryotes. An intriguing possibility is that the cheater segments or viruses may feature a limit to co-infection that differs from wild type ϕ6;

for instance, the limit may be increased in the derived viruses due to the benefit that results from co-infection with other genotypes. This idea is highly speculative, and relies on the untested assumption that the limit is at least partially controlled by virus genes.

VII. PARASITISM IN PLANT AND ANIMAL VIRUSES

Viruses of plants, humans, and other animals are very well-studied because they can be infectious agents of agricultural, medical, and veterinary importance. Most often these viruses are examined under artificial conditions (such as in laboratory tissue culture) that differ dramatically from their natural ecological settings. When viruses are propagated in laboratory environments where co-infection is common, the evolution of parasitic viruses is often observed. These cheating viruses tend to be shortened forms of the original virus and are generally known as defective-interfering (DI) particles (terminology coined by Huang and Baltimore, 1970). Here, 'defective' refers to their lack of essential genes or parts of the genome that were present in the parental virus. DIs 'interfere' specifically with co-infecting ordinary (helper) viruses by replicating at their expense, usually resulting in reduced titers of the ordinary viruses relative to their reproductive output when DIs are absent. Interestingly, recent evidence shows that DI-helper interactions can be mutualistic, at least in the case of an insect baculovirus that increases the pathogenicity of the DI-helper virus population (Lopez-Ferber *et al.*, 2003); future work will determine whether this one reported case is a rare outlier.

In essence, the vast majority of DIs are parasites of parasites, because they rely on one or more proteins synthesized by co-infecting helpers. Use of these intracellular products by DIs allows them to achieve high frequencies in virus populations because their shortened length affords an advantage in terms of rates of genome replication relative to helper viruses. Antagonistic DIs are particularly well-described in vesicular stomatitis virus (VSV), an RNA virus that is transmitted by several insects including mosquitoes, and which infects a wide variety of mammals. VSV has a rich history as a preferred model to study the molecular genetics and experimental evolution of RNA viruses (Holland *et al.*, 1991; Turner and Elena, 2000; Rose and Whitt, 2001). But when particles of VSV greatly outnumber their host cells in laboratory tissue-culture experiments, this produces strong selection for DIs (e.g., Li and Pattnaik, 1997; Whelan and Wertz, 1997). Figure 5 depicts one example of the interaction between ordinary (full-length) VSV and an antagonistic DI particle. A thorough description of the varied plant and animal systems where defective (shortened) viruses have been observed extends beyond the space available in this chapter; several recent papers

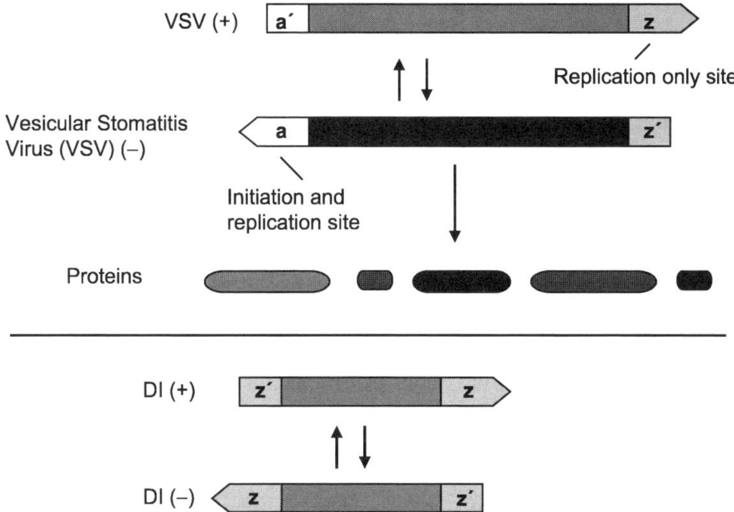

Figure 5 VSV is a single-stranded negative-sense RNA virus, which infects certain insects and mammals. Growth of VSV under high levels of co-infection in the laboratory strongly selects for parasitic defective-interfering (DI) particles. When VSV enters the cell it transcribes viral RNA to make monocistronic mRNAs to produce proteins. However, VSV also replicates by serving as the template for the complementary positive-strand genome, VSV(+), which is used to make more VSV(−). The 3′ terminus of VSV(−) contains an initiation site (**a**) for both transcription and replication, whereas the 3′ terminus of VSV(+) has a site (**z**) for only replication. The complementary sequences of **a** and **z** on the opposite strands are **a**′ and **z**′. DI particles of VSV are typically shorter and lack various protein-coding sequences. In this particular example, the DI particle has a **z** replication site at the 3′ terminus of both its negative and positive strands. Whereas VSV(−) allocates time for transcription, DI(−) does not and is effectively equal to DI(+). Thus, the DI RNA relies on the complete virus to provide proteins, but it has a higher replicative fitness than the VSV genome. Adapted from Turner, 2003.

describe DIs and other defective viruses in these systems (Holland, 1991; Vogt and Jackson, 1999).

Obviously, the long-term persistence of DIs depends on the continued presence of ordinary viruses. But persistence of the helpers is unlikely because DIs typically experience a replication advantage over full-length viruses and may easily out-compete their helpers. Elimination of the helpers results in extinction of the virus population because DIs cannot replicate on their own. Thus, the appearance of DIs in laboratory cultures is often viewed as a nuisance (Holland, 1991). The long-term dynamics of DI-helper virus interactions have received theoretical treatment (Nelson and Perelson, 1995; Frank, 2000) but have rarely been studied in empirical detail (but see e.g., Duhaut and Dimmock, 1998). Unfortunately, the appearance of DIs may be unavoidable in certain industrial microbiology applications (e.g., virus vaccine production), because the manufacturer of high virus titers (densities) is commercially desirable but can lead to high levels of DIs that reduce titers of ordinary viruses. Beyond the realization that DIs easily occur in laboratory and other artificial environments, there is a pressing need to establish whether DIs are also prevalent in natural populations of viruses. For DIs to flourish in the wild, co-infection must be common enough that viruses lacking essential genes can readily encounter their helpers. Evidence suggests that this condition may be fulfilled in certain systems, because helper-dependent (but non-DI) viruses can be isolated from wild populations of infected animals and plants (see Vogt and Jackson, 1999).

VIII. CONCLUDING REMARKS

As the aforementioned studies demonstrate, recent laboratory experiments with phage $\phi 6$ are highly valuable for elucidating seldom-studied aspects of phage ecology. Laboratory studies in the evolutionary ecology of viruses can be extremely informative, but even more can be gained by examining the ecology of viruses in the wild. Of course, the latter is extremely challenging due to the complexities and uncontrolled factors inherent to field studies. But worthwhile data may be obtained by combining the study of virus ecology with modern molecular techniques. For example, one can acquire natural samples of viruses, isolate viral DNA or RNA and conduct sequence analyses to examine genetic relatedness within and between samples from given locales (e.g., Paul and Kellog, 2000; Breitbart *et al.*, 2002). In addition, valuable data would be biotic factors such as the host genotype or species from which the virus was isolated, as well as abiotic factors such as the local climate or pH at time of isolation. The information can then provide a context for relating the impact of environment on estimates of genetic diversity in the natural samples.

Viruses are so abundant in natural systems that they can influence large-scale ecological processes, such as nutrient cycling and bacterial biodiversity in marine environments (Fuhrman, 1999). The role of parasitic or other interactions between viruses in these ecosystem-level phenomena is unknown. Many viruses are multipartite, featuring genomic segments that are packaged into separate virus particles (Knipe and Howley, 2001). Obviously, frequent co-infection of cells is required to evolve and maintain multipartite genomes, providing further indication that ecological interactions between viruses are not uncommon. For these reasons, it is easy to argue that much greater effort should be placed in examining virus ecology and in determining the influence of virus co-infection on host disease and the evolution of virus traits.

ACKNOWLEDGMENTS

I thank two anonymous reviewers for their insightful and valuable suggestions to improve the clarity of the text, and F. Cohan for inspiring the use of a literary quote by Swift.

REFERENCES

Antia, R., Nowak, M.A. and Anderson, R.M. (1996) Antigenic variation and the within-host dynamics of parasites. *Proc. Natl. Acad. Sci. USA* **93**, 985–989.

Axelrod, R. (1985) *The Evolution of Cooperation*. Basic Books, Boulder, CO.

Barton, N.H. and Charlesworth, B. (1998) Why sex and recombination? *Science* **281**, 1987–1990.

Bergh, O., Borsheim, K.Y., Bratbak, G. and Heldal, M. (1989) High abundance of viruses found in aquatic environments. *Nature* **340**, 467–468.

Breitbart, M., Salamon, P., Andresen, B., Mahaffy, J., Segall, A., Mead, D., Azam, F. and Rohwer, F. (2002) Genomic analysis of an uncultured marine viral community. *Proc. Natl. Acad. Sci. USA* **99**, 14250–14255.

Brown, S.P. (2001) Collective action in an RNA virus. *J. Evol. Biol.* **14**, 821–828.

Burch, C.I. and Chao, L. (1999) Evolution by small steps and rugged landscapes in the RNA virus phi-6. *Genetics* **151**, 921–927.

Cairns, J. Stent, G.S. and Watson, J.D. (1992) *Phage and the Origins of Molecular Biology*. Cold Spring Harbor Laboratory Press, Cold Spring Harbor, NY.

Chao, L. (1988) Evolution of sex in RNA viruses. *J. Theor. Biol.* **133**, 99–112.

Chao, L. (1990) Fitness of RNA virus decreased by Muller's ratchet. *Nature* **348**, 454–455.

Chao, L., Tran, T.T. and Tran, T.T. (1997) The advantage of sex in the RNA virus ϕ6. *Genetics* **147**, 953–959.

Chao, L., Hanley, K.A., Burch, C.L., Dahlberg, C. and Turner, P.E. (2000) Kin selection and parasite evolution: Higher and lower virulence with hard and soft selection. *Q. Rev. Biol.* **75**, 261–275.

Cohen, M.S. (1998) Sexually transmitted diseases enhance HIV transmission: No longer a hypothesis. *Lancet* **351**(Suppl. 3), 5–7.

Crick, F.H.C., Barnett, L., Brenner, S. and Watts-Tobin, R.J. (1961) General nature of the genetic code for proteins. *Nature* **192**, 1227–1232.

Dugatkin, L.A. (1997) *Cooperation among Animals: An Evolutionary Perspective*. Oxford University Press, Oxford.

Duhaut, S.D. and Dimmock, N.J. (1998) Heterologous protection of mice from a lethal human H1N1 influenza A virus infection by H3N8 equine defective interfering virus: Comparison of defective RNA sequences isolated from the DI inoculum and mouse lung. *Virology* **248**, 241–253.

Elena, S.F., Miralles, R. and Moya, A. (1997) Frequency-dependent selection in a mammalian RNA virus. *Evolution* **51**, 984–987.

Fisher, R.A. (1930) *The Genetical Theory of Natural Selection*. Oxford University Press, Oxford, UK.

Frank, S.A. (1992) A kin selection model for the evolution of virulence. *Proc. R. Soc. Lond. B* **250**, 195–197.

Frank, S.A. (1996) Models of parasite virulence. *Q. Rev. Biol.* **71**, 47–78.
Frank, S.A. (2000) Within-host spatial dynamics of viruses and defective interfering particles. *J. Theor. Biol.* **206**, 279–290.
Frank, S.A. (2002) *Immunology and Evolution of Infectious Disease*. Princeton University Press, Princeton, NJ.
Fuhrman, J.A. (1999) Marine viruses and their biogeochemical and ecological effects. *Nature* **399**, 541–548.
Hammond, J., Lecoq, H. and Raccah, B. (1999) Epidemiological risks from mixed virus infections and transgenic plants expressing viral genes. *Adv. Virus. Res.* **54**, 189–314.
Holland, J. (1991) Defective viral genomes. In: *Fundamental Virology* (Ed. by B. Fields and D. Knipe), 2nd ed., pp. 151–165. Raven Press, New York, NY.
Holland, J.J., de la Torre, J.C., Clarke, D.K. and Duarte, E. (1991) Quantitation of relative fitness and great adaptability of clonal populations of RNA viruses. *J. Virol.* **65**, 2960–2967.
Horiuchi, K. (1975) Genetic studies of RNA phages. In: *RNA Phages* (Ed. by N. D. Zinder), pp. 29–50. Cold Spring Harbor Laboratory, Cold Spring Harbor, NY.
Horwitz, J.P., Chua, J., Curby, R.J., Tomson, A.J., Da Rooge, M.A., Fisher, B.E., Mauricio, J. and Klundt, I. (1964) Substrates for cytochemical demonstration of enzyme activity I. Some substituted 3-inodoly1-β-D-glycopyranosides. *J. Med. Chem.* **7**, 574–575.
Huang, A.S. and Baltimore, D. (1970) Defective viral particles and viral disease processes. *Nature* **226**, 325–327.
Hurst, C.J. (Ed.) (2000) *Viral Ecology*. Academic Press, San Diego, CA.
Knipe, D.M. and Howley, P.M. (Eds.) (2001) In: *Fields Virology*, 4th ed. Lippincott Williams & Wilkins, Philadelphia, PA.
Kohler, R.E. (1994) *Lords of the Fly*. University of Chicago Press, Chicago, IL.
Lenski, R.E. (2002) Experimental evolution. In: *Encyclopedia of Evolution* (Ed. by M.D. Pagel), pp. 339–344. Oxford University Press, Oxford.
Levin, S. and Pimentel, D. (1981) Selection of intermediate rates of increase in parasite-host systems. *Am. Nat.* **117**, 308–315.
Lewontin, R.C. (1970) The units of selection. *Annu. Rev. Ecol. Syst.* **1**, 1–18.
Li, T. and Pattnaik, A.K. (1997) Replication signals in the genome of vesicular stomatitis virus and its defective interfering particles: Identification of a sequence element that enhances DI RNA replication. *Virology* **232**, 248–259.
Lopez-Ferber, M., Simon, O., Williams, T. and Caballero, P. (2003) Defective or effective? Mutualistic interactions between virus genotypes. *Proc. R. Soc. Lond. B* **270**, 2249–2255.
Lu, M. and Henning, U. (1994) Superinfection exclusion by T-even-type coliphages. *Trends Microbiol.* **2**, 137–139.
Luria, S.E. and Delbruck, M. (1943) Mutations of bacteria from virus sensitivity to virus resistance. *Genetics* **28**, 491–511.
Lythgoe, K.A. and Chao, L. (2003) Mechanisms of coexistence of a bacteria and a bacteriophage in a spatially homogeneous environment. *Ecol. Lett.* **6**, 326–334.
May, R.M. and Nowak, M.A. (1995) Coinfection and the evolution of parasite virulence. *Proc. R. Soc. Lond. B* **261**, 209–215.
Michod, R.E. and Levin, B.R. (1988) *The Evolution of Sex. An Examination of Current Ideas*. Sinauer, Sunderland, MA.
Mindich, L. (1999) Reverse genetics of dsRNA bacteriophage phi-6. *Adv. Virus Res.* **53**, 341–353.

Mindich, L., Sinclair, J.F., Levine, D. and Cohen, J. (1976) Genetic studies of temperature-sensitive and nonsense mutants of bacteriophage ϕ6. *Virology* **75**, 218–223.

Mindich, L., Qiao, X., Qiao, J., Onodera, S., Romantschuk, M. and Hoogstraten, D. (1999) Isolation of additional bacteriophages with genomes of segmented double-stranded RNA. *J. Bact.* **181**, 4505–4508.

Mosquera, J. and Adler, F.R. (1998) Evolution of virulence: A unified framework for coinfection and superinfection. *J. Theor. Biol.* **195**, 293–313.

Muller, H.J. (1932) Some genetic aspects of sex. *Am. Nat.* **66**, 118–138.

Nagy, P.D. and Simon, A.E. (1997) New insights into the mechanisms of RNA recombination. *Virology* **235**, 1–9.

Nee, S. (2000) Mutualism, parasitism, and competition in the evolution of coviruses. *Phil. Trans. R. Soc. Lond. B* **355**, 1607–1613.

Nee, S. and Maynard Smith, J. (1990) The evolutionary biology of molecular parasites. *Parasitology* **100**(Suppl. S5–S18).

Nelson, G.W. and Perelson, A.S. (1995) Modeling defective interfering virus therapy for AIDS: Conditions for DIV survival. *Math. Biosci.* **125**, 127–153.

Nowak, M.A. and May, R.M. (1994) Superinfection and the evolution of parasite virulence. *Proc. R. Soc. Lond. B* **255**, 81–89.

Nowak, M.A. and May, R.M. (2000) *Virus Dynamics: The Mathematical Foundations of Immunology and Virology*. Oxford University Press, Oxford.

Olkkonen, V.M. and Bamford, D.H. (1989) Quantitation of the adsorption and penetration stages of bacteriophage ϕ6 infection. *Virology* **171**, 229–238.

Onodera, S., Qiao, X., Gottlieb, P., Strassman, J., Frilander, M. and Mindich, L. (1993) RNA structure and heterologous recombination in the double-stranded RNA bacteriophage ϕ6. *J. Virol.* **67**, 4914–4922.

Paul, J.H. and Kellogg, C.A. (2000) Ecology of bacteriophages in nature. In: *Viral Ecology* (Ed. by C.J. Hurst), pp. 211–216. Academic Press, San Diego, CA.

Petney, T.N. and Andrews, R.H. (1998) Multiparasite communities in animals and humans: frequency, structure and pathogenic significance. *Int. J. Parasitol.* **28**, 377–393.

Popham, E.J. (1942) Further experimental studies on the selective action of predators. *Proc. Zool. Soc. Lond.* **112**, 105–117.

Pressing, J. and Reanney, D.C. (1984) Divided genomes and intrinsic noise. *J. Mol. Evol.* **20**, 135–146.

Qiao, J., Qiao, X., Sun, Y. and Mindich, L. (2003) Isolation and analysis of mutants of double-stranded-RNA bacteriophage ϕ6 with altered packaging specificity. *J. Bacteriol.* **185**, 4572–4577.

Read, A.F. and Taylor, L.H. (2000) Within-host ecology of infectious diseases: Patterns and consequences. In: *Molecular Epidemiology of Infectious Diseases* (Ed. by R.C.A. Thompson), pp. 59–75. Arnold, London.

Read, A.F. and Taylor, L.H. (2001) The ecology of genetically diverse infections. *Science* **292**, 1099–1102.

Rose, J.K. and Whitt, M.A. (2001) *Rhabdoviridae*: The viruses and their replication. In: *Fields Virology* (Ed. by D.M. Knipe and P.M. Howley), pp. 1221–1244. Lippincott Williams & Wilkins, Philadelphia, PA.

Rosenzweig, R.F., Sharp, R.R., Treves, D.S. and Adams, J. (1994) Microbial evolution in a simple unstructured environment: Genetic differentiation in *Escherichia coli*. *Genetics* **137**, 903–917.

Rozen, D.E. and Lenski, R.E. (2000) Long-term experimental evolution in *Escherichia coli*. VIII. Dynamics of a balanced polymorphism. *Am. Nat.* **155**, 24–35.

Sevilla, N., Ruiz-Jarabo, C.M., Gomez-Mariano, G., Baranowski, E. and Domingo, E. (1998) An RNA virus can adapt to the multiplicity of infection. *J. Gen. Virol.* **79**, 2971–2980.

Singh, I., Suomalainen, M., Varadarajan, S., Garoff, H. and Helenius, A. (1997) Multiple mechanisms for the inhibition of entry and uncoating of superinfecting Semliki Forest virus. *Virology* **231**, 59–71.

Sokal, R.R. and Rohlf, F.J. (1995) *Biometry*, 3rd ed. Freeman, San Francisco, CA.

Szathmary, E. (1993) Co-operation and defection: Playing the field in virus dynamics. *J. Theor. Biol.* **165**, 341–356.

Tsiamis, G., Mansfield, J.W., Hockenhull, R., Jackson, R.W., Sesma, A., Athanassopoulos, E., Bennett, M.A., Stevens, C., Vivian, A., Taylor, J.D. and Murillo, J. (2000) Cultivar-specific avirulence and virulence functions assigned to avrPphF in Pseudomonas syringae pv. phaseolicola, the cause of bean halo-blight disease. *EMBO J.* **19**, 1–11.

Turner, P.E. (2003) Searching for the advantages of virus sex. *Orig. Life Evol. Biosph.* **33**, 95–108.

Turner, P.E. and Chao, L. (1998) Sex and the evolution of intrahost competition in RNA virus ϕ6. *Genetics* **150**, 523–532.

Turner, P.E. and Chao, L. (1999) Prisoner's dilemma in an RNA virus. *Nature* **398**, 441–443.

Turner, P.E. and Chao, L. (2003) Escape from prisoner's dilemma in RNA phage ϕ6. *Am. Nat.* **161**, 497–505.

Turner, P.E. and Elena, S.F. (2000) Cost of host radiation in an RNA virus. *Genetics* **156**, 1465–1470.

Turner, P.E., Burch, C., Hanley, K. and Chao, L. (1999) Hybrid frequencies confirm limit to coinfection in the RNA bacteriophage ϕ6. *J. Virol.* **73**, 2420–2424.

Turner, P.E., Souza, V. and Lenski, R.E. (1996) Tests of ecological mechanisms promoting the stable coexistence of two bacterial genotypes. *Ecology* **77**, 2119–2129.

van Baalen, M. and Sabelis, M.W. (1995) The dynamics of multiple infection and the evolution of virulence. *Am. Nat.* **146**, 881–910.

Vidaver, K.A., Koski, R.K. and Van Etten, J.L. (1973) Bacteriophage ϕ6: A lipid-containing virus of. *Pseudomonas phaseolicola. J. Virol.* **11**, 799–805.

Vogt, P.K. and Jackson, A.O. (Eds) (1999) Current Topics in Microbiology & Immunology, Vol. 239: Satellites and defective viral RNAs. Springer-Verlag, New York.

West, S.A., Lively, C.M. and Read, A.F. (1999) A pluralist approach to sex and recombination. *J. Evol. Biol.* **12**, 1003–1012.

Whelan, S.P.J. and Wertz, G.W. (1997) Defective interfering particles of vesicular stomatitis virus: Functions of the genomic termini. *Sem. Virol.* **8**, 131–139.

Wommack, K.E. and Colwell, R.R. (2000) Virioplankton: Viruses in aquatic ecosystems. *Microbiol. Mol. Biol. R.* **64**, 69–114.

Constructing Nature: Laboratory Models as Necessary Tools for Investigating Complex Ecological Communities

MARC W. CADOTTE, JAMES A. DRAKE AND TADASHI FUKAMI

Ecological systems are arguably among the most complex of all systems found in nature at any scale of observation. The origin of this complexity arises not only from the stunning diversity of entities that comprise such systems (i.e., species, individuals and highly variable nature of interactions among these entities), but also because such systems exhibit non-equilibrium, non-linear, historically contingent, self-organizing behavior replete with a host of emergent properties and attendant noise. There are few constants or laws past thermodynamic constraints and simple mechanics, and here process and mechanism operate across vast scales of existence in a fashion that is not precisely replicable. The basic approach of ecologists has been brute-force observation and experimentation in specific settings, with the tacit hope that answers will emerge from large data collection. Laboratory experiments, of course, suffer from many of the same difficulties that plague field studies. At the same time, the simplifying assumptions of laboratory community studies permit explorations of intact communities—something never accomplished in the field (Drake *et al.*, 1996). But the lab is not enough. The general dynamic behaviors and phenomena observed in the lab provide fodder (in the form of potential and probable processes directing community structure) for field approaches.

ADVANCES IN ECOLOGICAL RESEARCH VOL. 37

0065-2504/05 $35.00
DOI: 10.1016/S0065-2504(04)37011-X

Early explorations into the structure of ecological reality were largely limited to description and dominated by natural historians, typified by the supposedly inductive methodologies of the great Victorian naturalists, such as Richard Owen and Charles Darwin. This, of course, was the essential first step, as one must know what building blocks exist before any consideration of composite structure can be envisioned. Nevertheless, brilliant insights into possible inner-workings of ecological systems, by largely non-inductive methods, began to emerge in the late 1800s (e.g., Camerano, 1880; Forbes, 1887), and coupled with mathematical application (e.g., Volterra, 1926; D'Ancona, 1954) rapid progress was made in establishing working foundations for understanding nature.

Concomitant with this foundation, a variety of laboratory studies were conducted as tests of emerging paradigms in this loose discipline (Gause, 1934; Park, 1962). Early in the development of ecology as a science, investigators realized that the experimental use of rapidly growing species in a precisely controlled and readily replicated environment was essential to answering many of the questions posed by theory (Lawton, 1998; Jessup *et al.*, 2004). Literally thousands of generations can be readily observed and tracked in the laboratory using appropriate organisms; a thousand generations of birds, on the other hand, would require many generations of scientists devoting their entire lives to this single endeavor. However, there are a few examples of data collection and analyses selflessly extending beyond a single research career (e.g., Schindler *et al.*, 2003). Interestingly, and in similar fashion to the physical sciences, scientists working in the laboratory, in the field, and with models, fight among themselves for primacy (Kohler, 2002).

Here, we argue that laboratory studies are not only valid investigative tools, but may also better capture the essential dynamics of ecological systems compared with field studies. Through exploring the use of microcosms we wish to illustrate the dynamic reality of ecological systems and examine the systems in which explicit tests of these realities are possible. Further, we argue that most field studies, while remaining the cornerstone of ecology, are inadequate to answer many fundamental ecological questions largely because of inappropriate time series, field-induced reductionism, and unwieldy scale issues (Lawler, 1998). Such studies provide little more than a snapshot of ecological reality, exposing day-to-day variance and uncovering proximate mechanism. Herein lies the dilemma; laboratory studies are artificial constructs and caricatures of reality (where parameters are imposed by the would-be observer), while field studies address the real world but with little hope of stepping past the workings of the presently observed state. We offer that rather than being antagonistic approaches, as is the current state, field and laboratory studies should be conducted in concert. Field and laboratory studies are inevitably linked by the ecological processes that bind all living organisms. However, insights into the workings of nature not only advance through direct observation but also through deductive and logical musings into ecological possibilities. Mathematical and conceptual models, we posit, make assumptions and predictions better tested, in the first instance, using controlled laboratory studies. For example, Tilman (1977) famously used algae in aquatic microcosms to test the predictions made by resource competition models (e.g., R^*). These concepts extended to and refined by natural grasslands provide validation of the power of a synergistic laboratory and field approach.

We proceed by examining various concepts in community ecology, which may be difficult to observe in natural systems and examine some laboratory studies that document the production of ecological structure in multi-species systems. Specifically, we will explore two aspects of ecology where laboratory studies are crucial for our understanding of ecological phenomena. The first aspect is parameter- and measurement-oriented, what must be measured and how should it be done to make observations in ecological studies. We ask whether we are adequately testing ecological theory. Secondly, we explore the interactive nature of ecological phenomena and our inability to reduce these processes to their constituent components. These two discussions implicitly address the relationship between reductionism and holism. Based on these, we will contrast the laboratory approach with the field and model approaches to identify the prescriptive role for laboratory microcosms in ecological research.

I. TIME, SCALE, AND OBSERVATION IN ECOLOGICAL SYSTEMS

"Ecology differs from other natural sciences in its emphasis on the primacy of direct observation" (Keller and Golley, 2000, pg. 10). Given this, ecologists must be critical of what they are observing and what these observations mean, especially as they pertain to understanding nature. Nature viewed through human eyes is just that—reality tuned and created by a particular neural net and observed at specific long wavelengths, while being framed by mental limitations and preconceived ideas about what should be observed. Ecologists employ both empirical and theoretical models to formalize the operation of nature. Properly employed empirical and theoretical models feed back to one another, generating a worldview. For reciprocal translation between empirical and theoretical claims, models and observations have to be temporally and phenomenologically equivalent (Hastings, 2004) or we will enter the domain of the canonical apple- and-orange metaphor. There is no need to invoke the age-old perceived tension between empiricism and rationalism, where 5th century BC contradictory philosophies of Heraclitus and Parmenides on the nature of change brought into question the reliability of observation (Lindberg, 1992). Instead, in ecology the ground seems ripe for a reciprocal relationship between the various methods of asking questions and obtaining knowledge about how nature works.

Ecological communities are composed of a vast array of variously interacting species, so much so that community investigations in the real world typically focus on something other than the community—generally, the set of visually prominent or taxonomically interesting species in some area (e.g., fishes, birds, invertebrates, lizards, trees and so on). We have argued elsewhere that erroneous conclusions about the nature of the community, and certainly its operation, are a necessary outcome of considering pieces of a larger structure (Drake *et al.*, 1996; Samuels and Drake, 1997). Just as the façade of a building reveals little of the internal structure capable of supporting that façade, analyses of pieces of communities have also led us astray and unfortunately such analyses form much of the present ecological paradigm. Herein lies one of the pre-eminent roles for laboratory experimentation—the ability to consider an intact and readily replicable ensemble of species. With microcosms, ecologists have the ability to construct communities as complex or simple as desired. In these systems, species composition, trophic structure and complexity are readily manipulated (e.g., Balčiunas and Lawler, 1995). For example, Fox and McGrady-Steed (2002) examined the role community complexity plays in system-level properties such as 'stability'. Several recent microcosm experiments attempted to account for all species at all trophic levels within the system, from bacteria to top predators (e.g., Balčiunas and Lawler, 1995; Morin, 1999; Diehl and Feißel, 2000; Fox 2002) or even manipulate multiple trophic levels (e.g., Naeem *et al.*, 1994), in order to better understand the role various phenomena play in community dynamics. Without doubt, powerful analyses such as these are all but impossible in the 'real world'. However, armed with concepts, dynamics and behaviors observed in the lab, the pieces of nature, which are intractable in the field, may yield their secrets.

Approaching the appropriate structure, however, is only one aspect of an adequate community characterization. These are generally long-lived structures where dynamics on a successional time scale must be invoked to understand observation. Even at simpler levels of organization, multigenerational data often reveal aspects of population dynamics not readily observed in short studies. In examining population dynamics of weedy annual plants, Freckleton and Watkinson (2002) noted that long-term data were required and that such data were surprisingly rare. When we start looking at basic interspecific interactions, the number of generations of competitors or predators and prey required to expose the essential dynamic nature of these interactions can be enormous (Table 1). Tilman's (1977) elegant studies of competition between diatoms employed the highly controllable environment of the chemostat. Given the duration of the study and organisms involved, Tilman likely observed somewhere between 15 and 30 generations before dynamics were definable. Similarly, Gause's (1934) paramount competition studies with protists spanned roughly 20 generations. Other relatively 'simple' experiments (Table 1) still required multiple generations to show relevant patterns. Luckinbill (1973) and Vandermeer (1969) both examined two-species dynamics (predator-prey in the former and competition in the later), and both required at least 10 generations to correctly parameterize dynamics. More recently, Holyoak and Lawler (1996) examined a two-species, predator-prey metapopulation and due to the complicating factor of habitat subdivision, they needed at least 25 generations to observe cyclic population dynamics. Similarly, Morin (1999) needed at least 50 generations to observe outcomes of two-species competition experiments.

Utida (1957) presented laboratory time-series data of a host-parasite system. The first five generations of his study show a crash in the parasite and an increase and subsequent crash in the host (Fig. 1). This length of time is readily scalable to, for example, five years worth of breeding bird data. However, the oscillatory nature of this interaction does not become apparent for at least 10 generations, and an observable damping of oscillations did not occur for at least 25 generations (Fig. 1) (Utida, 1957). Utida's (1957) experiment reveals that observations at three different temporal scales could potentially lead to three different conclusions about host-parasitoid dynamics. The need for long-term observations also seemed to strike Utida, when commenting on a paper he previously published on the same data. "This trend of convergence was not found in the data described in the previous paper, for the 25 generations provided too short a time in which to detect it" (Utida, 1957, pg. 443). Relatively long time series are necessary to establish the dynamic nature of a single mechanism, under highly controlled conditions.

However, do Utida's findings typify other systems? What about characterizing complex population dynamics, where the stochastic process potentially lead to apparently chaotic trajectories? The relatively few instances where long-term field data have been collected seems to largely support the notion that population dynamics are often non-linear (Kendall, 2002). In order to properly characterize non-linear dynamics in field populations, data spanning decades is required. Unfortunately, these types of data are rare (Freckleton and Watkinson, 2002). Even less common are situations where long-term demographic data are collected on more than one interacting species, and yet this type of data are regarded as quintessential for understanding non-linear dynamics (e.g., Krebs *et al.*, 1995; King and Schaffer, 2001). The idea of non-equilibrium, transient ecological systems has changed

Table 1 Examples of multispecies microcosm studies of ecological phenomena

Researchers	Organisms	Ecological phenomena	Time scale of result
Balčiunas and Lawler, 1995	Two ciliate prey and a single predator	Resource level, predation and prey composition on dynamics	Experiment ended on day 50 (50 or more generations of species) and some results not finalized (e.g., their Fig. 4—not clear if cyclic or extinction, and Fig. 5a—extinction trajectory?)
Dickerson and Robinson, 1985	19 species of algae, protists and rotifers	Richness-area relationships	About 20 weeks (more than 100 generations) needed for patterns
Diehl and Feißel, 2000	Bacterial basal level and ciliate consumer and omnivore	Resource level on three-level food chains with omnivory	Outcomes of experiments were not realized, in some cases, until 100 to 150 generations into the experiment
Drake, 1991	13 species of algae, protists, cladicerans and amphipods	Assembly sequence and community structure	Systems still dynamic from 60 to 160 days
Have, 1993	Two rotifers, one Ostracoda and 20 ciliates	Species-area relationships	Species-area relationships continue to change up to 90 days (i.e., 90 generations for most species)
Holyoak and Lawler, 1996	Single predator and prey pair	Predator and prey population dynamics across patchy landscape	At least 25 generations to see nature of cycles
Lawler, 1993	Two ciliate bacteriovores, one intermediate predator and top predator	Multi-trophic interactions and population dynamics	At least 50 generations for four species treatment

Lawler and Morin, 1993	Two bacteriovorous and one omnivorous ciliates and two Sarcodine predators	Food-web architecture and population dynamics	Depending on composition, up to 40 days needed (i.e., about 40 generations)
Luckinbill, 1973	Ciliate predator and prey	Predator-prey co-existence	At least 15 days generations to observe two full predator-prey oscillations
McCauley and Murdoch, 1990	Daphnia and algae	Resource and predator-prey dynamics	200 days to observe cycles
McGrady-Steed *et al.*, 1997	Algae, herbivore rotifers and protests and predators	Biodiversity and ecosystem predictability	Diversity stabilized at about 24 generations
Morin, 1999	Bacteria and two ciliate bacteriovores	Competition dynamics at differing resource concentrations	Two-species competition dynamics continued to change after 50 days
Petchey, 2000	Three ciliate bacteriovores and single predator	Community complexity and stability	Equilibrium around 16 generations
Robinson and Edgemon, 1989	28 species of algae and protists	Predation and invasibility of communities	Equilibrium after 15 weeks (about 100 generations)
Tilman, 1977	Two species of algae	Resource competition	Some cases more than 20 generations required
Utida, 1957	Bean weevil larvae and parasites	Host-parasite dynamics	10 to 25 generations to observe nature of cyclic patterns
Vandermeer, 1969	Three ciliates	Ciliate competition	Single populations took about 10 generations to reach carrying capacity
Weatherby *et al.*, 1998	63 combinations of six protists species	Persistent communities	Community persistence could not be assessed until 10 to 100 generations of constituent species

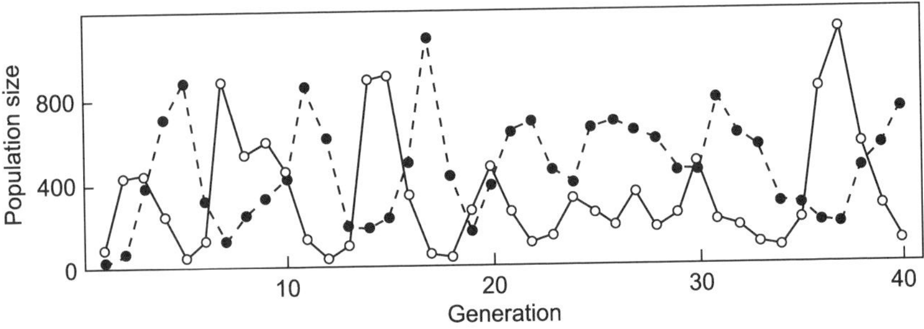

Figure 1 Fluctuations in population size in azuki bean weevil, *Callosobruchus chinensis* (open circles), and its larval parasite, *Neocatolaccus mamezophagus* (closed circles). Adapted from Utida (1957).

how we conceive of ecological dynamics (Hastings, 2004). However, laboratory studies offer the ability to manipulate parameters and test predictions about the nature of non-linear dynamics (Constantino *et al.*, 1995, 1997). Constantino *et al.* (1997) famously used *Tribolium* beetle laboratory populations to test mathematical theory predicting population transitions to chaotic dynamics. By manipulating mortality and recruitment rates they were able to predict when population dynamics changed from stable equilibrium to quasiperiodic cycles to chaos. To do a manipulative field experiment like that of Constantino *et al.* (1997), which integrates predictive models with real data would require decades of labor and vigilance. Probably the most famous case of data showing non-linear dynamics in natural populations (lynx-hare dynamics) (Krebs *et al.*, 1995) was a serendipitous study, via hunting records. Although immensely valuable and insightful, such a time series does not represent the rigorous methodology that is expected of modern experimental ecology.

The aforementioned examples examined ecological phenomena associated with relatively 'simple' situations with a few species, and yet required truly multigenerational observations. However, community ecology is often concerned with much more complex assemblages, where multiple processes may be interacting to produce the observed patterns. Such complex assemblages likely require even longer observational time series in order to adequately account for community dynamics. For example, Cadotte and Fukami (2005) needed from 50 to 100 generations of the organisms making up microcosm communities to adequately observe richness patterns in groups of communities (Fig. 2). In their experiment, they observed patterns of species richness at the metacommunity level, where the different treatments corresponded to differing levels of dispersal. From Fig. 2, it is readily apparent that although initial conditions were identical, one of the treatments (no dispersal) showed a different history compared to the other two

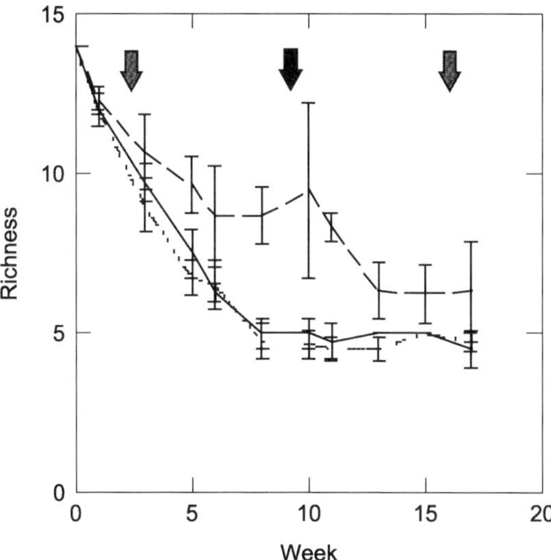

Figure 2 Change in landscape-level richness over time in three dispersal treatments (dashed line refers to no dispersal treatment). Adapted from Cadotte and Fukami (in review). The three arrows, discussed in text, refer to the fact that differing conclusions would be drawn if the system were sampled at these time periods.

treatments, despite the fact that the metacommunities end up at similar richness states. Singular observations at the three arrows would lead investigators to different conclusions about the role of dispersal.

In general, studies employing more complex community structures required many generations (usually 50 or more) to adequately account for community dynamics (Table 1). Microcosm studies investigating species-area relationships (Dickerson and Robinson, 1985; Have, 1993) required many generations (at least 90) to account for these patterns. In fact, Have (1993) showed that the nature of the species-area relationship quantitatively changed up to 90 days into the experiment, when the experiment was terminated. Studies examining the effect of resource level on species interactions (McCauley and Murdoch, 1990; Balčiunas and Lawler, 1995; Diehl and Feißel, 2000) all required more than 50 generations to make adequate observations. Invasion resistance is another concept where microcosms offer great opportunity in advancing our understanding. Robinson and Edgemon (1989) used 28 species of algae and protozoan to investigate invasion resistance, they needed about 100 generations to address invasion resistance in communities. Assembly history is an area of investigation in community ecology which can only be sufficiently addressed by microcosm studies (see more on this in section II.A). Drake (1991) investigated the process of community assembly on community structure and made observations for 60 to 160 days, where most of the organisms he used had a generation time of about a day. Several other studies, examining such things as food-web interactions and population dynamics (Lawler, 1993; Lawler and Morin, 1993; Diehl and Feißel, 2000), needed anywhere from 40 to 150 generations to observe crucial dynamics. Many of these studies reveal that there may be no *a priori* determination of the number of generations needed to make sufficient observations. A number of these results observed changes in dynamics or patterns at some later time in the experiment. Time-specific observations (such as at the arrows on Fig. 2), depending on the hypotheses being tested, would seriously restrict inferential space and possibly lead to erroneous conclusions. Perhaps even worse is the contention that the origins of observed dynamics require even more historical information that is unavailable by direct experimentation (Drake *et al.*, 1999). We are not suggesting that microbial systems are any less prone to exhibiting transient patterns after many generations, but that if transient behavior typifies natural systems (Hastings, 2004), we had better be able to observe them.

II. CONTINGENT STRUCTURE AND RECIPROCAL INTERACTIONS

Laboratory studies can address fundamental questions. Here we address two rapidly expanding foci of study in ecology: the interaction between historical contingency and ecological processes, and the interaction among processes operating at different spatial scales. The processes constituting these two foci do not appear to be understandable in terms of dichotomous states, which are often the typology used to conceptualize or parameterize ecological systems (Peters, 1991). Nor are these properties likely to be reducible to constituent components (e.g., Levins and Lewontin, 1985). The role for microcosms in examining intertwined and possibly non-reducible processes is enormous. Despite the fact that very sophisticated statistical models have been developed to tease apart avenues of causation in ecological systems [e.g., path analysis (Shipley, 2000)], field studies likely lack the temporal scale and manipulative control to investigate the nature of such processes.

A. Historical Contingency and Ecological Processes

There is no doubt that both historical contingency and local ecological dynamics are paramount processes in shaping communities. From an early point in time, interspecific interactions, such as competitive exclusion and predation, have been recognized as key factors in community dynamics (e.g., Gause, 1934; Huffaker, 1958; Chase and Leibold, 2003). More recently, understanding the contingent nature of community assembly has also been considered necessary for understanding community patterns (e.g., Drake, 1991; Drake *et al.*, 1999; Law and Morton, 1996). These two community-structuring influences (interspecific interactions and community assembly) are not readily separable, and understanding this interaction requires highly controlled and replicable experiments. We say that they are not readily separable because, although we can conceptualize a community where all possible species from the regional pool arrive at the same time (the community inception) and hence interspecific interactions are the primary structuring process, this is very unlikely to be the case in most systems. Therefore, most communities must have some history of assembly or continuing invasion where the arrival of a novel species initiates new sets of interspecific interactions (such as in community succession). The lack of independence stems from the idea that each arriving species, if it persists, interacts and changes some portion of the community into which it invades.

Two recent studies suffice to show the importance of the interactive nature of history and local processes. First, Warren *et al.* (2003) showed that, what they termed 'catalytic species', could not persist in either of two possible final community states, but yet switch the communities between these two states via their invasion. One of the two end communities could not be reassembled from the final species membership (aka Humpty-Dumpty community) (e.g., Luh and Pimm, 1993; Samuels and Drake, 1997), but required the presence of the catalytic species in the community's assembly history. Although the final communities appear to be structured by local interactions, assembly history appears to have a critical role in directing structure to alternative basins of attraction. The two antithetical models of community assembly—deterministic versus alternative stable states—appears to actually not be dichotomous but co-occurring processes.

A second recent study, by Fukami and Morin (2003), showed that productivity-diversity relationships, which are often thought to be caused by local dynamics, may depend upon the history of community assembly. They show that productivity-diversity relationships are greatly influenced by assembly history of protozoan invasions into microcosm communities. Previous to this study, Waide *et al.* (1999) summarized research into productivity-diversity patterns, and found that a number of different patterns are commonly found in nature. Fukami and Morin (2003) showed that depending on assembly sequence, linear, quadratic (convex and concave), and non-significant productivity-diversity patterns were all possible. This research reveals that community assembly processes may be very important for understanding extant ecological patterns.

B. The Interaction of Ecological Processes Operating at Different Spatial Scales

There have been a number of recent theoretical studies examining species co-existence as the product of ecological processes operating over larger spatial scales (e.g., Amarasekare and Nisbet, 2001; Shurin and Allen, 2001; Mouquet and Loreau, 2002). Early work by Horn and MacArthur (1972) and Levin (1974) showed that competitively inferior species could co-exist with competitively superior species given a spatial context. The mechanism for this co-existence is a colonization-competition tradeoff, where competitively inferior species are likely to be superior at colonizing new habitats. Since that time the idea that local-level processes (e.g., competition and predation) and larger-scale processes (i.e., dispersal) are important for shaping biodiversity patterns has been formalized (e.g., Srivastava, 1999). Community ecology research concerned with patterns at larger spatial scales, involving multiple local communities, is referred to as the metacommunity-level (more detail on the metacommunity will be discussed in the following text).

A number of studies view the effects of local and regional processes as discrete and in some cases as diametrically opposing forces. These processes may not be readily separable, in similar fashion to the interaction between historical contingency and local processes previously mentioned. Dispersal of propagules originates from within patches, which is likely related to the density of individuals. At the same time, dispersal of new individuals into a habitat likely alters local population dynamics and species interactions thereby revealing the integrated nature of these two processes (e.g., Levins and Lewontin, 1985). The research into the interaction between local and regional processes has not kept pace with theoretical advances. However, some recent work, utilizing both laboratory and field microcosms, have addressed the influence that dispersal has on structuring local communities and diversity patterns (Warren, 1996; Shurin, 2001; Kneitel and Miller, 2003). Important advances in this area have been made by Marcel Holyoak and colleagues who examined metapopulation dynamics of interacting species (e.g., Holyoak and Lawler, 1996; Amezcua and Holyoak, 2000; Holyoak, 2000). Although the role of dispersal on more complex assemblages is only just beginning, one of the greatest strengths of microcosms is the ability to control and quantify dispersal. Quantifying dispersal has been an incredible challenge in natural systems, and it is essential for testing models.

A frontier of modern community ecology is examining the processes and dynamics occurring at spatial scales larger than the local communities, and vast potential rests in this area of research (e.g., Maurer, 1999). This recent area of research has greatly benefited from advances made by microcosm experiments done both indoors and outdoors (Warren, 1996; Shurin, 2000, 2001; Kneitel and Miller, 2003; Cadotte and Fukami, 2005). We would define the metacommunity as not just immigration and emigration of individuals among local communities—which is important—but also the interconnectedness of community processes. This interconnectedness arises via the moving of competitive and predatory interactions, stabilizing and destabilizing units, chaotic attractors and so on among local communities. If local communities exhibit local chaotic attractors directing community patterns, then what happens when these local systems are interconnected? Do local attractors themselves converge, or do we get some higher-level, emergent phenomena at the metacommunity scale? These complex issues, in the immediacy, can only be addressed using highly controlled microcosm systems, where again, the proper time scales are crucial.

C. What These Examples Tell Us

These two examples of complex interrelations of ecological processes are not meant to convey the perception that microcosms can reduce ecological phenomena to their constituent processes. Rather, laboratory microcosms offer an important avenue towards observing and measuring these interactive, intertwined processes and emergent properties. In the first case (contingent history), Fukami and Morin (2003) reveal that local processes do create a productivity-diversity relationship, but the exact nature of these relationships is highly influenced by history. All natural communities have history. Understanding the influences of history requires ecologists to follow communities from their inception through many generations, something that few field systems would allow (Berlow, 1997; using sessile marine organisms is a notable exception). In the second example (processes at different spatial scales), microcosm studies are crucial because researchers can examine communities structured only by local processes and compare them to communities that include regional dynamics. Microcosms allow controlled experiments to understand how local and regional processes combine to structure local communities.

III. PROBLEMATIC FIELD STUDIES

Some questions in ecology are of fundamental importance, given the current expectations for ecologists to predict the consequences of human-caused environmental impacts (Lawton, 2000; Shrader-Frechette, 2001; Ehrlich, 2002). The importance of these questions are confounded by the ethical dilemmas of performing experiments on intact communities, or the fact that constructing and replicating human-caused change may be functionally impossible to engineer. Microcosms offer much to the field of invasion biology. Whereas there are obvious ethical concerns about purposefully introducing invaders into intact systems, as well as manipulating parameters such as richness of these intact communities, microcosms are able to overcome these ethical concerns (e.g., Robinson and Edgemon, 1989; McGrady-Steed *et al.*, 1997).

Similarly, the effects of global change are of great concern for many ecologists. Yet, answering the question of how global change may affect communities is very complex, and the experiments (including FACE sites) are expensive and, due to small sample sizes, statistically cumbersome. Microcosm investigations into global change (Petchey *et al.*, 1999) not only can overcome many logistical problems, but also could be used to inform what factors field studies should look at.

IV. THE ROLE OF MICROCOSMS IN COMMUNITY ECOLOGY

Gilpin *et al.* (1986) noted that laboratory microcosm experiments have a great potential to offer insight into ecological communities, but that lab experiments thus far were too simplistic. In the almost 20 years since the Gilpin *et al.* (1986) paper, there have been several important studies into the structure of communities using microcosms. Examples include: Drake (1991); Lawler (1993); Diehl and Feißel (2000); Holyoak (e.g., 2000); Fox and McGrady-Steed (2002); and most recently Warren *et al.* (2003). Present microcosm studies are examining much more complex assemblages and phenomena compared with earlier studies.

There is concern that energies devoted to laboratory studies may be better spent for advancing theoretical models, since laboratory microcosms are essentially messy models. This concern misses the many potential contributions, stark and subtle, that laboratory research offers (e.g., Lawton, 2000; Jessup *et al.*, 2004). True, laboratory systems are much simpler than real ones and use organisms which are themselves physically and behaviorally simpler than other organisms and, in a sense, follow many simplifying assumptions in general models (e.g., asexual reproduction, discrete community borders, stable trophic structure and so on). Gilpin *et al.* (1986) rightly noted that future lab studies would find that laboratory communities have inherent complexity. This complexity is at the heart of why laboratory communities are so useful to ecologists. This complexity indicates that assemblages of species may exhibit patterns that are on the one hand greater than the sum of their parts, and on the other may reveal patterns and dynamics not predicted by mathematical models (see more discussion in the following text on variability and stochasticity in microcosms).

One of the greatest strengths of microcosm systems is the ability to directly test theoretical models. Examples include Constantino *et al.* (1995, 1997) and Warren *et al.* (2003), both previously discussed, as well as experiments comparing predictive models of predator-prey interactions (Bohannan and Lenski, 1997; Kaunzinger and Morin, 1998). Experimental microcosms also refine theory by stipulating under what circumstances processes operate. For example, experimental microcosms show that the intermediate disturbance hypothesis operates as the well-known unimodal relationship in heterogeneous environments but as a linearly declining relationship in homogeneous environments (Buckling *et al.*, 2000).

Diamond (1986) rightly notes the important contributions that microcosm studies have had on fundamental principles of ecology. However, we disagree with his assessment that lab experiments are weak in terms of temporal and spatial extrapolations. A most recent reincarnation of the same idea can be found in the otherwise excellent article by Ricklefs (2004, p. 5). Their conceptualization of temporal scale seems to rest on 'real' time (measured in days), where laboratory studies typically run for less than one year. However, different organisms should be viewed under the auspices of generational time (Petersen and Hastings, 2001). As previously discussed, laboratory experiments may follow populations for many more generations than in many field studies. Diamond believes that "[natural experiments] are the sole method capable of studying the genetic changes after 10,000 years." Yet work by Lenski *et al.* (2003) on genetic changes over tens of thousands of generations in bacterial populations reveal immense insight into what laboratory experiments offer. This is not to say that bacterial population dynamics offer particular insights into, for example, marine mammals, but our general understanding of genetic change and ecological factors has been greatly advanced by such microcosm studies (e.g., Bohannan and Lenski, 1997; Lythgoe and Chao, 2003; Yoshida *et al.*, 2003).

As for spatial scale, Diamond (1986) and Ricklefs (2004) seemed to compare scale to human relative measurements, while we would argue that spatial scale should be relative to the scale of ecological processes of the organisms involved. Again as previously outlined, laboratory experiments can include habitat heterogeneity and larger spatial processes. For example, Ricklefs's (2004) hypothesis that diversity patterns are historically contingent over large scales can be explicitly tested by laboratory microcosms in ways that are not otherwise possible (Rainey and Travisano, 1998; Fukami and Morin, 2003). More importantly, the criticism that microcosms are too simple and too small a scale ignores that, with care, systems can be scaled down with appropriate processes and dynamics being maintained (Petersen and Hastings, 2001).

An important criticism is that microcosms limit variability and stochasticity, which may be critical for real communities. Again, we believe that microcosm studies can help ecologists understand the potential roles variability and stochasticity play in ecological communities. Environmental variability can be not only eliminated, but also induced (e.g., Petchey *et al.*, 2002) to see how variability interacts with other processes. More subtle is the role stochasticity plays in ecological processes. As noted earlier, laboratory systems can still exhibit chaotic fluctuations and exhibit stochastic and other complex or unexpected patterns. For example, the results of a recent experiment (Fig. 3; Cadotte, unpublished data, 2003) show that the inherent stochastic nature of population dynamics in monocultures of *Paramecium aurelia*. The four population trajectories correspond to identical initial conditions, and yet two of the four replicates reach a carrying capacity of about 600 individuals/ml in six days, while the other two replicates settle at $<$100 individuals/ml. These apparently stochastic processes may be the result of a number of mechanisms, from genotype differences to contamination of infectious agents or toxic microorganisms such as fungi. This is not to say that microcosms do not confer a high degree of repeatability (e.g., Holyoak and Lawler, 1996; Morin, 1999), but microcosm populations are not immune to uncontrolled stochasticity. Whereas natural systems may not allow ecologists to witness stochastic trajectories (rather, time-specific variation), microcosm communities allow ecologists to observe stochastic trajectories in controlled replicated experiments.

We see laboratory microcosms as the link between models and the real world. The reasons for this have been outlined throughout this chapter.

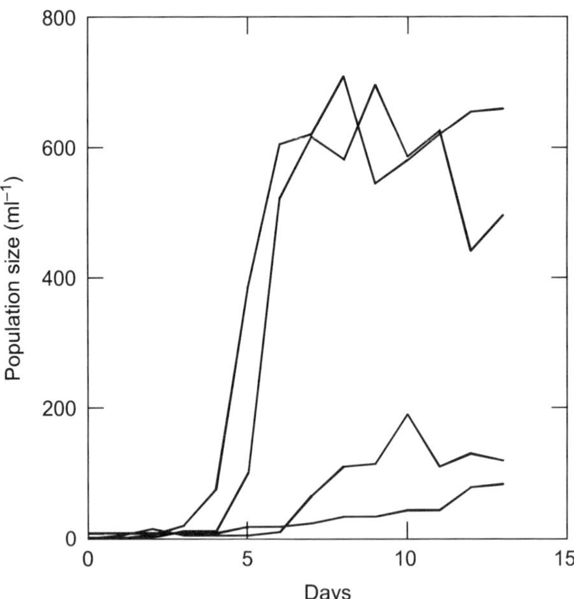

Figure 3 Population growth curves for four populations of *Paramecium aurelia* in identical conditions, showing the stochastic nature of population growth (Cadotte, unpublished data).

Basically, we are forwarding the notion that there may be a discontinuity between general models and observations based in the real world. This continuity comes in the form of scale of observation (temporal and spatial), elucidating non-independent and contingent processes, as well as observing complex, unexpected patterns.

V. ENDNOTE

Science advances by the generation of new theories, solving existing puzzles within the extant framework, and the largely non-epistemological need to explain more phenomena with fewer explanations. Scope, simplicity and fruitfulness are several measures of useful and growing research programmes (e.g., Brown, 2001), although we must also not forget the value of specificity in science (Motokawa, 1989; Simberloff, 2004). The ability to address important problems depends upon scientists being able to ask the correct questions and make precise and unequivocal observations.

In ecology, important problems abound. Field studies and conceptual models have helped ecologists clarify some of the critical problems, yet natural systems are teeming with complexity, so much so that there seem to be exceptions to any rule. Should we resign ourselves to accept that we may never be able to disentangle patterns in community ecology? Or that we may only be able to make generalizations at extreme macroscopic scales (Lawton, 1999)? We would argue that, in the face of insurmountable complexity, we redouble our efforts, using new tools and concepts, deconstruct and reconstruct communities and use our ability to construct functioning caricatures of nature in the laboratory. Microcosm communities are real communities, although not natural. They contain the same types of interacting species: competitors, predators and prey, mutualists and so on; undergo the same types of processes: species extinctions, succession, species sorting and so on; and show the same type of patterns: diversity-productivity patterns, community complexity and resultant stability, species-area relationships and so on. We argue then, given that microcosm communities confer experimental benefits, they are more useful for understanding some of the primary processes in community ecology than generally thought.

REFERENCES

Amarasekare, P. and Nisbet, R.M. (2001) Spatial heterogeneity, source-sink dynamics, and the local coexistence of competing species. *Am. Nat.* **158**, 572–584.

Amezcua, A.B. and Holyoak, M. (2000) Empirical evidence for predator-prey source-sink dynamics. *Ecology* **81**, 3087–3098.

Balčiunas, D. and Lawler, S.P. (1995) Effects of basal resources, predation, and alternative prey in microcosm food chains. *Ecology* **76**, 1327–1336.

Berlow, E.L. (1997) From canalization to contingency: Historical effects in a successional rocky intertidal community. *Ecol. Monogr.* **67**, 435–460.

Bohannan, B.J.M. and Lenski, R.E. (1997) Effect of resource enrichment on a chemostat community of bacteria and bacteriophage. *Ecology* **78**, 2303–2315.

Brown, J.R. (2001) *Who Rules in Science? An Opinionated Guide to the Wars.* Harvard University Press, Cambridge, MA.

Buckling, A., Kassen, R., Bell, G. and Rainey, P.B. (2000) Disturbance and diversity in experimental microcosms. *Nature* **408**, 961–965.

Cadotte, M.W. and Fukami, T. (2005) Dispersal spatial scale and species diversity in a hierarchically structural experimental landscape. *Ecology Letters* **8**, 548–557.

Camerano, L. (1880) Dell'equilibrio dei viventi merce la reciproca distruzione. *Accademia delle Scienze di Torino* **15**, 393–414. (translated in the cited source by C.M. Jacobi and J.E. Cohen (1994): On the equilibrium of living beings of reciprocal destruction). In: *Frontiers in Mathematical Biology* (Ed. by S.A. Levin), pp. 360–380. Springer-Verlag, Berlin, Germany.

Chase, J.M. and Leibold, M.A. (2003) *Ecological Niches: Linking Classical and Contemporary Approaches.* University of Chicago Press, Chicago, IL.

Constantino, R.F., Cushing, F.M., Dennis, B. and Desharnais, R.A. (1995) Experimentally induced transitions in the dynamic behavior of insect populations. *Nature* **375**, 227–230.

Constantino, R.F., Desharnais, R.A., Cushing, F.M. and Dennis, B. (1997) Chaotic dynamics in an insect population. *Science* **275**, 389–391.

D'Ancona, U. (1954) *Struggle for Existence.* E. J. Brill, Leiden, Germany.

Diamond, J. (1986) Overview: Laboratory experiments, field experiments, and natural experiments. In: *Community Ecology* (Ed. by J. Diamond and T.J. Case), pp. 3–22. Harper and Row, New York, NY.

Dickerson, J.E. and Robinson, J.V. (1985) Microcosms as islands: A test of the MacArthur-Wilson equilibrium theory. *Ecology* **66**, 966–980.

Diehl, S. and Feißel, M. (2000) Effects of enrichment on three-level food chains with omnivory. *Am. Nat.* **155**, 200–218.

Drake, J.A. (1991) Community assembly mechanics and the structure of an experimental species ensemble. *Am. Nat.* **137**, 1–26.

Drake, J.A., Hewitt, C.L., Huxel, G.R. and Kolasa, J. (1996) Diversity and higher levels of organization. In: *Biodiversity: A Biology of Numbers and Difference* (Ed. by K.J. Gaston), pp. 149–166. Blackwell Science, Cambridge, MA.

Drake, J.A., Zimmerman, C.R., Purucker, T. and Rojo, C. (1999) On the nature of the assembly trajectory. In: *Ecological Assembly Rules: Perspectives, Advances, Retreats* (Ed. by E. Weiher and P. Keddy), pp. 233–250. Cambridge University Press, New York, NY.

Ehrlich, P.R. (2002) Human natures, nature conservation, and environmental ethics. *Bioscience* **52**, 31–43.

Forbes, S.A. (1887) The lake as a microcosm. Ill. *Nat. Hist. Surv. Bull.* **15**, 537–550.

Fox, J.W. (2002) Testing a simple rule for dominance in resource competition. *Am. Nat.* **159**, 305–319.

Fox, J.W. and McGrady-Steed, J. (2002) Stability and complexity in microcosm communities. *J. Animal Ecology* **71**, 749–756.

Freckleton, R.P. and Watkinson, A.R. (2002) Are weed population dynamics chaotic? *J. Applied Ecology* **39**, 699–707.

Fukami, T. and Morin, P.J. (2003) Productivity-biodiversity relationships depend on the history of community assembly. *Nature* **424**, 423–426.

Gause, G.F. (1934) *The Struggle for Existence*. Dover, NYReprinted 1971.

Gilpin, M.E., Carpenter, M.P. and Pomerantz, M.J. (1986) The assembly of a laboratory community: Multispecies competition in *Drosophila*. In: *Community Ecology* (Ed. by J. Diamond and T.J. Case), pp. 23–40. Harper and Row, New York, NY.

Hastings, A. (2004) Transients: The key to long-term ecological understanding? *Trends Ecol. Evol.* **19**, 39–45.

Have, A. (1993) Effects of area and patchiness on species richness: An experimental archipelago of ciliate microcosms. *Oikos* **66**, 493–500.

Holyoak, M. (2000) Habitat patch arrangement and metapopulation persistence of predators and prey. *Am. Nat.* **156**, 378–389.

Holyoak, M. and Lawler, S.P. (1996) Persistence of an extinction-prone predator-prey interaction through metapopulation dynamics. *Ecology* **77**, 1867–1879.

Horn, H.S. and MacArthur, R.H. (1972) Competition among fugitive species in a harlequin environment. *Ecology* **53**, 749–752.

Huffaker, C.B. (1958) Experimental studies on predation: Dispersion factors and predator-prey oscillations. *Hilgardia* **27**, 343–383.

Jessup, C.M., Kassen, R., Forde, S.E., Kerr, B., Buckling, A., Rainey, P.B. and Bohannan, B.J.M. (2004) Big questions, small worlds: Microbial model systems in ecology. *Trends Ecol. Evol. Biol.* **19**, 187–197.

Kaunzinger, C.M.K. and Morin, P.J. (1998) Productivity controls food-chain properties in microbial communities. *Nature* **395**, 495–497.

Keller, D.R. and Golley, F.B. (2000) *The Philosophy of Ecology: From Science to Synthesis*. University of Georgia Press, Athens, GA.

Kendall, B.E. (2002) Chaos and cycles. In: *Encyclopedia of Global Environmental Change: Volume 2, The Earth System: Biological and Ecological Dimensions of Global Environmental Change* (Ed. by H.A. Mooney and J.G. Canadell), pp. 209–215. John Wiley and Sons, Chichester, UK.

King, A.A. and Schaffer, W.M. (2001) The geometry of a population cycle: A mechanistic model of snowshoe hare demography. *Ecology* **82**, 814–830.

Kneitel, J.M. and Miller, T.E. (2003) Dispersal rates affect species composition in metacommunities of *Sarracenia purpurea* inequilines. *Am. Nat.* **162**, 165–171.

Kohler, R.E. (2002) *Landscapes and Labscapes: Exploring the Lab-Field Border in Biology*. The University of Chicago Press, Chicago, IL.

Krebs, C.J., Boutin, S., Boonstra, R., Sinclair, A.R.E., Smith, J.N.M., Dale, M.R.T., Martin, K. and Turkington, R. (1995) Impact of food and predation on the snowshoe hare cycle. *Science* **269**, 1112–1115.

Law, R. and Morton, R.D. (1996) Permanence and the assembly of ecological communities. *Ecology* **77**, 762–775.

Lawler, S.P. (1993) Species richness, species composition and population dynamics of protists in experimental microcosms. *J. Animal Ecology* **62**, 711–719.

Lawler, S.P. (1998) Ecology in a bottle: Using microcosms to test theory. In: *Experimental Ecology: Issues and Perspectives* (Ed. by W.J. Resetarits, Jr. and J. Bernardo), pp. 236–253. Oxford University Press, New York, NY.

Lawler, S.P. and Morin, P.J. (1993) Food web architecture and population dynamics in laboratory microcosms of protists. *Am. Nat.* **141**, 675–686.

Lawton, J.H. (1998) Ecological experiments with model systems: The Ecotron facility in context. In: *Experimental Ecology: Issues and Perspectives* (Ed. by W.J. Resetarits, Jr. and J. Bernardo), pp. 170–182. Oxford University Press, New York, NY.

Lawton, J.H. (1999) Are there general laws in ecology? *Oikos* **84**, 177–192.
Lawton, J.H. (2000) *Community Ecology in a Changing World*. Ecology Institute, Oldendorf/Luhe, Germany.
Lenski, R.E., Winkworth, C.L. and Riley, M.A. (2003) Rates of DNA sequence evolution in experimental populations of *Escherichia coli* during 20,000 generations. *J. Molecular Evolution* **56**, 498–508.
Levin, S.A. (1974) Dispersion and population interactions. *Am. Nat.* **108**, 207–228.
Levins, R. and Lewontin, R. (1985) *The Dialectical Biologist*. Harvard University Press, Cambridge, MA.
Lindberg, D.C. (1992) *The Beginnings of Western Science*. University of Chicago Press, Chicago, IL.
Luckinbill, L.S. (1973) Coexistence in laboratory populations of *Paramecium Aurelia* and its predator *Didinium nasutum*. *Ecology* **54**, 1320–1327.
Luh, H.K. and Pimm, S.L. (1993) The assembly of ecological communities: A minimalist approach. *J. Animal Ecology* **62**, 749–765.
Lythgoe, K.A. and Chao, L. (2003) Mechanisms of coexistence of a bacteria and bacteriophage in a spatially homogeneous environment. *Ecol. Lett.* **6**, 326–334.
Maurer, B.A. (1999) *Untangling Ecological Complexity*. University of Chicago Press, Chicago, IL.
McCauley, E. and Murdoch, W.W. (1999) Predator-prey dynamics in environments rich and poor nutrients. *Nature* **343**, 455–457.
McGrady-Steed, J., Harris, P.M. and Morin, P.J. (1997) Biodiversity regulates ecosystem predictability. *Nature* **390**, 162–165.
Morin, P.J. (1999) Productivity, intraguild predation, and population dynamics in experimental food webs. *Ecology* **80**, 752–760.
Motokawa, T. (1989) Sushi science and hamburger science. *Perspect. Biol. Med.* **32**, 489–504.
Mouquet, N. and Loreau, M. (2002) Coexistence in metacommunities: The regional similarity hypothesis. *Am. Nat.* **159**, 420–426.
Naeem, S., Thompson, L.J., Lawler, S.P., Lawton, J.H. and Woodfin, R.M. (1994) Declining biodiversity can alter the performance of ecosystems. *Nature* **368**, 734–737.
Park, T. (1962) Beetles, competition, and populations. *Science* **138**, 1369–1375.
Petchey, O.L. (2000) Prey diversity, prey composition, and predator population dynamics in experimental microcosms. *J. Animal Ecology* **69**, 874–882.
Petchey, O.L., Casey, T., Jiang, L., McPhearson, P.T. and Price, J. (2002) Species richness, environmental fluctuations, and temporal change in total community biomass. *Oikos* **99**, 231–240.
Petchey, O.L., McPhearson, P.T., Casey, T.M. and Morin, P.J. (1999) Environmental warming alters food-web structure and ecosystem function. *Nature* **402**, 69–72.
Peters, R.H. (1991) *A Critique for Ecology*. Cambridge University Press. Cambridge, UK.
Petersen, J.E. and Hastings, A. (2001) Dimensional approaches to scaling experimental ecosystems: Designing mousetraps to catch elephants. *Am. Nat.* **157**, 324–333.
Rainey, P.B. and Travisano, M. (1998) Aadpative radiation in a heterogeneous environment. *Nature* **394**, 69–72.
Ricklefs, R.E. (2004) A comprehensive framework for global patterns in biodiversity. *Ecol. Lett.* **7**, 1–15.
Robinson, J.V. and Edgemon, M.A. (1989) The effect of predation on the structure and invasibility of assembled communities. *Oecologia* **79**, 150–157.

Samuels, C.L. and Drake, J.A. (1997) Divergent perspectives on community convergence. *Trends Ecol. Evol.* **12**, 427–432.

Schindler, D.E., Chang, G.C., Lubetkin, S., Abella, S.E.B. and Edmondson, W.T. (2003) Rarity and functional importance in a phytoplankton community. In: *The Importance of Species: Perspectives on Expendability and Triage* (Ed. by P. Kareiva and S.A. Levin), pp. 206–220. Princeton University Press, Princeton, NJ.

Shipley, B. (2000) *Cause and Correlation in Biology: A User's Guide to Path Analysis, Structural Equations and Causal Inference*. Cambridge University Press, New York, NY.

Shrader-Frechette, K. (2001) Non-indigenous species and ecological explanation. *Biol. Philosophy* **16**, 507–519.

Shurin, J.B. (2000) Dispersal limitation, invasion resistance, and the structure of pond zooplankton communities. *Ecology* **81**, 3074–3086.

Shurin, J.B. (2001) Interactive effects of predation and dispersal on zooplankton communities. *Ecology* **82**, 3404–3416.

Shurin, J.B. and Allen, E.G. (2001) Effects of competition, predation, and dispersal on species richness at local and regional levels. *Am. Nat.* **158**, 624–637.

Simberloff, D. (2004) Community ecology: Is it time to move on? *Am. Nat.* **163**, 787–799.

Srivastava, D.S. (1999) Using local-regional richness plots to test for species saturation: Pitfalls and potentials. *J. Animal Ecology* **68**, 1–17.

Tilman, D. (1977) Resource competition between planktonic algae: An experimental and theoretical approach. *Ecology* **58**, 338–348.

Utida, S. (1957) Cyclic fluctuations of population density intrinsic to the host-parasite system. *Ecology* **38**, 442–449.

Vandermeer, J.H. (1969) The competitive structure of communities: An experimental approach with protozoa. *Ecology* **50**, 362–371.

Volterra, V. (1926) Fluctuations in the abundance of a species considered mathematically. *Nature* **118**, 558–560.

Waide, R.B., Willig, M.R., Steiner, C.F., Mittelbach, G., Gough, L., Dodson, S.I., Juday, G.P. and Parmenter, R. (1999) The relationship between productivity and species richness. *Ann. Rev. Ecology Systematics* **30**, 257–300.

Warren, P.H. (1996) The effects of between-habitat dispersal rate on protist communities and metacommunities in microcosms at two spatial scale. *Oecologia* **105**, 132–140.

Warren, P.H., Law, R. and Weatherby, A.J. (2003) Mapping the assembly of protist communities in microcosms. *Ecology* **84**, 1001–1011.

Weatherby, A.J., Warren, P.H. and Law, R. (1998) Coexistence and collapse: An experimental investigation of the persistent communities of a protist species pool. *J. Animal Ecology* **67**, 554–566.

Yoshida, T., Jones, L.E., Ellner, S.P., Fussman, G.F. and Hairston, N.G., Jr. (2003) Rapid evolution drives ecological dynamics in a predator-prey system. *Nature* **424**, 303–306.

Index

Advances in Ecological Research Volume 1–37

Cumulative List of Titles